T0396418

INTERPRETING PHYSICS

BOSTON STUDIES IN THE PHILOSOPHY OF SCIENCE

VOLUME 289

For further volumes:
http://www.springer.com/series/5710

INTERPRETING PHYSICS

Language and the Classical/Quantum Divide

by

EDWARD MACKINNON

California State University East Bay, CA, USA

Edward MacKinnon
California State University East Bay
Manzanita Drive 2045
94611 Oakland
USA
emackinnon@comcast.net

ISSN 0068-0346
ISBN 978-94-007-2368-9 e-ISBN 978-94-007-2369-6
DOI 10.1007/978-94-007-2369-6
Springer Dordrecht Heidelberg London New York

Library of Congress Control Number: 2011939573

Printed on acid-free paper

Springer is part of Springer Science+Business Media (www.springer.com)

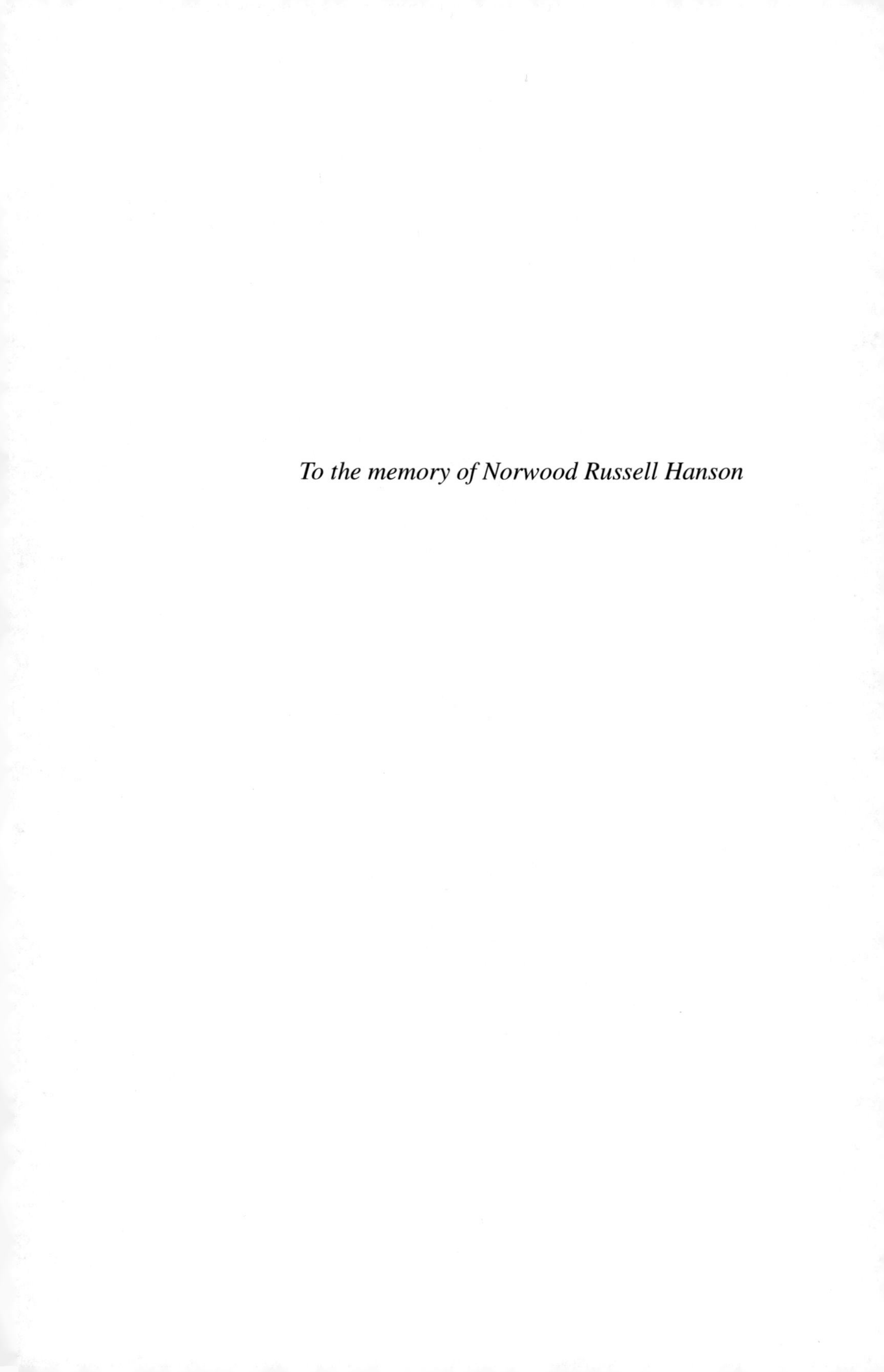

To the memory of Norwood Russell Hanson

Preface

It will be misjudged because it is misunderstood, and misunderstood because men choose to skim through the book and not to think through it, a disagreeable task, because the work is dry, obscure, opposed to all ordinary notions, and moreover long-winded.
Immanuel Kant, Prolegomea

This is a book about the role of language in physics, and especially about its role in the interpretation of quantum mechanics. Yet, it is utterly unlike any other book written on the interpretation of quantum mechanics. This introduction gives a preliminary account of the reasons for the novel approach.

There are different philosophical approaches to the interpretation of quantum mechanics and quantum field theory. Most share a common feature. Interpretation is essentially a matter of specifying the physical significance of mathematically formulated theories, or of models that relate to theories. The language in which these are expressed is not a critical factor and is not included in accounts of the way physics is interpreted. The truth of the theory being interpreted is determined by how well conclusions drawn from the theory accord with experimental observations.

In contemporary particle physics conclusions from a theory are *never* tested against observations. They are tested against inferences based on observations and a network of presuppositions supporting the inferential process of experimental physics. This inferential system relies on the highly developed language of physics. Typically an experimental begins with a descriptive account of the problem being considered, the equipment used, the scope of its valid applicability, and a consideration of related work. Within this general linguistic framework one can draw on other parts of physics: established facts, experimental results, and whatever theories seem appropriate. The enveloping matrix for these considerations is language, not a theory. Thus the interpretation of physics, rather than isolated physical theories, involves two separate, but related, inference systems. The formal one, deduction from axioms, is one that philosophers of science have explored in detail. The informal one, inferences in the developed dialect of physics, remains, to adopt Vico's famous phrase, *terra incognita*.

The separation of the two inferential systems is not a critical issue in classical, or pre-quantum, physics. The evolution of the language of physics and the co-evolution of mathematics and physics since the seventeenth century harmonized the relation between the two. The redevelopment of the foundations of mathematics at the beginning of the nineteenth century initiated a separation between mathematics and physics. However classical physicists, almost without exception, relied on the older physicalistic mathematics of Euler and Lagrange. The development of quantum physics in the mid 1920s led to a linguistic crisis. The familiar language of physics could be used to report an individual experiment without contradictions. However, an attempt to extend this reporting language to broader contexts could engender contradictions. Most physicists originally thought of this in terms of puzzling properties of subatomic particles. How could electrons and photons have both wave and particle properties? Niels Bohr treated this as a problem about concepts, not objects. His somewhat nebulous doctrine of complementarity provided an expedient for avoiding difficulties. He contended that the language of classical physics is an extension of ordinary language and that its usage in experimental contexts must meet the conditions for the unambiguous communication of experimental information. If experimenters limited the extended usage of key terms in classical physics to a particular experimental context and rejected further extensions to other experimental situations, then they avoided contradictions. After this practice was adopted the original linguistic crisis gradually slid into a collective oblivion.

Philosophers, and even some historians, of science have regularly castigated Bohr's complementarity interpretation of quantum mechanics as naive, ad hoc, arbitrary, outmoded, and even irrational. Most of these criticisms involved a serious misunderstanding. Bohr never intended an interpretation of quantum mechanics *as a theory*. We are, he contended, suspended in language. He treated mathematical formalisms as useful tools, not as theories to be interpreted. He struggled to clarify the basic concepts of physics and the limits of their valid applicability. These struggles supply a point of departure for an analysis of the role of language in quantum physics and for the interrelation between the reporting of experimental results and theories.

How could philosophers who have taken or survived the linguistic turn neglect such a fundamental issue as the role of language in physics? This will be treated in Chapter 1. In this preliminary orientation we can cite two types of reasons. The first is that Anglo-American philosophy, subsequent to Quine and Sellars, experienced a split between philosophers of science and analysts. Both groups avoided this issue for very different reasons. Philosophers of science have increasingly come to focus on theories as the primary products of advances in physics and to regard theories as mathematical structures that can support different physical interpretations. The mathematical structure, accordingly, must be able to stand on its own independent of particular interpretations. The same structures can be used by people speaking many different languages. The analysis and interpretation of such structures is a matter of logic and mathematics. Any analysis of usages proper to one particular language seems irrelevant. An analysis of the role of language in reporting experimental results and in the dialog between experimenters and theoreticians could be avoided by following the Logical Positivist's precedent of relegating experiments to the role of supplying facts that test theoretical predictions.

In this book I draw on methods developed within the analytic tradition. This tradition, however, has sharp limitations. Three such limitations especially concern us. The methods of analysis are essentially synchronic, centered on the way language now functions. A diachronic analysis is needed to understand how the language of physics developed. Here the hermeneutic wing of the phenomenological tradition offers some helpful precedents. Both traditions, however, share a common limitation. They do not consider the way in which a specialized extension of language can interrelate with and support a mathematical formalism used to explain physical reality. A reference to Kuhnian relativism often supplied a justification for this omission. If the history of physics is interpreted as a succession of theories, paradigms, research programs, or problem-solving methodologies, then currently accepted theories are as open to rejection and replacement as their predecessors. Why bother with the ontological claims or epistemological justifications of such ephemeral entities?

The third reason is that it is difficult to clarify and find methods of resolving the basic problems. I will use a bit of personal history to illustrate the nature of the difficulties. In the years 1963–1965 I was a post-doc in the Yale philosophy department. My Ph.D is in physics, not philosophy. Each term I sat in on N. R. Hanson's seminars and also had weekly discussions with him and two other students on problems in the history and philosophy of physics. What proved more instructive than the seminars was Hanson's own process of philosophizing. He believed that ordinary language analysis, where he was especially influenced by Wittgenstein, should be related to the philosophy of science. Yet, his attempt to do this systematically bogged down in the problem of cross-type inferences. He argued that there are no valid inferences from the necessary to the contingent, and vice versa. He also believed, long before it became fashionable, that philosophers of science should relate their work to serious historical studies. As he put it: History of science without philosophy of science is blind; philosophy of science without history of science is empty. Yet, he also argued that any attempt to establish a logical relation between the two disciplines entails a genetic fallacy.

Accepted philosophical methods supplied no basis for relating ordinary language analysis to philosophy of science, and no systematic way of relating philosophy of science to the history of science. Yet, the need remained. What Hanson actually did, with limited success, was to pick problems, chiefly in astronomy, particle physics, and aeronautics, involving the opposed fields, muddle through particular solutions, and hope that methodological enlightenment would emerge. My protracted efforts to interrelate the historical development of physics with the interpretation of quantum mechanics, and of ordinary language analysis with the formal interpretation of theories did not draw upon, or even bear any conscious relation, to Hanson's earlier efforts. A request to write a summary account of Hanson's work triggered the realization that my procrustean lucubrations were mimicking my mentor. Hence the dedication.

The problems Hanson faced have mutated in the ensuing years. Yet, ordinary language analysis, whether in an analytic or hermeneutic version, still remains separated from the philosophy of physics. When I retired from teaching I used my new freedom to begin a systematic exploration of the problem. I found that the only way I could make sense of the specialized extension of language to physics was to track the conceptual evolution involved. The earliest stages in this evolution were not motivated by physical goals. The historical analysis in Chapter 2 focuses on the evolution of a specialized language, rather than on ancient physics The intermediate stage involved the co-evolution of physical and mathematical concepts. This, in turn, led to the emergence of the three master theories of classical physics, mechanics, thermodynamics, and electrodynamics. Here we are focusing on the classical physics that both supports and conflicts with quantum physics.The final stage, the classical/quantum interface and a recognition of the fundamental role of quantum mechanics required a departure from normal philosophical practices. Rather than focusing on individual theories as units of interpretation I focus on the practice of physics and the dialog between experimenters and theoreticians. This required a detailed presentation of the physics involved.

The eventual result was a work that was obscure, opposed to all ordinary notions, and quite long. In the hope that philosophers might read, rather than skim through it, I shortened and streamlined it. As a method of shortening I appeal, whenever possible, to the work of historians of science and omit details. Second, I treated the technical physics in a series of specialized articles and present a fairly non-technical summary in the present work. Physics is the basic science. The treatment of other philosophical issues often a presupposes an interpretation of physics. The new interpretation of physics presented here changes the problematic status of these issues. The concluding chapter surveys this changing problematic.

In writing this book I have benefited from discussions with Harold Brown, Catherine Chevalley, James Cushing, Olivier Darrigol, Ravi Gomatam, Richard Griffiths, William Reuter, and anonymous referees who criticized this work or parts of it developed as independent articles. I am especially grateful to my wife and fellow philosopher, Barbara, for her patience and support in the the long protracted process of writing and rewriting.

Contents

Part II The Classical/Quantum Divide

Acronyms

aka	also known as
BKS	The 1924 Bohr-Kramers-Slater paper
B-S	The Bohr-Sommerfeld atomic theory
CH	Consistent Histories formulation of quantum mechanics
CMB	Cosmic Microwave Background Radiation
CMS	Compact Muon Solenoid
CP	Correspondence Principle
C-P	Classical Physics
EFT	Effective Field Theory
EM	Electrodynamics
EOL	Extended Ordinary Language
E-R	Emergence-Reduction
FAPP	For all practical purposes
G-H	Gell-Mann–Hartle project of developing a foundational quantum mechanics
LHC	Large hadron Collider
NRQM	Non-relativistic quantum mechanics
p_c	classical particle
p_t	theoretical particle
QM	Quantum Mechanics
QED	Quantum Electrodynamics
QFT	Quantum Field Theory
RQM	Relativistic quantum mechanics
T-T'	Thomson and Tait's *Treatise on Natural Philosophy*
Umwelt	The social-physical world of ordinary experience

Part I
The Language of Physics

Chapter 1
A Philosophical Overview

If it is true that there are but two kinds of people in the world—the logical positivists and the god-damned English professors—then I suppose that I am a logical positivist.
Clark Glymour, Theory and Evidence

1.1 Introduction

One of Donald Davidson's most influential articles begins:

> In sharing a language, in whatever sense this is required for communication, we share a picture of the world that must, in its large features, be true. It follows that in making manifest the large features of our language, we make manifest the large features of reality. (Davidson 1985, p. 199)

As he explicitly notes, Davidson is putting himself in the company of philosophers from Plato and Aristotle to Wittgenstein, Strawson, and Quine. He is not, however, putting himself in the company of contemporary philosophers of physics. For most of them an analysis of language plays no role in the interpretation of physics. Theories, rather than language, supply the basis for interpretation, especially for ontological issues. Interpretation is viewed as a relation between a mathematical formulation, viewed as the foundation of a theory, and the reality of which it is a theory. The principal goal of the present work is to restore an analysis of language to a central role in the interpretation of physics.

From the time of Galileo and Kepler physicists have attempted to replace this common sense ontology with a scientific world view. Though this has only been achieved in a piecemeal fashion, the limited success supplies a basis for linking basic ontological issues to fundamental theories and to anticipated unifying theories. However, the ontological assumptions implicit in ordinary language also play an essential role in supporting inferences. Physics, as reconstructed by some philosophers, can dispense with the role of language in the interpretation of theories. The physics of the physicists lacks this streamlined clarity. What theories are tested against is physical reality as reported by experimenters. Language is indispensable and, as we will see in more detail later, structures within the language of physics,

E. MacKinnon, *Interpreting Physics*, Boston Studies in the Philosophy of Science 289,
DOI 10.1007/978-94-007-2369-6_1,

play an essential role in supporting inferences. Here again, philosophers need not deny this but, as is their wont, could support the ultimate dismissal of language by distinguishing between fundamental theories and limited theories. Limited theories do not determine ontology; they presuppose it. When the question is raised of replacing the ontology implicit in ordinary language by an ontology that reflects and supports the scientific world view, then one must turn to fundamental theories. Such theories will be considered in much more detail in later chapters. A tradition in the philosophy of science has banked on anticipation of the form such ultimate theories should have. These anticipations shaped the projected path from an ordinary language starting point to the anticipated terminal. We will consider the main stream of this tradition.

1.2 Language and Logical Positivism

The logical positivists and their latter day successors, Quine and Sellars, recognized ordinary language as the starting point in any clarification of scientific epistemology. Progress, however, was seen as a matter of replacing, rather than relying on, ordinary language usage. In his autobiographical account, Rudolf Carnap summarized his basic orientation: "Chiefly because of Frege's influence, I was always deeply convinced of the superiority of a carefully constructed language and of its usefulness and even indispensability for the analysis of statements and concepts, both in philosophy and science" (Schilpp 1963, p. 28). Coffa's (1976) penetrating study of the underlying continuity in Carnap's thought brought out the peculiar status of language. In a humorous metaphor he related Carnap's work to the then popular mini-series, *Upstairs Downstairs*. Upstairs are the masters, the scientists, who speak about the world in whatever language they choose. Downstairs live the servants, whose task is to cleanse the house of knowledge and, occasionally, to give instructions to the masters. The analytic/synthetic distinction, so basic to early logical positivism, fared rather differently on these two levels. Upstairs, in scientific discourse about objects, there are analytic sentences, synthetic sentences, and many sentences that do not fit easily into either category, items that Coffa dubbed 'strange sentences'. These are sentences that have the form of empirical claims, but whose acceptance does not depend on empirical evidence. Examples are easy to multiply from the axioms of geometry to sentences like, "Physical bodies are in space". These are like Kant's synthetic a priori principles in straddling the analytic/synthetic distinction. They differ, however, in that they need not be principles of a science. They could be protocol sentences or sentences that constitute a framework for discourse: time is one-dimensional; space is three-dimensional; a thing is a complex of atoms; a thing is a complex of sense data. Friedman (1991, 2008) clarified the development of Carnap's position on theoretical terms. Within science a realistic language is normal, treating theoretical entities as real. External questions about the reference of theoretical terms are dismissed as pseudo-question. The point Friedman stresses, and on which I concur, is that Carnap was not trying to regiment or change physics. His theme was: Let physics be physics and dissolve the philosophical pseudo-questions.

A facile application of positivist principles might seem to require rejecting such strange sentences, since they are neither true by virtue of logical form nor open to empirical verification. Carnap did not reject them. His problem was to find downstairs equivalents of these upstairs sentences. This he worked out through detailed examples for sentences concerning: meaning, universals, and philosophical claims. His basic move was to reconstruct these strange sentences in the material mode as pseudo-object sentences in the formal mode. Thus, "Five is not a thing but a number" gets reconstructed as " 'Five' is not a thing-word, but a number-word."

When Carnap came to add pragmatics to his earlier accounts of the role of syntax and semantics (Carnap 1950, pp. 205–220) he used this addition to sharpen the difference between the task of the physicist and the task of the philosopher. Relying on the rather infelicitous term 'linguistic framework' Carnap distinguished between what is done in using and in analyzing a linguistic framework. One can use the familiar common sense framework of things with properties, or a linguistic framework in which one speaks of numbers, or linguistic frameworks concerned with atoms and molecules. "Let us grant to those who work in any special field of investigation the freedom to use any form of expression which seems useful to them; the work in the field will sooner or later lead to the elimination of those forms which have no useful function" (Ibid., p. 221). In reconstructing a linguistic framework Carnap used an 'internal/external' distinction as a way of dissolving ontological issues. Within the linguistic framework of numbers such statements as: "Five is a number" or "There are numbers" are analytic. Similarly for such statements as "There are things" in a common sense framework, or "There are molecules" in the linguistic framework of molecular physics. What of such questions as: "Are numbers real?" or "Are molecules real?" considered as external questions? Such questions, Carnap insisted, are misconstrued. The issue here is pragmatic, not ontological. It concerns a decision about accepting a framework, not the imposition of an ontology. Carnap insisted that the physicist should be free to introduce new entities and the terminology necessary to speak about them. The question, internal to science, of whether these postulated entities should be accepted as real is an empirical question to be answered by empirical means. The choice of appropriate means depends on the discipline. The physicist's method of deciding whether quarks are real differs from the historian's method of determining whether King Arthur was real. Carnap's goal was to answer external questions by an appropriate reconstruction of this empirical language. My goal is to understand this empirical language as it actually functions.

Within the main stream of logical positivism, concept analysis always relied on logical reconstruction. It gradually became clear, however, that such a formal analysis presupposed some sort of prior and independent clarification of an empirical language, or of language used in the material mode. Two examples of such preliminary informal analyses are Carnap's informal analysis of the meaning of 'probability' in both ordinary language and science (Carnap 1950, pp. 1–51, 161–191) and Reichenbach's analysis of the qualitative properties of time terms

(Reichenbach 1956, chap. 2).[1] However, the logical positivists never developed any account of the role of the language in use and the appropriate method of analysis.

The movement away from logical positivism had two principal directions. The historical wing, led by Thomas Kuhn, will be considered subsequently. We will focus on the reconstructive wing, led by W. V. O. Quine and Wilfrid Sellars. Each developed a strategy for beginning with the ordinary language that grounds the meanings of the terms used in discourse and advancing to scientific theories that explain physical reality in terms of its ultimate constituents. Neither man accommodated real physics. Rather they assumed that whatever the ultimate physics might be it could be recast in the language of first order predicate logic with or without identity. The anticipation that eschatological physics can be recast in the language of first order prepositional calculus no longer seems plausible. For this reason we will be more concerned with a holistic view of the scaffolding and its relation to ontology than with the unrealized goals.

Quine has an overall vision of the great sphere of knowledge, a vision first fleshed out in *Word and Object* and subsequently modified in many details, but never abandoned. We will begin with two distinctive points. The first is naturalized epistemology. Quine never tries to prove, or even question, the existence of a real world apart from a knowing subject. His purpose is to systematize, not justify, knowledge. The second is the unique form of holism, virtually unprecedented in the empirical tradition. For most empiricists who have taken the linguistic turn, some terms and sentences directly relate to reality: observation sentences, an observation language, names, rigid designators. Theories, at least reconstructed theories, are generally treated as detached entities connected to reality, or to an observation language, by some sort of correspondence rules. For Quine all knowledge is, or should be, of one piece. The great sphere of knowledge relates to reality only through the exciting of sensory nerve endings. All else is constructed. The familiar middle sized objects of ordinary experience, the gods of Homer, the particles of contemporary physics: all are posits of language. Effectively, all knowledge, in a Quinean perspective, functions the way theories do for more traditional empiricists.

Quine and Sellars reintroduced ontology in the empiricist tradition by extending Carnap's argument. Acceptance of a theory as fundamental and irreducible pragmatically entails acceptance of the entities postulated by the theory. Ontology is an inescapable aspect of general theories (Quine 1961, 1969, 1976, Davidson and Hintikka 1969, chap. 2). Quine never attempted to supply an ultimate ontology of physical reality. He assumed that such an ontology would come from the physical theory accepted as fundamental, on the pragmatic ground that there is no further court of appeal. The aspect that concerns us is the functional role ontological presuppositions play in supporting inferences.

---

[1] Salmon's (1979), survey article forcefully presents the reasons why such informal semantic analyses are needed prior to any formal reconstruction. Even Patrick Suppes (1979), dean of formal reconstructionists, cautions against excessive reliance on formalistic approaches to the interpretation of science (Ibid., pp. 16–27).

In the suppressed introduction to Word and Object he distinguished three different types of foundations (1966, p. 233). **Sense data** are evidentially fundamental; **physical particles** are naturally fundamental; **common sense** bodies are conceptually fundamental. When Quine speaks of foundations he simply means what we decide to accept as foundational in a particular way of ordering our knowledge. Bodies, in Quine's account, are not given by experience. They are posits of individual knowers and of language sharing communities. The ontology implicit in ordinary language is one of bodies with characteristic properties and activities sharing a spacetime framework. Quine's appraisal of such common sense realism is clear and consistent. The origins of this ontology are shrouded in prehistory (1960, p. 22). It is a remnant of muddy savagery (Davidson and Hintikka 1969, p. 133). It is vague and untidy, countenancing both too many objects and too much vagueness in scope It would be desirable to dispense with such ontological foundations in favor of the ontology specified by the precise referential apparatus of logic. Yet, Quine reluctantly admits, a commitment to bodies remains indispensable to shared discourse. "Bodies are the charter members of our ontology, let the subsequent elections and expulsions proceed as they may" (1973, p. 85). Physical objects are indispensable as the public common denominators of private sense experiences.

There is, for Quine, no way of avoiding the ontology of ordinary language when one is considering induction. To see why one must distinguish between the doctrine and the practice of induction. Quine despairs of an adequate solution to the doctrinal problems: "The Humean predicament is the human predicament" (Davidson and Hintikka 1969, p. 72). The practice of induction is based on the classification of objects into natural kinds. This is a notion foreign to Quine's preferred foundations. Membership in a set is a question of extensionality, not similarity. The notion of natural kinds based on similarity carries in its wake an epistemologically disreputable lot: essences, dispositions, subjunctive conditionals, singular causal statements. Yet, the urge to reform through elimination encounters something of a paradox. As long as the notion of natural kinds remains disreputable it is also indispensable; when it achieves respectability it also becomes dispensable. The notion of natural kinds is indispensable because the ordinary practice of inductive reasoning is based on a classification of objects into natural kinds. The notion becomes less entrenched when class membership is determined more by scientific criteria than by similarity. The notion of natural kinds should vanish when a fully explanatory theory is achieved.

The projected disappearance has not occurred, even in biology. Though most other aspects of biology have been transformed by the Darwinian revolution, pre-Darwinian taxonomies and the principles underlying their construction have remained largely unaltered. Evolution, rather than leading to a new systematics, has so far merely been granted the role of supplying an explanatory basis for the order already manifest in the existing taxonomies (de Queiroz 1988). Nor has it occurred in the discourse of physics. One speaks of how electrons, mercury atoms, or W^+ particles behave on the assumption that the entities sharing the same label are alike in basic respects. Our concern is not the dubious ontology of natural-kind terms, but the role that categorization and classification retain as basic inference-supporting structures. Wilfrid Sellars' overall strategy was one of anticipating the replacement of an ordinary language framework, the manifest image, by a radically different framework, the scientific image. The program never really worked. Sellars developed it in fragments tied together by strategy arguments, promissory notes, and the endless numbering of paragraphs and sample sentences. A major obstacle was Sellars' unyielding intellectual integrity. Any program of reductive materialism has difficulty explaining away persons, intentionality, intensions, and secondary qualities. Quine argued that if these cannot fit into the ultimate perspective then they are not important and may be neglected. Sellars took the opposite tack. The issues that do not fit the scientific image are the ones requiring the most detailed analysis. What we wish to appropriate is Sellars' clarification of the role of inference-supporting structures in ordinary language, or the manifest framework.

Most philosophers working within an ordinary language framework find it almost impossible to treat this as a conceptual system. If there is no alternative, and no way of getting out of it, then all we can do is explore an ordinary language framework from within. Sellars believed it could be replaced by a framework based on the theoretical entities postulated by the ultimate community of inquirers. Sellars was the only one who explored the manifest framework justifying the direct realism of the lived world and its extensions, and also stepped back and treated this framework as a conceptual representation.

When Sellars distinguishes the manifest and scientific images, his use of 'image' is not intended to signal a denial of the reality involved. As he explains it, it is more like Husserl's bracketing in transforming these images from ways of experiencing the world into objects of philosophical reflection and evaluation.[2] The manifest image is the framework in which man came to be aware of himself as man in the world. If we ignore the sexist terminology, then the manifest image is essentially common sense realism. It can support both philosophy and science. Sellars uses the blanket term, 'philosophia perennis' to include any philosophy that accepts the manifest image as real and seeks to refine its categorial framework. Aristotelianism, Marxism, and pragmatism qualify. What this framework does not support is replacement of its basic entities by postulated unobservable entities, the characteristic note of the scientific framework. The two frameworks should be thought of as complementary or parts of a stereoscopic vision.

Sellars conducted a campaign against the myth of the given, the idea that empirical knowledge rests on a foundation of non-inferential knowledge of matters of fact. This confuses the non-inferential experiencing of a sense-datum, e.g., hearing a $C^\sharp$ sound, and knowing in a non-inferential way that a sound is a $C^\sharp$. The latter requires extensive knowledge of the language used, musical notation, a recognition of circumstances as normal, and much more. Wittgenstein and Ryle led the movement showing that the meaningfulness of language is essentially public. Sellars accepted this and showed how one could graft on this publicly meaningful language a new dimension in which one could speak of private mental acts, states, and intentions.

---

[2] Sellars developed this in (1963b, pp. 1–40).

His clearest presentation of this development involved a myth, in the Platonic sense of a likely story. Sellars (1963a) presents an account (of how) a primitive semantic genius, Jones, develops a theory of mental states using models based on speech and on public objects. Jones postulates concepts as something entitatively different from the term that expresses them. But, the term supplies a model for the concept, considered as an inner word. Similarly he postulates judgments as inner mental acts, but models discourse about judgments on statements, or affirmed propositions. Thus within the Ryleian framework concepts and judgments are introduced as theoretical entities. However, a point basic to Sellars' realism, having good reasons for accepting a theory entails having good, essentially the same, reasons for accepting the entities postulated by the theory. Within the manifest framework Sellars is defending a direct realism. We directly perceive objects with shapes, sizes, colors, textures, and sometimes tastes and aromas. Sellars set this direct realism in a larger framework that effectively denies this realism any ultimate significance. Where Berkeley claimed that physical objects do not really exist as objects, but only as ideas, Sellars, who adapted Berkeley's strategy (Atkinson 1984), claims that the objects of the manifest framework do not really exist as objects in the world. There are no such things.

Again, comparison of Sellars with Husserl is enlightening. For Husserl the lived world is a world polarized into subject and objects and the basic objects are the familiar ones we perceive, use, interact with and discuss. It is not until we bracket the reality of this world that we realize that objects are constituted as objects only in human consciousness. Sellars adds that we constitute objects as objects by the imposition of a shared inherited categorial framework in which 'object' serves as an ultimate category. Since the meaningfulness of this categorial system is essentially public, we are situated in the logical space of a community using and shaped by a shared language. In the order of being concepts existing within individual minds are basic. In the order of knowing, or in semantic analysis, the language-sharing community and the public meaningfulness of language are basic. Sellars never related his anthropological fiction to cultural evolution. I will indicate such a relation in beginning the historical account.

This is the basic point I wish to adapt from Sellars, the difference in perspective between using the manifest image and it systematic extensions to represent the world we discuss, and studying this image as a representation while bracketing the question of the reality of the world represented. Sellars' position on material rules of inference is the point of immediate concern. Material rules of inference are illustrated by conclusions drawn on the basis of real world connections. "There is smoke; so there must be fire". "Keep clear. That dog is a pit bull". "The screen disappeared from the monitor because you hit the Escape key". Here one can always isolate any individual inference, interpret it as an enthymeme with a suppressed major like "All pit bulls are dangerous", and rely exclusively on formal, or content-independent, rules of inference. However, this formal perspective entails giving up on any analysis of scientific discourse within a language-sharing community. As Sellars saw it, from an internal perspective material rules of inference embody our consciousness of the lawfulness of things. The inferences are made on the basis of the way things in the world act and interact. Fire causes smoke. Pit bulls sometimes attack without provocation. One becomes computer proficient by learning how computers work. Such inferences structure every aspect of our normal lives. When material rules of inference are treated externally then one enters a world of Byzantine complexity.[3] A key point is that descriptive concepts imply laws. The laws are reflected by material inferences, In an external perspective laws of nature (or more properly, first order laws of nature) are interpreted as material rules of inference. In an internal framework these material inferences are rarely recognized as inferences.

1.3 Language and Contemporary Philosophy of Physics

Kuhn's *Structure* provided the first methodology for according historical studies a philosophical significance. The key point was the choice of a unit of explanation larger than individual theories but smaller than the language of science. Paradigm replacement was accorded the status of a conceptual revolution, because paradigms were regarded as the units within which scientific terms have definite meanings. Paradigm replacement leads to incommensurability. Subsequently many aspects were modified. Critics gave examples of conceptual continuity underlying paradigm replacements. Kuhn's later writings gradually blunted the sharp edge that Feyerabend and the earlier Kuhn had accorded incommensurability. Yet, replaceable paradigms remained the basic conceptual units. The alternative schemata modified the units of explanation and the process of replacement. Research programs need replacement when they become degenerate (Lakatos 1978). Problem-solving methodologies require replacement when they no longer provide a basis for solving pressing problems (Laudan 1977). There was less stress on problems of meaning. Yet, none of these presented the language of physics as the basic framework for treating the related problems of conceptual change and underlying continuity. I will not argue the point here. The only effective answer is to show in historical detail how the language of physics plays this role.

Many philosophers of physics regarded paradigms, research programs, and problem-solving methodologies as fuzzy, poorly defined units incapable of supporting an inference supporting system. They came to focus on theories. As Van Fraassen explained:

> Philosophy of science has focused on theories as the main product of science, more or less in the way philosophy of art has focused on works of art, and philosophy of mathematics on arithmetic, abstract algebra, set theory, and so forth. Scientific activity is understood as productive, and its success is measured by the success of the theories produced. (Van Fraassen 1991, p. 1).

---

[3] Sellars basic treatments of material rules of inference are in his "Some Reflections on Language Games", Sellars (1963c, pp. 321–358, 1958, 1962); "Counterfactuals, Dispositions, and the Causal Modalities"; and "Time and the World Order" which treats the relation between 'thing frameworks' and 'event frameworks' and argues the conceptual priority of the former. Brown (2006, chap. 5) presents a systematic revision of the types of inferences treated.

Contemporary philosophy of physics may be negatively characterized by the absence of any consideration of the role of language analysis. A negative reason for this is the difficulties some philosophers encountered in relating linguistic analysis to theories. Mary Hesse's network model handles the introduction of new quantitative concepts by analyzing the compromises and accommodations involved in fitting a concept into a network of concepts while simultaneously adjusting it to precise mathematical expressions. Individual concepts can be accommodated. Theories are more difficult. Hesse concludes that statements ascribing a real existence to theoretical entities be accorded a probability of 0.[4] Peter Achinstein (1963, 1968) started with ordinary language examples, such as a doctor explaining stomach problems, brought out distinctive aspects of 'explains' such as its illocutionary force and the role of emphasis and then extended this to science. While this analysis could handle expressions of the form, "A explains X to B", it had grave difficulties accommodating expressions of the form, "The Bohr theory of the atom explains the mathematical form of spectral laws". I will be relying on a modified version of Hesse's network model later. Now we simply consider the methodological point. When functioning, rather than reconstructed, physics is made an object of study and treated as a system, then it should have a unified system of inferences (Kuhn 1962). If this is the informal method of linguistic analysis, then this, rather than deductively unified theories, supplies the basis for interpretation. The problem then is accommodating the role of theories.

No one denies informal analysis a preliminary role in analyzing experiments, setting up theories, or reporting historical developments. However, when the issue is one of interpreting the ontological significance of physics, then one seems to be confronted by a dilemma. Either take language as the ultimate unit of explanation with informal methods as fundamental and theories relegated to the role of subsidiary inference mechanisms, or take fundamental theories as basic and disregard the role of language. I can only see one way through the horns of this dilemma. That is to turn to history, to trace the way the language used in physics evolved to support the quantification of properties, mathematical inferences, and the development of theories.

1.4 Methods of Analysis

In Glymour's dichotomy it might seem that the methods of the god-damned English professors could be applied to the discourse of physics. It is a spoken language and must meet necessary conditions for meaningful communication and unambiguous reference. Yet, almost no one applies such methods to physical discourse. One reason for this is the obvious split between analysis and the philosophy of science. In

[4] See Hesse (1974, 1980). A more detailed summary of her position is given in my (1979, pp. 504–510). The somewhat negative appraisal given there will be revised in the course of the present work.

the background, there are some significant differences between 'ordinary language' and the idiolect of physics.

Analysis centers on a detailed examination of the accepted usage of problematic words in different contexts. The norm is the fluent native speaker or, in practice, what I and my friends would say. This is highly dependent on the particular language involved. The idiolect of physics is essentially the same in all cultures in which modern physics is taught and practiced. There are notorious difficulties in translating literature, poetry, even daily news, as well as such disciplines as sociology and theology into languages with different cultural traditions. There is little difficulty in translating physics from English to German, Russian, Arabic, or Japanese. The computer-assisted translation programs produce howlers epitomized by the English⇒Russian⇒English translation of "The spirit is willing, but the flesh is weak" into "The vodka is OK, but the meat is rotten". Such programs work quite well in translating physics. The reason for the difference is clear. The language of physical discourse has evolved into a highly structured idiolect. If an emerging nation wishes to teach and practice physics, it must assimilate the language of physical discourse. This is not merely a question of introducing new terms. It involves modifying the spoken language so that it incorporates the established structures basic to physical discourse. Similarly, the student becomes a practicing physicist only by assimilating a highly structured way of speaking and representing reality.

The most distinctive feature of the language of physics is the coupling of physical accounts to mathematical formulations. Realtors, accountants, bankers, and traders also have specialized extensions of language in which numbers play a prominent role. They do not, however, exhibit a similar dependence on deductive, or formalizable systems, except for simple arithmetic. Even when physicists are not explicitly using mathematical formulations they cannot use such terms as 'mass', 'energy', 'charge', 'entropy', and many other less familiar terms, in a critically acceptable way without a working knowledge of the role these terms play in laws and theories. This supplies a rigid, highly structured constraint on usage. Terms in ordinary language usage are also constrained, but not by such formal systems.

It is hardly necessary to argue that the standard methods of analysis are not easily adaptable to the discourse of physics, for no one believes they are. The more difficult problem is to find enough common ground to adapt any methods. We begin with three claims that are so basic that no one really disputes them. First, the language of physical discourse is a spoken language. Logic, formal systems, and mathematics are not spoken languages, i.e., they are not really languages at all. If there are certain basic conditions that must be met for any spoken language to serve as a vehicle for unambiguous reference and communication, then physical discourse must meet those conditions. Second, the language of physical discourse is Indo-European in it structure. This is not to deny that the Greeks borrowed from the Egyptians and Babylonians or that people speaking non Indo-European languages have made many significant contributions to physics. Our initial concern is with basic structures in the discourse of physics. These stemmed from Greece and Europe and were expressed in inflectional languages. The contributions of others were embedded in these structures. In non Indo-European languages, the language of physics was introduced as a linguistic parasite that exploits and adapts structures in the host to meet its own needs. Hence, our initial concern with ancient Greece, medieval Europe and the archeology of scientific language. Third, the language of physical discourse is extended ordinary language (or EOL). The language of physical discourse and standard Indo-European languages share a common core. The problem is one of clarifying what this is and how it developed.

1.4.1 Semantics and Ontology

Ordinary language analysis involves an interpretative perspective that differs significantly from that proper to the interpretation of theories. The difference of immediate concern is the relation between semantics and ontology. Theory analysis is foundational in orientation. In a formal system the truth status of conclusions depends on the truth status of axioms. Ordinary language analysis involves a surface level and a depth level, or various depth levels. One might try to parallel the theory type of analysis by arguing from claims to implicit presuppositions and on to a correspondence between basic categories that are implicit in claims accepted as true and the ontology this presupposes. The result is common-sense realism as a philosophical position. Ordinary language usage, however, does not lend much support to this pattern of inference. Consider a simplistic ordinary language analysis.

The shirt I am now wearing is yellow. (S1)

This might seem to support a translation into more ontological terms.

This shirt has the property of being yellow (S2)

The ontology may be given an Aristotelian formulation

Color is a property of extended material objects. (S3)

Within the broadly Aristotelian framework that characterized medieval philosophy (S3) was simply accepted as true. The objectivity of secondary qualities has been debated since Galileo and Locke made it an issue. I am not concerned with defending a position, but with the role of presuppositions. It is consistent, and in fact now common, to accept S1 as true (or false, depending on the shirt's color), while rejecting S3 as an ontological claim. S2 has an intermediate status. If it is simply intended as a paraphrase of S1 then it has the same truth status as S1. If it is taken as an abbreviated formula for a position on color realism, e.g., the claim that something in reality correlates with the experienced similarity structure of colors, then its acceptance reasonably depends on the arguments supporting it.[5] The key point here is that presuppositions in language do not function semantically the way axioms do in theories. Ordinary language usage is, at least in principle, anti-foundational. Truth claims are a function of what Wittgenstein called 'surface grammar'. Accepting claims as true need not entail accepting as true an explicit formulation of its presuppositions and, a fortiori, an ontological interpretation of the presuppositions. The relation is more complex and more flexible.

To parallel the shirt example we will consider a significant example of the discourse between experimenters and theoreticians. James Franck and Gustav Hertz began a series of experiments in 1913 to measure ionization potentials of different gases. Others, performing similar experiments, had obtained results different from those of Franck and Hertz (1967). Franck and Hertz were not merely trying to get more accurate results than others. They were using experiments as a way of implementing the new quantum theory of Planck and Einstein. J. S. Townsend, who had pioneered in the experimental investigation of ionizing gases through collision processes, interpreted his results in accord with the classical principle of energy conservation. An electron passing through a tenuous gas experiences a small energy change with each inelastic collision. Franck and Hertz accepted the new quantum assumption that energy transfer is an all or nothing affair. In an inelastic collision an electron ionizes a gas molecule through the transfer of a discrete quantum of energy. If the collision is inelastic then, according to their interpretation of the quantum postulate, there should be no energy transfer. Franck and Hertz, accordingly, devised an apparatus that could measure the onset of ionization. They also tried to establish the validity of their results by setting up the apparatus in a way that excluded the chief sources of experimental error and allowed more accurate results than any previously obtained. Their apparatus was immersed in a paraffin bath that was continuously heated. Within the apparatus they had a partial vacuum. A drop of mercury in the bottom became partially vaporized and supplied a tenuous gas of mercury atoms. At the center of the system was a platinum wire, D, that

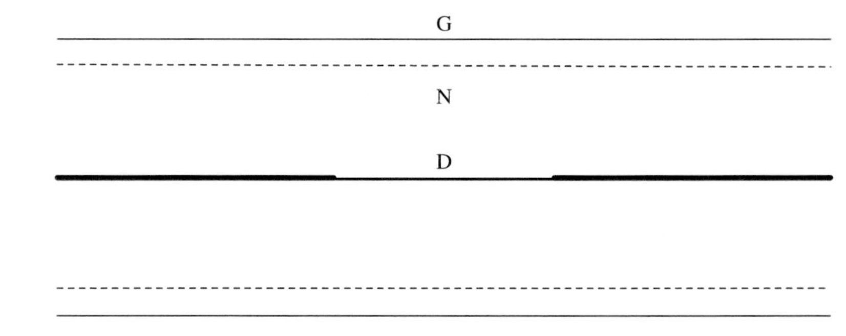

Fig. 1.1 The Franck-Hertz Experiment

[5] See Norris (2002) for discussions of the objectivity of conceptually structured properties and Giere (2006, chap. 2) for a perspectival reinterpretation of the problem.

can be brought to incandescence by a current. The electrons emitted can collide with mercury atoms and lose energy. The outer part of the cylindrical apparatus was a platinum foil,G, that could record electrons. A fine cylindrical wire mesh, N, 4 cm. form the inside the platinum foil supplied a retarding potential. As long as the accelerating potential between N and G is less than the retarding potential at N, no electrons are recorded. As the potential is increased there is a rise in current and then a sudden drop. They interpreted this drop as the onset of ionization. A further increase in the ionization potential produces an increase in current and then another drop. Franck and Hertz attributed this drop to the energy loss electrons experienced when they ionized mercury atoms and used their plots of ionization energy to determine the energy of ionization. Subsequently Bohr reinterpreted these drops as energy loss consequent upon raising the mercury's outside electron to a higher level. Subsequent developments, epitomized by 'wave-particle duality' called into question the assumption that electrons are simply particles traveling in trajectories. To parallel the previous example consider the three claims:

Electrons emitted from the heated wire collided with mercury atoms. (F1)

This strongly supports the contention

Electrons travel in trajectories. (F2)

This can be given an ontological interpretation:

Sharp spatio-temporal localization is a property of electrons. (F3)

Statement F1 had to be accepted as true to interpret the experimental results. Statement F3 is not now accepted as true, when considered as an ontological claim. The well-known problems associated with wave-particle duality, superposition of states, and the dependence of observed results on the questions posed by the measurement apparatus militate against that. Statement F2 has an in-between status, depending on how it is interpreted. Interpreted literally, it cannot be accepted as a descriptive account of what happens objectively. This might suggest simply considering F2 to be false. However, in interpreting the experiments it is treated as true. This is not simply an experimental simplification of a theoretical complexity. As Cartwright (1983, p. 172) has argued, similar claims are operative in the design and execution of particle accelerators. They are also operative in the normal functioning of particle detectors.

Normal scientific discourse, like any form of normal language, presupposes the acceptance of a vast collection of claims as true. It need not presuppose that ontological interpretations of the presuppositions of discourse are true. Yet, an ontological interpretation of such presuppositions is implicit in normal reasoning. The acceptance of statements like (S3) and (F3) as true greatly simplifies discourse. To clarify this oscillation between semantics and ontology and the functional role of ontological presuppositions it might help to consider a less problematic example, one where analysis had played a significant role.

Consider the discussion in the last quarter century on the issue of rights. Do women have the same rights as men? Do fetuses have rights, such as the right to life? Should Gays and Lesbians be accorded full rights to housing, employment, and public expression of their politics and preferences? Does a carrier of AIDS of HIV have a right to normal social relations, or does the community have the right to restrict the carrier's freedom of action? What of merely potential human beings? Do our practices of contaminating the water and polluting the air violate the rights of future generations? Do animals have rights? Do the children of illegal aliens have a right to public education?

To get at the presuppositions underlying these arguments we can make an initial distinction between conceptual and ontological modes of argumentation. The conceptual issues hinge on word-word connections. 'Right' has complex relations to other concepts. A right may be exercised, enjoyed, given, claimed, demanded, asserted, insisted upon, secured, waived, or surrendered. A right may be related to or contrasted with a duty, an obligation, a privilege, a power, or a liability. The interconnection of these concepts is not extensional or definitional. One gets at these connections by analyzing the way members of this family of terms are used in different contexts.

The process of argumentation is considerably simplified when one switches from a pure conceptual analysis to a more ontological mode based on word-world connections. Let's begin in a simplistic fashion by assuming that only beings who are capable of exercising, claiming, or surrendering a right, and are appropriate subjects for the predication of duties, obligations, and claims, are the possessors of rights. This approach rather obviously favors the status quo, which ordinary language usage reflects. Thus, animals, embryos, and future generations cannot assert, claim, defend, or surrender a right to life. Those interested in changing the status quo tend to look for a different basis for rights. We will briefly consider two such bases.[6]

The first is the traditional one. Only a person can be said to have rights in the full sense just as only a person can be said to have obligations, duties, etc. To settle the issue of whether a being has rights we must first determine whether or not the being in question counts as a person. Opponents may agree on this while disagreeing on the grounds for personhood. The hypothesis favored by many right-to-life advocates is that an ovum becomes a person the moment that it is fertilized, because that is the moment when the soul is infused. Some implications of this acceptance are immediately clear. A fetus is the subject of a right to life; an animal is not. Those who reject this dualism, but still accept being a person as the ground of rights, look for a different person-making factor, such as rationality or acceptance into a community of persons. This is less ontological. Yet it is still an ontological argument if one is claiming that possessing a certain characteristic or family of characteristics

[6] This is a summary of material treated in White (1984, chap. 6).

is either a necessary or a sufficient condition for the presence of the capability of possessing rights, or of being the kind of being that can have the right in question.

The second approach is to consider some other property besides personhood as the basis for rights. Rationality, for example, is often claimed as the basis for a strict right to life or for genuinely human rights. Again, one way of testing this hypothesis is by judging the acceptability of the conclusions it suggests. Does it imply that imbeciles, the irremediably comatose, and the truly senile are not the bearers of rights? This criterion seems too strict. It excludes those who, most feel, should be included. Those who want to change current laws and practices generally advocate other criteria, such as the capacity for sensation and/or experiencing pain. This seems too loose. It does not accord with the material rules of inference linking 'right' to 'exercise', 'enjoy', 'give', 'surrender', 'claim', 'demand', 'assert', 'duty', 'obligation', 'privilege', 'power', and further related terms.

We have not gone far enough to settle any interesting ethical issues. Nevertheless, we have gone far enough to illustrate the conceptual simplification induced by a reliance on material rules of inference. Consider the on-going dialog just sketched from two perspectives. The first is that of a participant. A participant presupposes a lived world and a categorial organization of this world into such different types of entities as plants, animals, humans, men and women, embryos and adults, straights and gays, with such different properties as life, consciousness, sentience, rationality, and variant sexual preferences. These are not treated as presuppositions of argumentation, but as factual truths about the world. Opponents may challenge some of the factual claims, but still argue from within the same general framework. It is only against this background that one can debate disputed facts and relations.

Now consider the same debates from the perspective of a detached critic. The world as it exists objectively does not make inferences. We do. Our inferences are based on the world as represented. We must have a shared language-based representation of a world in which we function as moral agents to make inferences and dialog possible. The ontology that supports our processes of making and criticizing inferences is the ontology implicit and functioning in a shared representations of reality. Any attempt to make explicit the representation of reality implicit in this ethical discourse would hopelessly complicate the ongoing dialog. Normal dialog is possible only when we isolate issues, which require arguments and invite criticism, from background facts, which are accepted as determined by the world rather than the debaters. Critical reflection on such dialog suggests that the background facts are presuppositions that are accepted as non-controversial.

A somewhat similar appraisal relates to our present concerns. Through a process that *Pickering* dubbed 'retroactive realism' hypotheses that become accepted soon acquire the status of facts. For a simple example consider the historical sequence of hypotheses that were eventually accepted as factual. The atom has a small nucleus. The nucleus contains electrons. The nucleus consists exclusively of protons and neutrons. Nucleons consist of quarks and gluons. This retroactive realism greatly simplifies routine inferences for the reasons that Shapere (1983) has clarified. In a scientific inquiry, one may call into question any particular hypotheses or presumed fact. However, any attempt to call all background presuppositions into question paralyzes the process of inquiry. The normal practice is to presume accepted facts, unless there is a specific reason for doubt. Thus, Franck and Hertz assumed that a heated wire emits electrons, that mercury atoms constitute a gas, that there are collisions between the electrons and some mercury atoms and that these collisions are of two kinds. In an elastic collision the energy of the electron is not changed. In an inelastic collision, the electron loses energy to the mercury atom. All of these facts had the status of hypotheses a generation before their experiment.

The transition from tentative hypothesis to fact greatly simplifies and streamlines the process of informal inference. This simplification, however, obscures an analysis of the role of informal inference. A further complication comes from the reversal of this process. Factual presuppositions that supply a platform for inferences may eventually be rejected as false. A simple example is the presumption that nuclei must contain electrons since they emit electrons. A more complex assumption is that the Aristotelian categorical system: substance, quantity, quality, relation, etc., supplies a factual basis for understanding the way objects exist objectively. This supplied an inference-supporting structure in the medieval developments leading to the scientific revolution. Another implicit inference supporting structure is the metaphorical extension of language.

The fundamental categorial system of the language used plays a role similar, in some respects, to the foundations of a theory. As a bit of background information we will first consider what some linguists, psychologists, and philosophers have said about the role of categorial frameworks. This does not supply a foundation for the historical analysis. It may, however, help to situate the analysis in a more critical historical perspective. Categorization is a basic feature of all natural languages (Rosch 1999). It provides maximum information with the least cognitive effort. By coding an object or event as a member of a class, one has a basis for drawing further conclusions concerning normal properties and activities. For such inferences to have a reasonable degree of accuracy the categorial structure must map significant features of the perceived world. In normal usage, including language learning, the categorial system is anchored in basic objects: chair, table, man, bird, cow. These are typically learned by prototypes, rather than definitions or necessary and sufficient conditions. This basis leads to fuzzy concepts with overlapping border. Ordinary usage handles this by various hedges (Lakoff and Johnson 1980). While standard objects anchor a categorial system, subordinate levels emphasize differentiation, a kitchen chair as opposed to a living room chair. Superordinate levels play a distinctive role in classifications of events. By 'event' we refer to activities that are routinely categorized as distinct units. Thus the students Rosch tested listed their normal daily activities through items such as: getting dressed (putting clothes on), having breakfast (eating food), traveling to the university (using a vehicle), and attending classes. Here, superordinate terms supply the normal means of categorizing activities. Such higher order activities as the classification of objects in terms of distinctive properties presupposes the establishment of the categorial system.

We will consider the way categorial systems structure and support different types of inferences. Jackendoff, more than anyone else, has taken up the study of the relation between linguistics and conceptual structures. In common with the M. I. T.

School, he accepts the thesis that basic syntactical structures are innate. Acceptance of this hypothesis is not essential to acceptance of his further claims. His basic contention is: "There is a single level of mental representation, conceptual structure, at which linguistic, sensory, and motor information are compatible" (Jackendoff 1983, p. 17). The core of this conceptual structure is an implicit ontology. Linguistics, he insists, is a much better guide to this core ontology than logic. To make sense out of the way reference functions in ordinary language one needs more categories beyond the constants, predicates, and variables of logic. Chomsky's X-Bar theory of grammatical categories leads to the conclusion that all major lexical categories (noun, verb, adjective, adverb, and proposition) admit of essentially the same range of types and modifications. Accordingly, Jackendoff can use syntactical generalizations among categories as evidence for parallel semantic generalizations. This leads to the conclusion that the basic ontological categories implicit in our conceptual structure are: things, properties, events, places, and paths. Events, places, and paths are basic categories in the context of human action in a lived world, not in a Quinean classification of the furniture of the world. Dubbing these 'ontological' effectively puts them in a theoretical context that is not a feature of ordinary language.

Ordinary language is also richer in inference supporting structures, such as ways of embedding one conceptual constituent within another, and principles of combination. A fundamental inference-supporting mechanism is categorization. In standard logic "a is a D (or Da)" is generally explained extensionally; a is included in the extension of D. Ordinary language categorization is not limited to this extensional mode (See Lakoff and Johnson 1980, Part I). Actions and events, all the major ontological categories, as well as things, admit of categorization. Jackendoff proposes handling this through tokens (for individuals) and types (for categories), with metalinguistic rules 'is an instance of' leading from types to tokens and 'is exemplified by' leading from tokens to types. Regardless of whether one accepts's metalinguistic rules, the basic underlying point is one that others have pointed out and attempted to systematize. We derive our ordinary language universals from individual instances of them by fuzzy predicates (Zadeh 1965) or 'resemblance class predicates' (Körner 1970), not through the formulation of necessary and sufficient conditions. Since types and tokens are characterized by the same semantic markers, inferences from one to the other are almost automatic, and often highly uncritical. Induction may be the despair of philosophers; it is a facile practice of the normal language user.

We will not rely on the details of Jackendoff's account of categorization. It is cited here to make a point that is not dependent on the acceptance of a particular linguistic theory. Ordinary language categorization is richer, more complex, and less organized than critically developed systematic accounts. It supplies the normal framework for ordering reality and making inferences. The point of departure for the distinctive language of physics is a more systematic categorization that supplies a more transparent vehicle for inferences. An initial problem, accordingly, is to understand how this emerged from an implicit ordinary language categorization. This brings up the issue of gradual changes in language, or in specialized extensions of language.

We have been speaking language as if it were something fixed. Language obviously changes. Neologisms, such as 'dotcom', 'email', 'snail-mail', 'hippie', and the new meanings attached to such older terms as 'gay', 'chauvinist', 'bug', 'virus', and 'rock' attest to growth and change. They do not, however, clarify the issue of how deeply these changes penetrate. Does the conceptual core of language also change? The analytic tradition, unfortunately, offers little guidance on such questions. The hermeneutic wing of the phenomenological movement is more helpful in treating texts, traditions, and externally induced changes. For these reasons we will freely mix analytic and phenomenological methods and slight the traditional opposition.

Our immediate concern, however, is not with such philosophical analyses, but with the extension of language to new domains. This often involves the invention of new terms, 'isotope', 'neutrino', 'ferromagnetism', etc. The more interesting issue concerns the extended or novel use of familiar terms. A brief consideration of some standard ways of doing this leads to the philosophical issue with which our survey commences. Figures of speech, especially metaphor, synecdoche, and metonymy, supply familiar means of extending standard usages. The more problematic issue concerns the implicit distinction between normal usage, presumably factual, and extended usage, presumably metaphorical or non-factual. Consider the statement that is used in routine traffic reports.

A jack-knifed big-rig is blocking the express lane. (M1)

Every descriptive and referential term in (M1) began its automotive career as a metaphor, extending a familiar term to a new domain. Yet (M1) is not a metaphorical expression. It is a factual report.

The role of such tropes in extending language to new domains is best illustrated by the most rapid technological transformation in human history, computerization. One can easily find synecdoches, such as 'chips' for 'computers', metonymy, as in 'the silicon industry', and some self-conscious use of metaphorical terms, such as 'bug', 'virus', 'mouse', 'menu' and 'piracy'. Our concern is with extended usages of established terms that have become so accepted that they routinely function in factual reports, rather than metaphorical descriptions. A partial list of such terms, omitting single quotes, would include: program, statement, declaration, drive, method, hardware, software, access, icon, operator, browse, button, platform, shell, file, folder, desktop, docking, domain, download, upload, library, filter, install, tool, diagnostics, password, view, profile, class, extension, record, buffer, pointer, tree, queue, code, recycle, object, field, property, inheritance, encapsulate, polymorphic, compile, interpret, loop, string, bit, mode, port. There are many linguistic and philosophical accounts of metaphors.[7] The focus of their concern is with non-literal

[7] I have been chiefly influenced by the linguistic accounts of Lakoff and Johnson (1980), Pinker (2007) and the philosophical account of Davidson (1985). A survey of other accounts may be found in Engstrom (1951).

meaning (common) or standard meaning and non-standard usage (Davidson) of metaphors. The terms just listed no longer function as metaphors in computerese. They are used to refer, describe, and give factual reports.

Other extensions are less obvious, because more shrouded in history. Their clarification requires conceptual archeology. Such considerations raise a philosophical problem first articulated by Nietzsche. We tend to distinguish a fundamental literal level of language, and a non-literal or metaphorical level, and assume that that the relation between language and reality explains the fundamental level, while extended levels are parasitic on the fundamental level. All is interpretation, Nietzsche insisted. The level we consider fundamental is an accidental byproduct of our temporal location. When the Greeks of the classical period came to recognize the role of language they too sought for a level of language that cuts nature at the joints. The peculiar historical outcome of this effort shaped the core language of physics.

The ideal way to present this development is to begin with the ordinary pre-philosophical language of ancient Greece and trace the linguistic extensions and developed structures that supplied the potential for the emergence of science. I doubt if such an analysis is possible. I am certain that I cannot provide it. A plausible substitute for a point of departure relies on two assumptions. First, the ordinary language from which the language of physics gradually emerged must have the basis features that any language needs to function as a vehicle of interpersonal communication. Second, there were higher order conceptual structures, such as totemism and mythology, that supplied frameworks for inferences.

The common sense view of reality, reflected in and transmitted through ordinary language, can be systematized as: descriptive metaphysics (Strawson 1959, 1992), the manifest image (Sellars), the natural standpoint (Husserl). We are not interested in the systematizations, but in the common core they share, a core that was presumably present and operative in early Greek culture. At the root of ordinary language is a subject/object distinction. The world is a collection of spatio-temporal objects with properties interconnected by various types of causal relations. Among objects, I, the speaker, have a unique status. My bodily presence anchors the space-time framework and makes unambiguous reference possible. I also ascribe to myself properties, such as life, consciousness, sensation, and thought. Yet, the terms I use to refer to myself as the conscious subject of experiences are terms I learned through the use others made of them. Any self-ascription of mentalistic predicates is logically incoherent unless it also entails other-ascription. I, the speaker, function in a physical and social world. Davidson's triangulation indicates the basis of their interrelation.

Davidson's gradual abandonment of an extensional theory of 'true' led to a critical rethinking of the interrelation of truth, language, interpretation, and ontology (Davidson 1986, 1990). I will summarize the overview presented in the concluding Essay of his latest book (2001, Essay 14). Philosophers have been traditionally concerned with three different types of knowledge: of my own mind; of the world; and of other minds. The varied attempts to reduce some of these forms to the one taken as basic have all proved abortive. Davidson's method of interrelating them hinges on his notion of radical interpretation. My attempt to interpret the speech of another person relies on the functional assumption that she has a basic coherence in her intentions, beliefs, and utterances. Interpreting her speech on the most basic level involves assuming that she holds an utterance true and intends to be understood. The source of the concept of truth is interpersonal communication. Without a shared language there is no way to distinguish what is the case from what is thought to be the case. I also assume that by and large she responds to the same features of the world that I do. Without this sharing in common stimuli thought and speech have no real content. The three different types of knowledge are related by triangulation. I can draw a baseline between my mind and another mind only if we can both line up the same aspects of reality. Knowledge of other minds and knowledge of the world are mutually dependent. Davidson categorizes as the third dogma of empiricism the distinction between schema and content, a distinction manifested in attempts to speak of some content which is independent of and prior to the linguistic ordering we impose on it. For the second we rely on a citation: "Communication, and the knowledge of other minds that it presupposes, is the basis of our concept of objectivity, our recognition of a distinction between false and true beliefs" (Ibid., p. 217).

Our ordinary language picture of reality is not a theory. It is a shared vehicle of communication involving a representation of ourselves as agents in the world and members of a community of agents, and of tools and terms for identifying objects, events, paths, and properties. Extensions and applications may be erroneous. There can be factual mistakes, false beliefs, incorrect usages, invalid inferences, and various inconsistencies. But, the designation of some practice as anomalous is only meaningful against a background of established practices that set the norms. Our description of reality and reality as described are interrelated, not in a vicious circle, but in a developing spiral. Perduring objects with properties, relations, and locations supply the basic furniture of the lived world and the properties that count as basic relate to our activity in the world. The conception of causality that emerges is quite opposed to the Hume-Mill account of causality in terms of regular succession. Regularity is one criterion used in assessing causality. However, the notion of causality is related to a wide range of concepts of things, qualities, actions, interactions, and intervention in the regular course of nature. The further we get from human agency and human intervention in the regular course of nature the more diminished this notion of causality becomes. Our ordinary language is geared to life and action in the world, not to a detached contemplation of objects and structures. A spoken language is a public vehicle that an individual assimilates by accommodation to the social and physical world. This interrelation of the individual speaker, the community, and the natural world imposes necessary conditions that any spoken language must fulfill. This is the gist of Davidson's contention that it is not possible to replace the basic conceptual scheme of ordinary language by a conceptual system that is not embedded in a natural language nor translatable into our language.

This sets the general perspective for the historical analysis that follows. Our basic concern is with the historical process through which the language of physics developed. This is usually done, in a whiggish fashion, by focusing on the development of mathematics, physics, and astronomy in the ancient world, or by tracing the gradual emergence of ideas that now have a foundational role in science such as atomism, as in my earlier survey (MacKinnon 1982), or of evolution. Such approaches effectively detach the 'real' development of science from the hurly-burly of the lived world. This is not an option when one is concerned with the modification of ordinary language. This language is geared to individuals functioning in a social and physical environment. The changes begin with people attempting to understand their place in such a lived world, not with a detached contemplation of nature. The early stages in the emergence of the language of physics were the outcome of projects and processes that did not intend physics as a goal.

References

Achinstein, Peter. 1963. *The Nature of Explanation*. New York: Oxford University Press.

Achinstein, Peter. 1968. *Concepts of Science*. Baltimore, MA: The Johns Hopkins Press.

Atkinson, Nancy. 1984. *Dissertation: Sentience and Stuff*. Pasadena, CA: Claremont College.

Brown, Harold I. 2006. *Conceptual Systems*. London; New York, NY: Routledge.

Carnap, Rudolf. 1950. Empiricism, Semantics, and Ontology. *Revue International de Philosophie, 4*, 20.

Cartwright, Nancy. 1983. *How the Laws of Physics Lie*. Oxford: Clarendon Press.

Coffa, J. Alberto. 1976. Carnap's Sprachanschauung Circa 1932. In Fred Suppe, and Peter Asquith (eds.), *PSA, 1976* (pp. 205–241). East Lansing, MI: Philosophy of Science Association.

Davidson, Donald. 1985. *Inquiries into Truth & Interpretation*. Oxford: Clarendon Press.

Davidson, Donald. 1986. A Coherence Theory of Truth and Knowledge. In Ernest LePore (ed.), *Truth and Interpretation: Persxpectives on the Philosophy of Donald Davidson* (pp. 307–319). Oxford: Basil Blackwell.

Davidson, Donald. 1990. The Structure and Content of Truth. *Journal of Philosophy, 87*, 279–328.

Davidson, Donald. 2001. *Subjective, Intersubjective, Objective*. Oxford: Clarendon Press.

Davidson, Donald, and Jaako Hintikka. 1969. *Words and Objections: Essays on the Work of W. V. Quine*. Dordrecht: D. Reidel.

de Queiroz, Kevin. 1988. Semantics and the Darwinian Revolution *Philosophy of Science, 55*, 238–259.

Engstrom, Anders. 1951. Hintikka and Sandu on Metaphor. *Philosophia, 28*, 391–410.

Franck, J., and G. Hertz. 1967. On the Excitation of the 2536 A Merc. reson. Line by Elec. Collis. In D. Ter Haar (ed.), *The Old Quantum Theory* (pp. 160–166). London: Pergamon Press.

Friedman, Michael. 1991. The Re-evaluation of Logical Empiricism. *Journal of Philosophy, 88*, 505–519.

Friedman, Michael. 2008. Carnap on Theoretical Terms: Structuralism Without Metaphysics. Address to the Bay Area Philosophy of Science Colloquium.

Giere, Ronald N. (ed.) 2006. *Scientific Perspectivism*. Chicago, IL: University of Chicago Press.

Hesse, Mary. 1974. *The Structure of Scientific Inference*. Berkeley, CA: University of California Press.

Hesse, Mary. 1980. *Revolutions and Reconstructions in the Philosophy of Science*. Bloomington, IN: Indiana University Press.

Jackendoff, Ray. 1983. *Semantics and Cognition*. Cambridge, MA: MIT Press.

Körner, S. 1970. *Categorical Frameworks*. New York, NY: Barnes & Noble.

Kuhn, Thomas S. 1962. *The Structure of Scientific Revolutions*. Chicago, IL: University of Chicago Press.

Lakatos, Imre. 1978. *The Methodology of Scientific Research Programmes*. Cambridge: Cambridge University Press.

Lakoff, George, and Mark Johnson. 1980. *Metaphors We Live By*. Chicago, IL: The University of Chicago Press.

Laudan, Larry. 1977. *Progress and Its Problems: Towards a Theory of Scientific Growth*. Berkeley, CA: University of California.
MacKinnon, Edward. 1979. Scientific Realism: The New Debates. *Philosophy of Science, 46*, 501–532.
MacKinnon, Edward. 1982. *Scientific Explanation and Atomic Physics*. Chicago, IL: University of Chicago Press.
Norris, Christopher. 2002. Ambiguities of the Third Way: Realism, Anti-realism, and Response-dependence. *The Philosophical Forum, 33*, 1–38.
Pinker, Steven. 2007. *The Stuff of Thought Language as a Window into Human Nature*. New York, NY: Viking.
Quine, Willard van Orman. 1960. *Word & Object*. Cambridge, MA: MIT Press.
Quine, Willard van Orman. 1961. On What There Is. In Willard van Orman Quine (ed.), *From a Logical Point of View* (pp. 1–19). New York, NY: Harper & Row.
Quine, Willard Van Orman. (ed.) 1966. *The Ways of Paradox and Other Essays*. New York, NY: Random House.
Quine, Willard van Orman. 1969. *Ontological Relativity and Other Essays*. New York, NY: Columbia University Press.
Quine, Willard van Orman. 1973. *The Roots of Reference: The Paul Carus Lectures*. La Salle, IL: Open Court.
Quine, Willard van Orman. 1976. Whither Physical Objects. In R. S. Cohen (ed.), *Essays in Memory of Imre Lakatos* (pp. 497–504). Dordrecht: D. Reidel.
Quine, Willard van Orman, Lewis Edwin Hahn, and Paul Arthur Schilpp. 1986. *The Philosophy of W.V. Quine*. La Salle, IL: Open Court.
Reichenbach, Hans. 1956. *The Direction of Time*. Berkeley, CA: University of California Press.
Rosch, Eleanor. 1999. Principls of Categorization. In Eric Margolis and Stephan Laurence (eds.), *Concepts: Core Readings*. Cambridge: MIT Press.
Salmon, Wesley. 1979. Informal Analytic Approaches to the Philosophy of Science. In P. Asquith, and H. Kyburg (eds.), *Current Research in the Philosophy of Science* (pp. 3–15). East Lansing, MI: Philosophy of Science Association.
Schilpp, P. A. 1963. *The Philosophy of Rudolf Carnap*. La Salle, IL: Open Court.
Sellars, Wilfrid. 1958. Counterfactuals, Dispositions, and the Causal Modalities. In Herbert Feigl, Michael Scriven, and Grover Maxwell (eds.), *Minnesota Studies in the Philosophy of Science, Vol. II. Concepts, Theories and the Mind-Body Problem*. Minneapolis, MN: University of Minnesota Press.
Sellars, Wilfrid. 1962. Time and the World Order. In Herbert Feigl, and Grover Maxwell (eds.), *Minnesota Studies in the Philosophy of Science: Vol. III*. Minneapolis, MN: University of Minnesota Press.
Sellars, Wilfrid. 1963a. Empiricism and the Philosophy of Mind. In Wilfrid Sellars (ed.), *Science, Perception and Reality* (pp. 127–196). London: Routledge and Kegan Paul.
Sellars, Wilfrid. 1963b. Philosophy and the Scientific Image of Man. In Willfrid Sellars (ed.), *Science Perception and Reality* (pp. 1–40).
Sellars, Wilfrid. 1963c. *Some Reflections on Language Games*. London: Routledge and Kegan Paul.
Shapere, Dudley. 1983. *Reason and the Search for Knowledge*. Dordrecht-Holland: D. Reidel.
Strawson, Peter. 1959. *Individuals: An Essay in Descriptive Metaphysics*. London: Methuen.
Strawson, Peter. 1992. *Analysis and Metaphysics: An Introduction to Philosophy*. Oxford: Oxford University Press.
Suppes, Patrick. 1979. The Role of Formal Methods in the Philosophy of Science. In Peter D. Asquith, and Henry E. Kyburg (eds.), *Current Research in the Philosophy of Science* (pp. 16–27). East Lansing, MI: Philosophy of Science Association.
Van Fraassen, Bas. 1991. *Quantum Mechanics: An Empiricist View*. Oxford: Clarendon Press.
White, Alan. 1984. *Rights*. Oxford: Clarendon Press.
Zadeh, L. 1965. Fuzzy Sets. *Information and Control, 8*, 338–353.

Chapter 2
From Categories to Quantitative Concepts

> *Just as geographers crowd on to the outer edges of their maps the parts of the earth which elude their knowledge, with explanatory notes that "What lies beyond is sandy desert without water and full of wild beasts", or "blind marsh", or "Scythian cold", or "frozen sea", so in the writing of my Parallel Lives, now that I have traversed those periods of time which are accessible to probable reasoning and which afford a basis for a history dealing with facts, I might well say of the earlier periods: "What lies beyond is full of marvels and unreality, a land of poets and fabulists, of doubt and obscurity".*
> *Plutarch, Life of Theseus*

In the next three chapters, I will be exploiting history for the purposes of a philosophical reconstruction. The materials chosen may seem arbitrary, or even bizarre. For this reason I will begin with a preliminary orientation indicating the purpose behind the selections. Standard historical accounts pick out two periods that played a formative role in the emergence of science: Greece of the golden age, when philosopher-scientists developed the idea of a rational explanation of natural events; and the late Renaissance, when scientists wedded quantitative concepts to mathematical accounts. I do not dispute this, but I wish to put it in a different perspective. My concern is with the gradual formation of an Extended Ordinary Language (EOL) shared by theoreticians and experimenters. The speculative hypotheses and bold theories that initiate scientific advances draw on the potentialities that language makes accessible. Theories, in turn, enrich EOL by residues, terms, usages, and structures that become a part of the language of scientific discourse. To focus on this language in a historical reconstruction means to enter the great stream of the history of science as something of a bottom feeder. My immediate concern is not with the brave new theories, but with the language that supplied the potentiality for such theories.

The two formative periods just indicated were preceded by changes in linguistic usages and structures that supplied the potentiality the intellectual revolutions actualized. These changes were not occasioned by the science that did not yet exist, but by other concerns. More particularly, the new ideals of Greek rationality emerged

E. MacKinnon, *Interpreting Physics*, Boston Studies in the Philosophy of Science 289,
DOI 10.1007/978-94-007-2369-6_2, © Springer Science+Business Media B.V. 2012

from a demythologizing of earlier frameworks of discourse. The problem that preoccupied the seminal thinkers was one of accounting for human actions and decisions in terms of internal principles, rather than through the whim of a god or the jealousy of a goddess. In doing this, they gradually separated explanations of human actions from explanations of other organisms and material bodies. The framework that emerged for explaining activities of persons supplied something of a paradigm for explaining the activities of bodies. Two key notions that emerged were nature and necessity.

Medieval scholastics inherited the idea of explanations through nature and necessity, not merely from the revival of Aristotelianism, but also from the way these concepts had been used to express Christian doctrine in terminology shaped by Greek philosophy. Too much stress on nature and necessity seemed to limit the power of God. A reaction set in gradually leading to a quantitative spatio-temporal description of bodies and an explanation of their activities in terms of external forces, rather than internal natures. In treating these developments, I am primarily concerned with their effect in modifying and extending the conceptual core of ordinary language. This too presents problems. To speak of a conceptual core found in different languages seems to imply what Davidson has dubbed the third dogma of empiricism. This is a sharp scheme/content distinction with the idea that there is some neutral core independent of any particular language and that different conceptual schemes express this same core.

However, one may take a somewhat different tack. The indispensable task of any spoken language is to accommodate the lived world. Regardless of the particular form this lived world takes, the language used must accommodate basic human needs. It must have some means for speaking of food, survival, procreation, basic bodily parts and functions, parents, children, siblings, mates. It must have some classification of familiar plants and animals, of night and day, up and down, forward and backward. Anthropological studies indicate that basic first order categories are similar as one goes from language to other unrelated languages. Different cultures distinguish dogs from cats, men from women, children from adults, sickness from health, rain from snow. However, there are wide divergences on higher order classes. The basic reason for this is that higher order classifications are underdetermined by perception and perceptual reports. This underdetermination gives higher order systems a flexibility that plays a role in shaping these systems into vehicles for reasoning.

2.1 Early Developments

Levi-Strauss's accounts of totemic classifications and kinship relations have won widespread acceptance, even from critics who reject his structuralism as too sweeping and too a priori. These illustrate the most primitive embedded examples of conceptual superstructures that supply vehicles for inference. The use of animal species as a conceptual tool for dividing some tribe or group into exogamous moieties is widespread. So too are the inferences it supports. If I am a Fox and you are a Bear then we are in different groups, may have different functions, and could be allowed mates from each other's group. This can be extended to other totalities that allow division into units manifesting some correlations and some elemental opposition. Similarly, kinship systems supply prototypes of hierarchical structures that can be extended to other domains. Here Levi-Strauss stresses the relationship of alliances between families more than the particular units that support these relations.[1]

In later mythological cultures, the actions of gods supplied a basis for explaining the basic structure of the natural and social orders. Primitive Egyptian creation accounts picture the god, Atum, standing in the fertile fields emerging as the muddy overflow of the Nile recedes, masturbating, and spreading his semen to generate the Ennead of nine lesser gods and all forms of life. Later accounts transformed this from a physical act, mediated by semen, to an intellectual act mediated by speaking a command.

> His Ennead is before him in (the form of) teeth and lips. That is (the equivalent of) the semen and hands of Atum. Wheras the Ennead of Atum came into being by his semen and fingers, the Ennead (of Pttah), however, is the teeth and lips in this mouth, which pronounced the name of everything, from which Shu and Tefnut came forth, and which was the fashioner of the Ennead. (Pritchard 1973, pp. 1–2)

The Mesopotamian Enuma Elish depicts the ordering of the cosmos as a consequence of the victory of Marduk, the young male god, over Tiamat, the old earth mother figure, an account which reflects the way the Patriarchical nomadic ancestors of these Mesopotamians crushed the primitive agricultural communities where the earth-mother fertility cults reigned (Pritchard 1973, pp. 31–39). Though these cultures are not our concern, it is not really possible to appraise the significance of demythologizing without some appreciation of how pervasive and all-embracing mythological thinking was. In his multi-volume study of mythology, Joseph Campbell brings out four basic functions of living mythology.[2]

- It awakens and maintains in the individual a religious sense of awe and humility in the face of ultimate mysteries.
- It renders an account of the cosmos that accommodates both the sense of awe and the experiences of a culture.
- It validates and supports a social order, chiefly through the semi-mythological role assigned cultural heroes, leaders, and priests.
- It guides the individual, harmonizing him or her into the order of nature/culture through initiations, participation, liturgical reenactments of mysteries, and a sense of belonging.

Demythologizing is not merely a matter of questioning stories formerly accepted uncritically. It involves a reorientation of society and of the way members of the society understand themselves and their world, and of how they cope with birth and death, success and tragedy.

---

[1] See Levi-Strauss (1962, 1966, 1969). For a critical appraisal, see Leach (1974).

[2] This summary is from Campbell (1959–1968, Vol. I, pp. 518–523, Vol. IV, pp. 609–624).

2.2 From Myth to Philosophy

G. S. Kirk, the scholarly editor of the Pre-Socratic fragments, claimed: "In a sense the history of Greek culture is the history of its attitudes to myth; no other important western civilization has been so controlled by a developed mythical tradition."[3] As he sees it, the real age of mythological thinking is long before the time of Homer and Hesiod. In both of these authors one finds a mixture of personification, allegory, and speculative myth that characterize a beginning of the transition from mythopoetic to rational thought. The gods themselves were highly anthropomorphic, with very human failings. Unlike the Enuma Elish, the gods of Greek myths did not confront men through brute force and senseless terror, but left them free to develop. Yet, a loose mythological framework still supplied the only coherent basis for relating the individual to nature/culture.

The aspect of this demythologizing that concerns us is the gradual evolution of a language in which one could speak of human activities in terms of inner sources and of events in nature as manifestations of a rational order. The transformation in the way of speaking of human action in terms of inner sources is most clearly seen in Greek drama, since major works survive. Greek drama originated with the northern cult of Dionysus and its crude revelries. The goat songs (the original meaning of tragedy) gradually expanded to include a chorus, one actor, two or more actors. The hero generally had an ambiguous role in the hierarchical ordering of god—man—beast, either involved in bestial action (incest, matricide, parricide, uncontrollable rage), or aspiring to god-like powers. Marriage and sacrifice, normal ordering elements of society, are fused in a series of perverted rites. Instead of an Apollonian contemplation of a serene objective order, the tragedies celebrate desperate attempts to discover or impose some order on the irrational compulsions below the surface actions of heroes and cultures. The overriding need the tragedies met was to find some principle making the catastrophic sufferings recorded in legends and reenacted in history appear as part of an intelligible order.

Originally, the inner necessity of the unfolding actions came from the wrath of some god, or from a cultural principle, such as murder requires revenge. This compelling necessity can be reconciled to the abiding Greek desire for self-mastery only if the hero accepts some responsibility for the tragedy that befalls him. What emerged was a split-level account, an abiding necessity and a tragic flaw. In Aeschylus the actors portray idealized prototypes of tragic heroes while the chorus reflects the decrees of the gods in terms of lessons to be learned: *the evil doer must suffer*, reflecting human responsibility; and *through suffering comes knowledge*, knowledge of the plans of the gods. Aeschylus was a traditionalist, clearly reacting against the criticism gradually transforming polytheism into monotheism and atheism. He probably held the enlightened opinion that the gods are real, though most of the stories about them lie. His younger, contemporary, Sophocles, changed the emphasis from the doom portended to the character who endured it. The traditional myth supplied

[3] Kirk (1970, p. 250). Kirk and Raven (1962, chap. 1).

an external framework, one due to the gods and communicated through prophets of the gods. This external necessity is matched by an internal necessity, stemming from the very excellence of the hero or heroine. Antigone must defy an impious law. Oedipus must read the riddle of the plague.

Euripides used, but eventually abandoned, these mythological frameworks.[4] He experienced the protracted chaos of the Peloponnesian war and its atrocities. It seems that, probably through the influence of Xenophanes, he gradually developed a very critical stance towards the anthropomorphic accounts of gods and nature. His final play, the Bacchae, depicted the destructive power of religious frenzy. The necessity which made the sequence of tragic actions intelligible came from within. In the Prologue to Hippolytus, Aphrodite informs the audience that she will ruin chaste Hippolytus by causing his stepmother to love him and so provoke the wrath of his father. Yet, Aphrodite is an ambiguous figure. As a woman, she is something of a caricature ("I shall not let the thought of her suffering stop me from punishing my enemies to my heart 's content"). As a personification, she symbolizes an irresistible inner compulsion. Phaedra vividly embodies a woman unsuccessfully struggling to overcome her passionate nature. The necessity is internal; the decree of the goddess an analogical extension of this inner necessity. Medea, a woman driven by rage, cries out that her fury is stronger than her better counsels.

Segal (1986, chaps. 2 and 3) explains how as myths functioned through literary forms they came to comprise a vast system of symbols, verbal, visual, and religious, a second-order semiotic system. As a system, they embody a network of logical relations, a mega-text that operates through specific narratives. Greek tragedy faced the disintegration of the cosmic, social, and psychological order without losing an underlying sense of coherence. The citizen-spectators, who witnessed the social and psychological orders perverted and turned against themselves, were presented with an opportunity to grasp the role and significance of the proper order. The shift from a mythological to a naturalistic perspective was a protracted wrenching effort mediated by the transformation of myths into symbolic systems that supplied inferential vehicles.

A framework for describing important human actions and explaining them through inner principles had been initiated. Necessity, determined by inner principles, functions within an overall order rendering a sequence of actions intelligible. Actions and sufferings of ordinary humans could be understood by analogy with idealized counterparts. Yet, there remained the problem of developing a terminology in which one could speak of the inner sources of human acts and choices. Homer never depicted his heroes as making decisions, and apparently had no such concept. The wrath of Achilles is an unexplained brute fact. In Aeschylus's *Suppliants*, King Pelasgus must choose between the protection of his city and the rightful demands of the oppressed maidens. This is the first clear presentation of a conscious decision in Greek literature. Without the requisite verbal tools people lacked conceptual access to inner states and processes.

---

[4] This interpretation of Euripides draws heavily on Conacher (1967) and also on Segal (1986).

The development of such access hinged on metaphor as a tool for extending language and linguistic structures.[5] This characterized a late stage in Greek poetry. Homer made a sparing use of similes, but no use of metaphors. In his similes, there is no identification. Each item must be seen in its own bright particularity:

> In the lead, as he came on, he took the spear-thrust squarely in the chest beside the nipple on the right side; piercing him, the bronze point issued by the shoulder blade, and in the dust he reeled and fell. A poplar growing in bottom lands, in a great meadow, smooth-trunked, high up to its sheath of boughs, will fall before the chariot-builder's ax of shining iron—timber that he marked for warping into chariot tire rims—and, seasoning, it lies beside the river. So vanquished by the god-reared Aias lay Simoeisios Anthemides. (*Iliad*, Book IV, lines 493–509, Fitzgerald translation)

Homer had three terms for what eventually became 'soul', but each was understood as an extension of a bodily organ. The *psyche* is the breath of life; the *thymus* is the organ of internal motions; and the *nous* is the inner absorber of the images seen with untiring eyes. There were no terms to refer to inner states or acts or decisions. The lyric poets also accepted the idea that love comes from Aphrodite. Yet, when Sappho described the pain of rejection and the frustration of unrequited love she was using behavioral manifestations to describe the inner state of an individual woman. Metaphor was the tool for linguistic extension. The simplest metaphors involved the function of a concrete noun. A more ambiguous extension came from adjectives ascribing properties. Pindar used adjectival metaphors for proportions, such as seeking what is brightest and best rather than what is less so. This scheme of proportionality was adapted by Heraclitus and later mathematicians. The poet's 'is as' became the mathematicians 'equals'. A listing of properties served to systematize and ultimately to delineate an individual. In such schemes, understanding depends on speaking of the unfamiliar in terms of the relatively familiar. Socrates could not be compared to anyone—and this made him incomprehensible.

Verbs provided metaphorical access to action. The process begins with typical verbs characterizing human actions and extends them to natural processes. Water runs; the wind blows. Later people speak of the run of a poem and the course of a speech. This eventually extended to talk about knowledge. The early idea, expressed in Hesiod is that men have seen little and therefore know little. The gods have seen much. The physician Alcmaeon, a disciple of Pythagoras, was the first to make a clear distinction between perception and understanding. The process of thought was spoken of as a road, or by adapting action terminology as the course of a speech or the run of a poem. Formal inference, the idea that one thought entails another, begins its career with Solon. To explain how unlawful gain is quickly attended by misfortune he compared it to Zeus watching men and nature. Through this mythic language, Solon sees an order: in the sequence of states in nature; in human fate; and in a man's ability to bring his thoughts into a connected order.

---

[5] The best source for this extension of classical Greek language through metaphor remains Snell (1960).

The early philosophers used mythological language, but are clearly breaking with mythological accounts. Gods had been the ultimate sources of activities. Thales said there are gods in all things. The Proem of Parmenides speaks of goddesses and chariots leading to the heavens. But, this is stage setting for a philosophical argument. Empedocles adapted Homeric similes to give an analogical account of processes. Just as a lamp lets out light and keeps out the wind, so the eye lets out the subtle fire while keeping in the vital fluids. The metaphorical extensions that were accepted and incorporated into later discourse no longer appear as metaphors. 'ἀρχή' a 'beginning' became a 'principle'. 'ἀιτία', an 'accusation' became a 'cause'. 'ἀναγκη' a 'force' became 'necessity'. The metaphorical extensions that did not become embedded still stand out as metaphors: Empedocles's appeal to 'love' and 'strife' and Heraclitus's invocation of 'logos' as organizing principles. When 'earth', 'air', 'fire' and 'water' refer to elements, they serve as metaphorical extensions of ordinary terms.

The final metaphorical extension to be considered is the abstract use of concrete terms. Here the Greek definite article played a unique role. When Homer spoke of a horse, it was always of some particular horse. When philosophers came to speak of *the* horse, it was neither a particular nor a collective, but a universal term. Combining the definite article with nouns derived from adjectives and verbs allowed one to speak about something beyond material things, the good, motion in general. Myth and logic are not separated by a sharp temporal division, but by the analogical use of mythical frameworks This metaphorical use of terms served as a bridge to an impersonal view of nature.

Plato and especially Aristotle represent the culmination of these trends. Both expressed the opinion that the old myths had some truth in what they were saying, but not in the way they said it. Such a distinction fits a conscious attempt to transform a mythological framework. It is hard not to see a parallel between the split level of Athenian tragedies, the onstage protagonists depicting significant events while the offstage chorus representing the gods explains the necessity these events embody, and the Platonic doctrine that the transient ephemeral beings of experience are to be explained through eternal necessary forms that exist in some separated realm. Aristotle, if Jaeger is to believed, began his career as a young Platonist with a systematic study of ancient myths and religions, while initiating the astral religion that attributed intelligent souls to stars and planets, and concluded his career, as an exile, with the statement, "The more solitary and isolated I am, the more I have come to love myths".[6]

Aristotle knew the myths and respected their role. The term 'myth' had not yet received the sense 'lie' that came from later opponents of paganism. Yet, he clearly recognized them as myths, and so could consciously change the framework of mythical discourse. Aristotle, both the man and the system, is the most thoroughly

[6] Jaeger and Robinson (1934, chap. 6, and p. 321) for the citation. For an evaluation of Jaeger's reconstruction see Grene (1963, chap. 1).

studied figure in the history of philosophy. I have nothing new to contribute to Aristotelian scholarship. However, I would like to set his works in a different perspective. Our concern is with the conceptual core of empirical discourse. Earlier we surveyed some philosophical positions, while sliding over the question of whose language we are discussing. My contention is that what we have been calling a conceptual core emerged as the functioning core of a potentially explanatory empirical science only in and through the writings of Aristotle. Before developing this, I should consider a preliminary question. Which Aristotle am I referring to, the actual historical figure with the restless exploring mind, or the tidy encyclopedist manifested in the works collected and edited long after his death? It is primarily the latter, for it is Aristotle's system that plays a crucial role in the development being considered.

The subject/object distinction, which had been blurred in mythical thought, now emerges in full clarity. Plato celebrated the discovery of the soul, the inner intelligible principle of thought and action, and speculated about its existence before birth and subsequent to death. Aristotle dissected the soul, the inner form of any living being, cataloged and ordered its activities and potentialities. He made a clear distinction between human actions, explained through the processing of sensations and decisions explained through motives and habits; and the actions of inorganic beings, explained through natures and causes. Aristotle shared the common Greek position that explanations must rest on what is permanent and unchanging. He was the first to locate the source of this necessity within perceptible bodies themselves, rather than through something separate, decrees of gods, forms, the logos, atoms.

Another significant point concerns the linguistic orientation of Aristotle's thought. It is necessary to situate this both historically and with relation to the two extremes of pure a priori reasoning and empirical investigation. A comparison with Plato brings out the historical aspect. Plato had an abiding concern with problems of language and meaning, from the early dialogs where Socrates tries to ferret out the true definitions of piety, or shape, or love, to the late dialogs where Socrates as midwife assists Theatetus in delivering a definition of knowledge. Yet, this linguistic concern is subordinate to the idea that true understanding comes from intuitive knowledge of the forms, something beyond the limits of language. The philosopher who returns to the cave cannot find the terms to communicate what he has seen.

Aristotle initiated the systematic study of language and logic, of metaphors and tragedy. We will focus on one of his early works, the *Categories*, which played a formative role in the evolution of both physics and biology. His list of ten categories is presented in slightly different forms in *Categories* 4(1b25–2a10) and in *Metaphysics* 5(1017a7–1018a20) (See Hacking 2001, sect. 3.3). Neither account presents any justification of these ten categories as basic. Aristotle's *Categories* is generally treated as the introductory part of his *Organon*, a somewhat fragmentary collection of lecture notes in which Aristotle treats the logic of terms prior to his more mature work on syllogisms, inference, and axiomatic systems. W. Mann (2000) has recently argued that the *Categories* should be interpreted as the discovery of things as things. I would modify this to the claim that in the *Categories* Aristotle discovered a way of making individual things subjects of science.[7] Before considering them we should avert to an important point Mann makes. Aristotle's *Categories* has not been interpreted as a breakthrough chiefly because the basic points made now seem obvious. Since the treatise offers virtually no supporting arguments, the impression is given that these points were obvious even in Aristotle's day. Its revolutionary nature appears only when the doctrine presented is situated in the problematic of the late Academy, where accounts of predication (or *κατηγορία*) supplied the tool for criticizing the ontological primacy of forms and tackling the foremost problem of the day, making change intelligible.

The striking new claim is that among entities in general (*τα ὀντα*) concrete individual things are the really real, the fundamental entities (*ὀυσίαι*). Though this term is generally translated 'substance', this translation effectively imposes Aristotle's later metaphysics, rather than the problematic term he shared with Plato. As suggested by Hacking, Aristotle's earlier use of this term will be translated 'what a thing is' or 'whatness'. The crucial citation is (Categories, 2b 5–6): "Thus everything except primary whatness is either predicated of primary whatness, or is present in them, and if these last did not exist it would be impossible for anything else to exist."

The doctrine is fairly clear. Its justification is quite obscure. Since we are concerned with the historical role of the doctrine, we will consider Aristotle's justification only to the degree helpful in clarifying the doctrine. Aristotle's analysis of predication was concerned with ontological priority. Unfortunately, his analysis is notoriously innocent of any use/mention distinction. So, it is not always clear when he is speaking about terms and when about the things terms signify.[8] His early position was given in his *On Interpretation.* Objects cause concepts and words express concepts. Though all men do not have the same words they do have the same basic concepts. Later in his "De Anima" he gave a theoretical justification for this apparent naiveté. The active intellect has the power of abstracting forms and impressing them on the passive intellect so that the form of the knowing is the form of the known. Acceptance of such views allowed Aristotle to get at reality by analyzing the way we speak about reality. Instead of focusing on such contemporary concerns, it is better to begin by situating the doctrine of the categories between Plato's late dialogs and Aristotle's later metaphysics. For Plato scientific knowledge (episteme, not modern science) must be certain and of what is. Forms, rather than changeable beings, fit the requirement. Changeable beings were understood in terms of one quality replacing another, as heat replacing cold or fire either retreating or perishing in the presence of cold (Phaedo, 103). Concrete individuals were conceived of, in Sellars's apt phrase, as leaky bundles of particulars.

---

[7] I am also relying on Sellars' article, "Aristotle's Metaphysics: An Interpretation" (in Sellars 1967, Vol. 1) and, with considerable reservations, on Anscombe and Geach (1961, chap. 1).

[8] This is not a simple whiggish criticism. Stoic logicians, extending Aristotle's work, were aware of semantic problems and made a distinction between: the signified (what is said), signifying, and the thing itself.

Aristotle used an analysis of predication to get at the reality of things. Effectively he treated concrete individuals as falling into one of three classes: heaps, natural units, and artifacts. Heaps fit the Platonic treatment of individuals as leaky bundles of particulars. Aristotle considered natural units to be objects of scientific knowledge, and in fact devoted much of his career to studying them. He constantly relies on the analogy between natural units and artifacts. A piece of leather becomes a sandal because of an imposed form and a purpose. His doctrine of forms came later. In the *Categories* he is concerned with getting at concrete units by analyzing predication. That aspect, however, should be situated in the context of the Socrates of Plato's dialogs, unceasingly searching for the true definitions of justice, piety, and other virtues.

Aristotle's initial discussion introduces two types of distinctions in talking about terms. The first set, 'equivocal', 'univocal' and 'derivative' (e.g., gambler, runner) depends crucially on features of language. The second set, 'said of' and 'present in' do not manifest the same dependence on the terms used (and so are more concerned with things than terms). They differ in transitivity of predication. Thus (as in Greek, omitting indefinite articles) if "Socrates is man" and "Man is animal" then "Socrates is animal" and the definitions of 'man' and 'animal' can be predicated of Socrates. Affirming "Socrates is pale", however, does not imply that the definition of 'pale white' applies to Socrates. It applies to colors. Putting the two distinctions together yields a fourfold classification. (a) Some items can be both **in** something as subject, and also **said of** something; (b) other items are **in** but not said of something; some items can only be **said of** something as subject but are not **in** anything; finally some items can neither be **said of** anything nor be **in** anything. The last class yields the items Aristotle treats as basic. The other classes involve problematic features concerning predication.

To get at these basic items we consider the role of definitions. A definition is a phrase signifying a thing's essence (*Topics* 101b38). It is an answer to the question "What is an X?" Plato thought that there could be no definition of sensible things (*Metaphysics*, 987b). For Aristotle, individual things are what they are because they are beings of a certain kind (*Topics*, 125b 37–39). This kind could be defined by a genus and specific difference. Their designation was through univocal terms that his logic required. Thus, the primary instances of 'things' are natural units. They were now open to scientific study.

Aristotle eventually realized that his analysis was seriously incomplete. When he returned to the task in the lectures later put together as *Metaphysics* his overriding concern was with being as being.

> And indeed the question which was raised of old and is raised now and always, and is always the subject of doubt, viz. What being is, is just the question, what is substance?.... And so we also must consider chiefly and primarily and almost exclusively what that is which *is* in *this* sense. (*Metaphysics*, 1028b 1–6, italics in McKeon)

His metaphysical account is a theory of being as composite of matter, form, and potency that seeks to make change intelligible. This doctrine applies primarily to substantial beings. The net result is that the things that count as subjects of scientific investigation are things belonging to types that admit of definition, at least in principle. Scientific investigation and demonstration is primarily concerned with things categorized as substantial units, which are characterized by their quantity, quality, relation, action, passion, plus further categories, and which can remain substantial units while undergoing accidental changes.

The Aristotelian categorial system, though rooted in basic features of the Greek language, was intended as a means of accounting for reality as it exists objectively. Aristotle treated categories both as 'kinds of predicates' (γένη των κατηγορίων) and 'kinds of being' (γένη των όυτωυ). The boundary is fuzzy. Yet, one clear separation constitutes a basic requirement of Aristotelianism. Some properties are essential for natural kind objects to be natural kinds, regardless of the language in which these properties are expressed. In making individual objects the units to be explained and in striving to make basic explanations independent of the features of any particular language, Aristotle initiated the language of physics.

Aristotle divided the axioms of any science into two types: common axioms, known through intuition; and proper axioms, characterizing a particular science (*Posterior Analytics*, I, 10). Aristotle saw physics, or the philosophy of nature, as the general part of a science, which had studies of plants, animals, humans, and to some degree, celestial bodies, as special topics. Empirical investigation was needed for these special studies, not for the principles they had in common. Aristotle's version of empiricism is manifested chiefly in his biological writings, not in his physics.[9] His physics relied heavily on linguistic analysis. Time, Timaeus explains, is the image of eternity moving according to number. In the Platonic tradition time is always properly understood as a stretched out reflection of eternity. Aristotle develops his definition of time by analyzing the way we speak about time, just as he gets at change by analyzing the way we speak of something becoming something new. Both begin with the situated language user. Similarly, he developed his doctrine of space by working out from the localization of the individual (Physics, Book 4).

For comparison it helps to return to the argument Strawson gave. From the problem of identifying an absent particular and re-identifying it as the same particular he derived the necessity of a sharp subject/object distinction and the contention that the basic particulars had to be things with properties existing in a common space-time framework anchored, for purposes of reference, by the subject's bodily presence. Neither this argument, nor the special type of common sense realism it supports would have made philosophical sense to any philosopher before Aristotle. Ancient people could, to be sure, refer unambiguously to absent particulars identified relative to other particulars. Telemachos tells the nursemaid Eurycleis to find and hide the bow of Odysseus, the one he left behind when he went to Troy. Yet, in a mythological framework all explanatory references ultimately vanish in a mythological realm. For Plato explanatory references terminated in something

[9] The role of physics in Aristotle's system of the world is treated in Solmsen (1960). The biological orientation of Aristotle's thought is emphasized in Grene (1963) and also, in a form that is perhaps too functional, in Randall (1960).

timeless and imperceptible: the good, the true, the soul as it existed before its union with the body. For Aristotle what exist in a primary sense are individual substances. Reference, whether for identification or explanation, leads to individual substances and to a hierarchically ordered network of individuals sharing a common space-time framework. It was only much later, when Aristotelianism had been assimilated into enlightened discourse that a residue of Aristotle's position could emerge as accepted common sense realism. Aristotle is the founding father of the empirical discourse that made our science possible.

Aristotle, however, did not develop or support a mathematical physics. Mathematics, in his view, is based on abstraction from physical objects (Physics 193b), not from a contemplation of pure forms. From units one abstracts numbers; from shapes one abstracts forms. These abstractions lead to arithmetic and geometry, a study of forms as abstracted from, rather than existing in, material bodies. Though mathematics did not play any direct role in Aristotle's physics it could play a subordinate role, especially in applied physics, i.e., optics, harmonics, astronomy, and mechanics (Physics 194a), where mathematics was invoked to give the reason for the established facts. Because of Zeno's paradoxes, Aristotle was also concerned with the applicability of numbers to the continuum and its physical manifestations, space, motion, and time.

With these transformations stemming from Aristotle and Greek mathematics it was possible to get away from the immediacy of the world as experienced and speak of physical reality in a more abstract fashion in terms of geometrical forms, causal relations, essences, and laws of nature. There is, however, one linguistic practice that physics requires and which the classical world never achieved. That is some method of speaking about qualities in a quantitative fashion. As Bochner summarized it:

> And yet, from whatever reasons, The Greeks never penetrated to the insight that an effective manner of pursuing mechanics, physics, and other science is to articulate qualitative attributes by quantitative magnitudes, and then to represent connecting rules and laws by mathematical formulas and relations in order that the application of mathematical procedures to the formulas and relations may lead to further explications and developments. (Bochner 1966, p. 31)

Bochner is certainly correct. Neither the classical Greeks nor their Hellenistic and Arabic successors developed a quantitative science of qualities. The reasons for this shortcoming are neither as arbitrary nor as impenetrable as the citation suggests. Assigning numbers to quantities would not have much use unless the assignments could support further inferences. Consider the type of simple example that an Alexandrian scientist could easily handle.

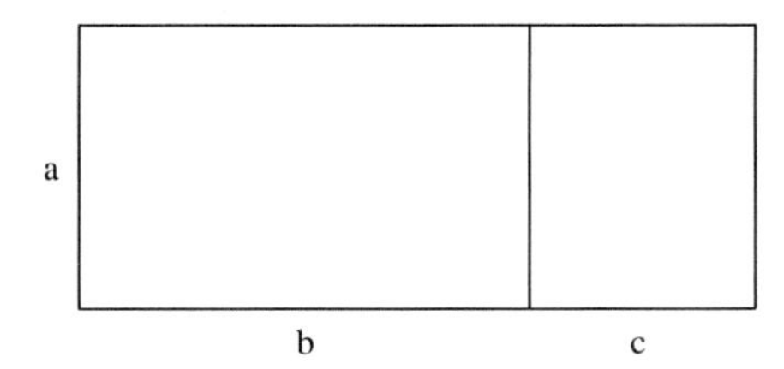

He could determine the overall area either by multiplying the height by the combined widths or determine the area of each rectangle and adding the two areas. The two methods yield the same results because $a \times (b + c) = (a \times b) + (a \times c)$. In this formula 'a', 'b', and 'c' could stand for numbers with '+' and '×' standing for addition and multiplication; or 'a', 'b', and 'c' could stand for lengths with '×' and '+' representing ways of forming areas and concatenating units. There was an implicit structural isomorphism between the representation of lengths, areas, and volumes and the number system. Arithmetic could be used to make inferences about areas because operations performed on numbers parallel operations performed on areas. No way had been developed to assign numbers to qualities, such as heat or color, in a way that supported inferences.

In Alexandrian science and its tributaries, geometry was the tool for scientific explanations. Following the Platonic tradition (Philebus 55a) they distinguished *logistic*, concerned with applications of numbers to practical affairs, from *arithmetic*, concerned with properties of numbers. The claim has frequently been made that there is a social explanation for the failure of Greek and Alexandrian scientists to develop a mathematical physics. The intellectuals who worked with their heads and studied arithmetic were separate from and superior to the laborers and merchants who worked with their hands and applied arithmetic to numbering and counting. I find this a radically insufficient explanation. Archimedes, the greatest of the ancient mathematicians, frequently applied his skills to practical affairs, such as levers and pulleys, the properties of fluids and floating bodies, or determining whether King Hiero's crown was pure gold. In his *Sand Reckoner* he demonstrated an awesome skill with applied mathematics. To show that he could express a number greater than the number of pebbles that could fill the universe he estimated the size of the universe in the classical geocentric version and also in the much vaster heliocentric version of Eratosthenes, where the absence of stellar parallax was explained by assuming vast distances for the fixed stars. To express an arbitrarily large number Archimedes divided numbers into classes of myriads and myriad-myriads, where a myriad is 10,000 and determined how many classes were needed. He incidentally set the stage for logarithms by noting that addition of classes corresponded to multiplication of numbers.[10] The last great Alexandrian mathematician, Heron, knew and used the ancient Babylonian approximate calculations and applied mathematics to surveying, to mechanical devices, to military weapons such as catapults, and even to the design of children's toys.

The necessary prerequisite to applying numbers to qualities in a way that supported inferences was a recognition, at least implicit and functional, of a partial isomorphism between arithmetical structures and structures implicit in a quantitative treatment of qualities. There was no such conceptual structure for qualities. Numbers were generally treated in a way that obscured the logical structures needed. In the Pythagorian tradition numbers were classified as even and odd, and then into evenly even (powers of two), evenly odd, and oddly even, with a further distinction

[10] In this summary I am relying on Boyer (1968, chaps. 5–11) and Kline (1972, chaps. 2–8).

into prime, composite and perfect numbers. This supported the popular trend seeking properties of particular numbers that were associated with particular qualities, rather than the properties of numbers as a system.

Consider two influential attempts to exploit the properties of numbers. Martianus Capella (c. 430 A.D.) wrote a treatise that became one of the most popular text books of the middle ages. In it, he explains properties of numbers that make them suitable for a physical interpretation:

> Three is the first odd number, and must be regarded as perfect. It is the first to admit of a beginning, a middle, and an end, and it associates a central mean with the initial and final extremes, with equal intervals of separation. The number three represents the Fates and the sisterly Graces; and a certain Virgin who, as they say, "is the ruler of heaven and hell," is identified with this number. Further indication of its perfection is that the number begets the perfect numbers six and nine. Another token of its respect is that prayers and libations are offered three times. Concepts of three have three aspects; consequently, divinations are expressed in threes. The number three also represents the perfection of the universe... (cited from Lindberg 1992, p. 146)

The association between numbers and properties was thought to hinge on distinctive properties of particular numbers. Thus, Philo Judaeus explained that 6 was the appropriate number of days for creation because of the perfection of the number. It is the only number that is the sum of its proper divisors.[11] However, one needs an ordered collection of numbers to fit different degrees of heat. For this we turn to Geber (Al Jabir, pp. 760–815), the greatest of the alchemists. He held that the number, 17, is the universal number power, for it has the property of being the sum of 1, 3, 5, and 8. Now consider the magic square of the first nine integers.

4 9 2
3 5 7
8 1 6

This is magic because the sum of its digits in any direction is 15. The lower left-hand corner contains a square of 1, 3, 5, and 8, the aliquot parts of 17. Removing this square leaves the gnomon

4 9 2
    7
    6

which sums to 28, or 2×17. Since the Arabic alphabet has 28 letters, this is a strong indication that the analysis is on the right track. Corresponding to the four elements are the four elemental qualities, hot, dry, cold, and wet. For the number system to fit, each quality must have 7 degrees, making 28 qualitative possibilities, which can be arranged in a matrix with 4 rows and 4 columns, each column having 7 units. To apply this to heat one should note that the greatest number of consonants in the name of any metal is 4. In Arabic, as in Hebrew, only consonants appear as letters. So, one assigns numbers to the consonants and determines the relative hotness of

[11] There are many more numbers with this property, e.g. 28, 496, 8218.

lead by combining the number of the consonants and their position in the matrix. This gives the external constitution, presumably the proportion of qualities in the metal as it appears. To get the internal constitution one needs manipulations that are more complicated.[12]

No one knew how to describe physical properties and formulate numerical relations so that the numbers fit the properties in a way that supported inferring statements about physical properties on the basis of numerical relations. No successful quantitative science of qualities could be developed until these features became linguistically accessible. Here again we have a protracted process of muddling through by people who, in the initial stages, neither anticipated nor intended the goal of a quantitative physics. We will imitate their example by muddling through another obscure development.

2.3 From Philosophy of Nature to Mechanics

Medieval Aristotelian philosophy of nature supplied the original matrix from which a quantitative physics emerged. What I intend to present here is not one more summary of this development but a consideration of the role this played in the gradual extension and transformation of the language used to speak about physical bodies and their properties, especially local motion. This language was gradually molded into a form that supported conceptual structures isomorphic to mathematical structures. For our purposes, we will be primarily concerned with two stages that can be roughly separated. The first is the transition from a strict Aristotelian account where changes are explained in terms of natures and causes to a perspective where local motion is described in a manner that supports some sort of mathematical representation. The second is the transformation of this conceptual philosophy of nature into a more mathematical and empirical science.

We begin with some general considerations. Medieval philosophers and theologians did a considerable amount or work that would now be labeled conceptual analysis. However, these philosophers, especially the Aristotelians, put this analysis in ontological terms and embedded it in a theological perspective.[13] A proper appreciation of the modifications requires some understanding of the general perspective. The prevailing accounts of sensation and cognition were based on the Aristotelian doctrines of matter and form, act and potency, and the four causes (See Smith 1981). In vision the sensible form of the object seen, its color and shape, are impressed on the medium, the air made transparent by light, and through the medium is impressed on the eye. The analogy then used was a seal on wax. The form impressed on the eye is thus causally determined by the thing known. The sensible in act is the sensed in act. Then the brain, which Galen had made the organ

[12] This is taken from Harré (1961, pp. 11–14).

[13] For general surveys of the development of science in this period see Crombie (1959) and Lindberg (1992). For the embedding of science in a theological perspective see Lindberg (1987).

of thought, abstracts perceptible forms. A rational, rather than a merely animal, soul is required for the abstraction of intelligible forms. This is not a myth of the given. It was a developed theory of how the form of the thing known together with the cognitive power of the mind makes a concept of an object correspond to the form of the object known. Conceptualists and Nominalists, who will be considered later, had some reservations. The Aristotelians realists had no difficulty with accepting the doctrine of Aristotle's "On Interpretation" that different languages have different terms. But, these different terms express the same basic concepts because the objects known determine the concepts of the objects.

Aristotle's Categories were well known even before the systematic translation of Aristotle's works in the late twelfth and early thirteenth century. Porphyry, a disciple of Plotinus, wrote a commentary on the Categories, the Isagoge. This had been translated into Latin, Syrian, Arabic, and Armenian and served as a staple text for the early arguments between Nominalists and Realists. The reliance on the familiar Porphyrian trees as a basis for specifying the nature and properties of things was a disputed issue. The only definition that seemed to capture an essence was 'rational animal'. The basic categories, however, were accepted as something determined by the nature of reality. This put analysis in ontological terms. Since the basic categories as well as concepts of particular objects are determined by the reality known, one can analyze objects by analyzing concepts of objects. This linguistic analysis was carried on in a theological perspective. The later reactions of the Scientific Revolution and the Enlightenment periods tended to regard theology as a restricting repressive force. To see how a theological perspective could lend positive support to a distinctively medieval type of conceptual analysis we turn to the most Aristotelian medieval thinker, Thomas Aquinas.

Aquinas thought of the method of philosophy as a process of analysis and synthesis, a terminology stemming from Pappus not Aristotle, repeated at different levels. Analysis begins with the confused knowledge of ordinary experience and the language in which it is described (the *magis nota quoad nos*) and seeks to understand the beings of ordinary experience, material beings, by a resolution into principles of being (the *magis nota quoad se*). The principles sought were the inner constitutive principles of matter and form (together with the privation of form needed to explain change) and the extrinsic principles of efficient causality, or agency, and final causality, or purpose. After such principles have been attained, one can form a synthesis in which the first principles of explanation express the ultimate results of the process of analysis. Thus, explanation in the synthetic mode has an ordering (Aquinas refers to it as a sapiential ordering) which is the reverse of that proper to the analytic mode.

This analysis and synthesis is part of the philosophy of nature, or physics in the Aristotelian sense, and is something distinct from and methodologically prior to metaphysics. It is a study of beings as changeable, but not of being as being. Metaphysics, in Thomas's view, begins with the judgment of separation, the conclusion that not all beings are material beings. Such a conclusion, in a systematized Aristotelianism, comes at the end of the *Physics*, based on arguments that the first mover cannot be a material being. If one concludes that 'being' does not necessarily mean 'material being' then one must confront the question: What is being as being? This, in the Aristotelian-Thomistic conception, launches metaphysics as a distinct science concerned with being as it exists. Our concern is not with this metaphysics, but with the philosophy of nature that precedes it. To see how philosophy of nature supplies a basis for conceptual analysis, however, we have to consider the way in which Aquinas interrelated logic, metaphysics, and the philosophy of nature.

Aquinas spoke of logic as a formal science. Yet, the term 'formal' did not have the meaning for him that it has for contemporary logicians. For Aquinas both metaphysics and logic have all of being as their scope. As Schmidt summarizes the difference: "While metaphysics studies the things presented to it in reality and studies them in themselves, logic studies rather the intellectual views or intentions which reason, in looking at these things, forms in the mind" (Schmidt 1966, p. 45). For Aquinas logic is formal in the sense that it is concerned with the formal principles of things and the concepts through which they are represented. The representation is, in his view, caused by the reality represented. Our concern is with the concepts used and the methods of analyzing them. Through the mediation of concepts, everything comes together in being and its essential attributes. Such concepts can be studied: either insofar as they express the reality of a thing, in which case one has metaphysics; or insofar as the mind has a certain way of relating one thing to another, in which case one has a science of the rules of predication. Thomas calls this logic, though much of this activity would now be classified as linguistic analysis. Though indispensable, this conceptual analysis is not an end in itself. It is a tool functioning in the protracted process of analysis and synthesis. The metaphysical analysis, according to this program, terminates in the ultimate inner constitutive principles of being, essence and existence, and the ultimate extrinsic principle, God as the first cause and final end of all being. In the synthetic mode, or when following the theological order, one begins with God as first cause (*Summa Theologiae, Pars Prima*) and final end (*Pars Secunda*) of all beings. This combination of a theological perspective and metaphysical realism supplied a framework for the interpretation of language.

God is truth. Things are true, ontologically, by virtue of their conformity to the divine mind, or the exemplary ideas in accord with which they were created. The human mind possesses truth by virtue of its conformity to things and through them to the divine mind. This conformity is expressed only in a judgment, which is an affirmation of the conformity (or lack of conformity) between the inner word, a concept, and that of which it is stated. An affirmed proposition involves a double composition. First, there is the composition of subject and predicate forming the proposition, or complex inner word. Second, there is the composition of the proposition and its affirmation, which should be based on a reflective grasp of evidence sufficient to warrant assent. Thus, any affirmed proposition is implicitly a truth claim. One can make this explicit by quoting the proposition and predicating 'true' of it. Here Aquinas anticipated the assertive-redundancy analysis of 'true'. "But 'true' and 'false' add nothing to the significance of assertoric propositions; for there is the same significance in 'Socrates runs' and 'it is true that Socrates runs' and in 'Socrates is not running' and 'it is false that Socrates is running.' "[14] Thus, in the order of synthesis, the Thomistic explanation of truth begins with God and terminates with a clarification of the syntactical features (relation of subject and predicate in a proposition), semantic roles (relation of subject and predicate to their supposit), and performative aspects (what one does with words in actually affirming rather than merely considering a proposition) of propositions accepted as true. In the order of analysis, one begins with this sort of conceptual analysis.

Our concern is with the philosophy of nature and the type of conceptual analysis it supported. Methodologically, the philosophy of nature did not presuppose metaphysics; it preceded it. This analysis, however, was conducted in the context of presuppositions which we would regard as metaphysical. Two such presuppositions are particularly significant. First, the universe was created by God in accord with a plan, his own nature as imitable. For Plato and Aristotle matter as such is unintelligible. It becomes intelligible only through forms imposed on it. Against the conceptual backdrop of creation, all matter must have some intelligibility. As Aquinas put it:

> In those things which are apprehended by all a definite order is found. For that which is first apprehended is being, so that the understanding of being is included in everything else that anyone apprehends. Accordingly, the first indemonstrable principle is that nothing can be affirmed and denied at the same time. This is based on the intelligibility of being and nonbeing; and all other principles are based on this one. (Aquinas, *Summa Theologiae*, III, a. 2, c)

This ontologically grounded semantic realism set limits to the applicability of language, something first made explicit by Moses Maimonides. Here again we have a semantic problem, how to handle contradictions, set in a theological context. Hillel initiated a way of treating apparent contradictions in the Torah in a holistic context. His concern was not with contradictory beliefs, but with conflicting practices. If the eve of Passover falls on a Sabbath, how does one reconcile keeping the Sabbath holy and sacrificing the Paschal lamb? He interpreted the Torah, not as a collection of separate laws and prescriptions, but as an encompassing unity, a guide for practical living. His famous seven rules find guidelines for resolving conflicts within the Torah itself.

It may well be that no legacy in human history has exhibited as much sophisticated concern with the resolution of contradictions as the Talmudic tradition. Moses Maimonides, the foremost Talmudic scholar of his era, brought the techniques of this tradition to bear on the problem of reconciling Biblical traditions with Aristotelian philosophy. Maimonides accepted as true a scriptural tradition that attributed to God such qualities as goodness, compassion, and wisdom. He also accepted as true the expanded Aristotelian doctrine that the first mover, which he identified with God, is an utterly simple being with no composition of any sort, one

[14] The Citation is from Bochenski (1961, p. 183). The doctrine briefly summarized is taken from Aquinas's *De Veritate*.

whose act of existence is identical with his own self-contemplation. The conjunction of the two traditions led to contradictions.

To make sense out of the way these were treated we must briefly consider the type of medieval hermeneutics that was developed by Maimonides and adapted by Aquinas. Both accepted Biblical revelation as the highest form of truth, but gave its claims a very Aristotelian interpretation. The hermeneutic principle was that Moses (considered the author of the Pentateuch) knew the real truth (or thought as an Aristotelian), but accommodated his speech to meet the needs of simple uneducated people. The Torah, Maimonides insisted, speaks the language of man. To understand what Moses really meant one must recast his sayings in their proto-Aristotelian form.[15] Maimonides reinterpreted any class of attributes that presupposed composition. This led to locutions like, "God is wise, but lacks the virtue of wisdom". This seems to countenance contradictions. Actually, Maimonides was attempting to eliminate contradictions even at the price of introducing a profound agnosticism. To say God is wise but lacks the virtue of wisdom is a way of making a claim about God while indicating that the normal presuppositions of attributive or descriptive claims, objects with properties in a shared space-time framework, do not hold in discourse about God. What then does it mean to claim that God is wise? Maimonides replies: "In like manner, the terms Wisdom, Power, Will, and Life are applied to God and to other things by way of perfect homonymity, admitting of no comparison whatsoever... there is, in no way or sense, anything common to the attributes predicated of God and those used in reference to ourselves" (Guide, Part I, chap. lvi). 'Perfect homonyms' implies that 'wise' in "God is wise" and "Socrates is wise" is like 'club' in "The murder weapon is a club" and "The Poetry Appreciation Society is a club." Aquinas substituted analogical predication for perfect homonyms. Yet, he matched Maimonides' agnosticism concerning knowledge of God. After a detailed treatment of the attributes of God Aquinas concluded: "We cannot grasp what God is, but only what he is not and how other things are related to him" (*Summa Contra Gentiles*, Book I, chap. 30, #4).

Ordinary language usage shapes, and is shaped by, the lived world. Scholastic Latin, however, was not an ordinary language. It was the language of religious rituals and the vehicle of instruction in the new universities. It was not spoken in homes, the market place, or used for ordinary life. Apart from a few Prioresses, women, for example, did not master Latin. No one assimilated Latin as a first language. This meant that scholastic Latin had a unique detachment from the normal factors shaping ordinary language usage.

Here an anecdote may help to clarify the peculiar nature of this detachment. Until the Second Vatican Council, in the early 1960s, scholastic Latin remained the language used for instruction in international Catholic seminaries. At the end of this period Father Frederick Copleston, the noted historian of philosophy, taught a course, in Latin, at the Roman Pontifical Gregorian University on the later

---

[15] This hermeneutic principle is developed in Maimonides (1963) Guide for the Perplexed, Part I, chap. xxxiii, and in Aquinas's Summa Theologiae, I, q. 68, a. 2.

philosophy of Wittgenstein. A few months after the course was over I had an opportunity to discuss it with him. He claimed that the course was an unmitigated disaster, the worst failure of his teaching career. The basic reason for this was that students master scholastic Latin by learning explicit definitions for all basic terms and then use these terms in accord with their defined meanings. The idea that a clarification of meaning, as dependent on usage, led to the dissolution of philosophical problems never filtered through the relatively opaque medium of scholastic Latin. The feature that is disadvantageous in the long run can be advantageous in the short run. This is especially true when we go from the context of students memorizing definitions to creative thinkers establishing a tradition. The detachment of the language of instruction from the constraints of normal usage allowed it to become more systematic and more flexible. Also, the rise of universities provided institutions in which grown men could spend their days discussing abstract issues of no practical significance.

The linguistic flexibility rooted in the detachment of scholastic Latin shaped the treatment of the Aristotelian categories. The categories are generally presented in the ordered list: substance, quantity, quality, relation, place, time, situation, state, action, and passion. When interpreted in the light of a doctrine of concept abstraction the first three categories embody an internal conceptual ordering proper to material beings, while the remaining categories represent external or relational properties of being. Substance, quantity, and quality are distinguished by the relation they have to matter. The order in which they are listed represents a conceptual ordering, which is the reverse of the perceptual ordering.

To make this more concrete consider the predicate 'red'. This cannot be used in a literal sense as a predicate proper to an immaterial being, such as an idea, or an unextended being, such as a point. A quality, such as color, presupposes the quantity, extension. This in turn presupposes a substance that is extended and colored. This led to a distinction between sensible matter and intelligible matter. Intelligible matter is not matter that thinks, but matter as understood. Since substance, quantity, quality, and relation represents an *ordered* list one can consider matter as extended while prescinding from further determinations, such as color or location. Matter cannot be perceived without color, but it can be understood without considering color. This notion of intelligible matter, in turn, was regarded as supplying a foundation for mathematics. A consideration of extended substances as units supplied the foundation for arithmetic, while a consideration of extended substances as having shapes supplied a basis for geometry. For Aquinas and most of his contemporaries, this exhausted mathematics.

This justification of mathematics was manifestly inadequate even to the data that Aquinas accepted. Thus, he had to conclude that it is impossible to number the angels, since they are not extended substances. However, he also held that there were less than an infinite number of angels and that they could be named. This suggests a method of counting. Another apparent limitation, stemming from the interpretation of the order of the categories as embodying a conceptual necessity, is that one could not seem to speak about the quantity of a quality. This inverts the proper ordering of the categories.

Here again, Aristotelian philosophy seemed to contradict the higher truth of theology. One's rank in heaven, according to accepted teaching, depends upon the degree of grace, or charity, that one has at the moment of death. Dante's Divine Comedy vividly illustrates the different degrees assigned in hell, purgatory, and heaven. Since grace is a quality, albeit a supernatural one, comparing degrees of grace is comparing quantities of qualities. Accordingly, a way had to be found to discuss the quantity of a quality. Thomas Aquinas followed the old tradition of assigning numbers to things based on an affinity between particular things and properties of individual numbers. This occurs regularly in the accounts of why it is fitting to have ten commandments, seven sacraments, twelve apostles, three theological virtues, and four cardinal virtues. Yet, he also went beyond such decorative arguments. He seems to have been the first medieval Aristotelian to give a coherent account of the way in which quantitative determinations can be given to qualities. Aquinas distinguished between quantity per se, or bulk quantity, and quantity *per accidens*, or virtual quantity (Aquinas, Summa Theologiae, I, q. 42, a.1, ad 1). Virtual quantity, or the quantity of a quality, can have magnitude by reason of the subject in which it inheres, as a bigger wall has more whiteness than a smaller one, or it can have magnitude by reason of the effect of its form. The first effect of a form is a way of existing, e.g., as human. The secondary effect of a form is shown through its action on objects. A comparison of relative effects serves as a measure of virtual quantity. Thus, one with greater strength can lift heavier rocks.

Measurement did not rest on the modern idea of a unit of measure or even a practice of measuring things. In medieval thought the Platonic idea that the perfect form is the measure for any being that participates in that form was reinforced by the scriptural statement that in creating the world God disposed all things in number, weight, and measure. Thus, the true measure of grace, for St. Thomas, is the grace of Christ in which man participates. Even when measurement is separated from a doctrine of participation and treated in terms of numbers, as in Aristotle's treatment of time, the constraint is that everything must be measured "by some one thing homogeneous with it, units by a unit, horses by a horse, and similarly times by some definite time" (Aristotle, Physics, 223b14). Applying quantities to qualities broke this Aristotelian constraint. This new idea of the quantity of a quality was the pivot leading from the Aristotelian philosophy of nature to a mathematical physics.

This history has been treated in detail elsewhere.[16] In summaries, this is often done by presenting the aspects leading to Newtonian physics in a manner intelligible to a modern audience. Here I wish to do the opposite, to bring out the complexities and confusion involved in developing an account of properties of matter and motion that admitted of a mathematical representation. The idea of the quantity of a quality matured into a doctrine of the intensification and remission of qualities. This led to a nest of conceptual problems. Does the quality itself change, the degree of participation in a quality, or does one quality replace another? If the quality changes

---

[16] An excellent brief summary is given in Lindberg (1992, chap. 12). More detailed treatments are given in Clagett (1959), Crombie (1959, Vol. 2, sect. I), Dijksterhuis (1961[1950], pp. 126–222).

by addition, rather than replacement, how is the addition of qualities to be understood? The nominalism, spearheaded by William of Ockham, led to a de-emphasis on the ontological aspects of this discussion. Instead of asking how intensification and remission of a quality takes place he sought a criterion allowing one to predicate 'strong' or 'weak' of the qualities a thing has.

The mathematization of this came chiefly from the 'Calculators' of the Merton school and later from the Parisian school. What mathematics did they have?[17] In the twelfth century, three new elements were introduced and gradually assimilated: Hindu-Arabic arithmetic with its superior notation, Euclidean geometry, and the algebra in the first (of three) parts of al-Khwarizmi's treatise. This part ended with the rule of three, or how to infer a fourth number on the basis of three. Thus, if eight cost five, how much do eleven cost? It was a verbal algebra. No symbols were used even for numbers. Jordanus Nemorarius in the thirteenth century first introduced these, in a very limited way. This treatment of proportions was gradually fused with Euclidean geometry. Euclid, more an organizer than an originator, had two distinct theories of ratios and proportions. The one in Book VII, stemming from Pythagoras, was limited to integers. The one in Book V, stemming from Eudoxos, treated continuous magnitudes. The medievals who adapted this lacked Eudoxos's concern with incommensurables and existence theorems. From these elements, they fashioned a conceptual tool that could treat intensification of quantities through a kind of verbal algebra.

To grasp the conceptual problems it is important to change perspectives. Today we would express the rule of three in a ratio which we would symbolize as $A/B = C/D$. The Merton calculators would symbolize this rule as as $(A,B) = (C,D)$. What is significant is not the change in format, but the interpretation given to it. The terms we treat as denominators were understood as parts of a system of classification which admitted of groups and sub-groups. If A is a multiple of B then (A,B) is a *multiple ratio*. If A contains B once with a remainder of 1 then (A,B) is a *superparticular ratio*. This admits of different kinds. (3,2) is a *sesquialterate ratio*; (4,3) is a *sesquitertian ratio*. If A contains B once with a remainder greater than one, then (A,B) is a *superpartient ratio*, which also admits of sub-groups. If A contains B more than once, then one has the general categories of *multiple superparticular ratios* and *multiple superpartient ratios*. In this way, Euclid's system of proportions gradually became assimilated to the arithmetic of fractions. This was extended to ratios of ratios, but still using terms and categories rather than mathematical symbols for numbers. Thus a limited verbal algebra was developed for expressing proportions for quantities of qualities.

This schematization was given a kind of state-space representation by Nicole Oresme at the University of Paris. In interpreting this it is important to realize the

[17] Most general histories of mathematics have little on this. Kline (1972) devotes a chapter 10 to this problem, but gives a rather whiggish account. A good survey, which I rely on here, is Mahoney (1987).

very abstract level at which the physical-mathematical correspondence is found. I will present Oresme's explanation of this and then comment on it.

> Every measurable thing except numbers is to be imagined in the manner of continuous quantity. Therefore for the mensuration of such a thing, it is necessary that points, lines, and surfaces, or their properties be imagined. For in them (i.e., the geometrical entities), as the philosopher has it, measure or ratio is initially found, while in other things it is recognized by similarity as they are being referred by the intellect to them (i.e., to geometrical entities). Although indivisible points, or lines, are non-existent, still it is necessary to feign them (*oportet ea mathematice fingere*) mathematically for the measures of things and for the understanding of their ratios. Therefore, every intensity which can be acquired successively ought to be imagined by a straight line perpendicularly erected on some point of the space or subject of the intensible thing, e.g., a quality. For whatever ratio is found to exist between intensity and intensity, in relating intensities of the same kind, a similar ratio is found to exist between line and line, and vice versa. (From Clagett's translation, Clagett 1966, pp. 165–167)

Consider a body and a variable quality, such as motion or heat. Represent the quality by a base line (eventually an x-coordinate) and the intensity by a perpendicular (or y-coordinate). 'Measurement', as Oresme uses this term, does not presuppose any unit or method of measurement. The length of the lines representing intensities has no absolute significance. What counts are the ratios. If the intensity doubles then the length of the perpendicular line should double. The initial assignments are always arbitrary, e.g. "Let the intensity of a quality have a value of four". The correspondence occurs at the level or ratios. As the intensity of a quality changes, the ratio of line lengths represents the ratio of the changing intensities. The mathematics of proportions handled ratios.

The men who developed these systems were, by our occupational categories, logicians, not physicists. They developed logical systems that admitted of mathematical representation and which might, incidentally, admit of physical examples. The term 'motion' still meant change in a quality with local motion gradually emerging as the most significant type and 'velocity' a term for the intensity of local motion. Wallace (1981) has clarified the basic systematizations used. I will use his notation. A uniform motion (U) is one with a constant velocity (v), while a difform motion (D) has changing velocity. Motion may be uniform or difform in two different way: with respect to space (U(x) or D(x)) depending on whether all the parts of a body move with the same velocity. Difform motion is of various kinds. Motion that is difform with respect to the parts of the object moved may be uniformly difform D(x), in the sense that there is a uniform spatial variation in the velocity of the various parts of the body, or difformly difform D(x) if there is no such uniformity, or it may be uniformly difform, D(t) or difformly difform, D(t) with respect to time. This allows of various schema for treating velocity. I will outline the one that Oresme used:

U(x) . U(t) (1)
U(x) . D(t) (2)
D(x) . U(t) (3)
D(x) . D(t) (4)

The standard example of (3) is a rotating wheel, or, for Oresme, the heavens, as circling around the earth or the sun (a hypothesis he considered). The different parts move at different velocities, but do not change in time. An example of (2) is a falling stone. All the parts move at the same velocity, but the overall velocity increases. This is not as good an example, since it is not really uniformly difform. The increase in velocity gradually stops, especially for light bodies.

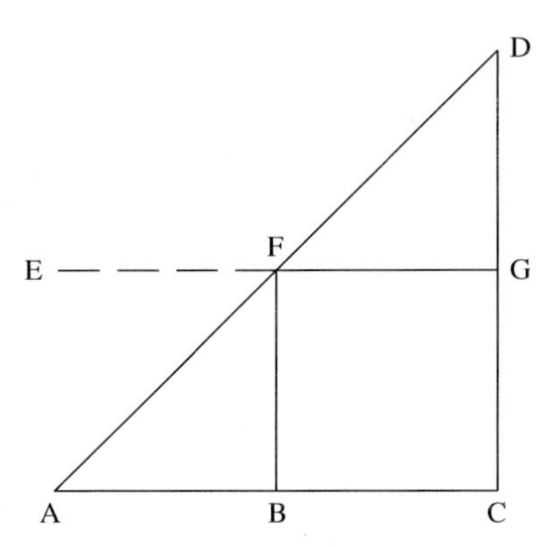

Either (2) or (3) may be represented by a diagram which illustrates the Merton theorem on uniformly difform motion. AC is the time axis, while BF and CD represent intensities of uniformly difform motion. The quantity of motion is represented by the area under the line representing motion. By comparing the area of the rectangle, AEGC, with the triangle, ACD, it is clear that the total motion of a uniformly difform motion is equal to the total motion of a uniform motion with half the terminal velocity. It also follows that the total motion in the second interval is three times the motion in the first interval.

This abstract approach raised questions that could not be handled in the received Aristotelian terminology. Thus, half the terminal velocity meant the instantaneous velocity at the point, F. This seems to involve vanishingly small distances and times. The development of this path led to Newton's theory of fluxions, or differential calculus. The notion of quantity of motion was still obscure. However, it was represented by the area under a line. For uniformly difform motion the areas involved triangles and rectangles, something these medieval logicians could handle. However, for difformly difform motion the quantity of motion was the area under an irregular curve. Archimedes had treated some such areas by his method of exhaustion. His works were not yet available. The development of this path led to the integral calculus of Leibniz and his account of functions. For Leibniz a function expressed the relation of a dependent variable to an independent variable. The ultimate independent variables are space and time. Functions, so defined, express the relations represented in Oresme's diagrams.

Before moving on, we should reflect on the significance of these developments. Christian Wolff introduced the term 'ontology', based on ὀν for a study of being as being. The developments considered gradually moved form a study of being as being to a systematization of the properties of beings. It could be labeled *idiontology*, by adapting the term, ἴδιον', which Aristotle used for a distinctive property. In place of an ontology based on Aristotle's ideas on substances and natures it stems from Aristotle's *Categories*, where a primitive whatness supports the properties systematized. The conceptualization of properties that emerged supported structures isomorphic to mathematical structures. However, both the conceptual structures and the mathematics were still radically underdeveloped. The subordination of the philosophy of nature to metaphysics/theology obscured the fact that idiontology was emerging as a new philosophical unit.

In the sixteenth century, Italy became the center for the development of mechanics, and of a mechanics that became more clearly related to dynamics. There were three somewhat separate traditions. The first, an academic tradition, was an extension of the work of the Calculators. Paul of Venice studied at Oxford and then, on his return, taught Mertonian ideas (Wallace 1981, Part II). Here the treatment of motion was caught up in the traditional clash between nominalists and realists. The realists were opposed to any simple identification of motion with a quantitative ratio. As Paul of Venice put it: "Motion is not a ratio, because a ratio is only a relative accident, whereas motion is an absolute accident" (Cited from Wallace, p. 68). This concern influenced the dynamic tradition that Galileo redeveloped. The nominalists focused on the new mathematical treatment and slighted the realistic significance of their formalism. The other two traditions, developed apart from the universities, were strongly influenced by the publication, in 1453 and later, of Moerbeke's translation of some of Archimedes' works. The Northern group, Tartaglia, Cardano, and Benedetti, was very concerned with applications of mechanics to ballistics. The group in central Italy, Commandind, Ubaldo, and Baldi, were more mathematically oriented (Drake and Drabkin 1969).

Aristotelian natural philosophy was still taught at the universities. After the fall of Constantinople (1453) some Greek scholars had moved to Italy and taught Greek literature and philosophy in the Academies sponsored by some noble families. This led to a type of Neo-Platonism opposed to the Aristotelianism taught in the universities. The Neo-Platonists especially studied Plato's Timaeus, where the astronomer, Timaeus, presents an extended explanation (29e–92c) of the order of the universe, beginning with the Demiurge fashioning preexistent matter in accord with mathematical forms and proportions and terminating with an explanation of illness and disease resulting from a lack of proportion of the four elements. A fusion of this idea of the universe as an embodied mathematical system with the Biblical account of creation supported the idea that God created the world in accord with ideal mathematical forms. The further fusion of these rather nebulous ideas with the mathematical treatment of quantities of qualities led to the conclusion that in coming to know things through their proper mathematical forms human knowledge matched divine knowledge.

In the Preface to his *Mysterium Cosmographicum*, Kepler wrote: "The ideas of quantities have been in the mind of God from eternity, they are God himself; they are therefore also present as archetypes in all minds created in God's likeness" (Cited from Koestler 1960, p. 65). In a letter to a friend he said:

> For what is there in the human mind besides figures and magnitudes. It is only these which we can apprehend in the right way, and if piety allows us to say so, our understanding is in this respect of the same kind as the divine, at least as far as we are able to grasp something of it in our mortal life. Only fools fear that we make man godlike in doing so; for the divine counsels are impenetrable, but not his material creation. (Citation from Baumgardt 1951, p. 50)

Galileo did not share Kepler's mathematical mysticism (See Cassirer 1967). Yet, he assumed a similar relation between quantitative forms as archetypes in the mind of God and men:

> I say that human wisdom understands some propositions as perfectly and is as absolutely certain thereof, as Nature herself; and such are the pure mathematical sciences, to wit, Geometry and Arithmetic. In these Divine Wisdom knows infinitely more propositions, because it knows them all, but I believe that the knowledge of these few comprehended by human understanding equals the Divine. (Galileo Galilei 1953, p. 114)

Galileo revived the distinction, which Democritos had adumbrated, between primary and secondary qualities.[18] Only primary qualities have objective reality, and these are the qualities that can be represented mathematically.

As these citations indicate, physical explanations were still functioning in a theological context. There was, however, the beginning of a shift from a theological perspective to an observer-centered perspective. Early in the fifteenth century, the Florentine architect and engineer, Filippo Brunelleschi, developed the basic laws of linear perspective, reportedly by painting a copy of part of the cathedral on top of its mirror image. Massaccio, della Francesca and others transformed painting by making perspective basic. Leon Batitista Alberti codified the rules of perspective in his book, Della pittura (1436), with a vanishing point and a horizon, both determined by the position of the observer. This is a geometrical representation of space. In linear perspective, the two dimensional representation of a three-dimensional space is thought of as a projection on a two dimensional surface of light rays traveling from the source to the eye of the observer, rather than the flat space of medieval painters. The space represented is a Euclidean homogeneous space organized from the standpoint of an outside viewer. In spite of strenuous opposition this new way of organizing representations of reality from the perspective of an outside observer rapidly spread to other fields. Classical French drama respects the 'Aristotelian' dramatic unities of an integrated story completed in one day at one locale. Aristotle had only insisted on unity of action. The 'classical Aristotelian doctrine' was articulated by sixteenth century Italian critics influenced by perspective. The dramatic action should be presented from the perspective of an observer.

Perspective spread to physics when Kepler, influenced by Dürer's perspectival methods as well as Galileo's account of his telescope, showed, in his *Dioptrice*, how a correct geometrical analysis of light rays explained vision. The theory it replaced, Aristotle's doctrine of transmitted images received as impressed sensible species, was never able to account for the fact that distant objects look smaller. Descartes's

[18] The scattered texts in which Galileo uses this distinction are collected in Burtt (1954, pp. 75–78).

La Dioptrique extended Kepler's work by giving a correct law of refraction. He explained different colors in terms of light producing different pressures on the eyeball. Perspective entered mathematics with Descartes' analytic geometry and the representation of bodies through coordinates in Euclidean space. Most analyses of this focus on the fact that the geometry is Euclidean, rather than on the portrayal of space from the perspective of an outside observer. The idea of the detached observer regarding physical reality from an external viewpoint culminates in Descartes' *Discourse on Method* and *Meditations*.[19] This detached-observer view of reality was gradually transformed into the notion of classical objectivity that Husserl sharply criticized.

Regardless of whether gunnery practice or perspective was the primary factor relating the observer to the described motion, the final result is clear. In place of the abstract ratios of the Calculators, the new methods begin with a three dimensional space, which supplies a framework for the measurement of motion. The bodily presence of the subject anchors this framework. Galileo extended this through his development and use of the telescope, describing in precise detail the positions of the Medicean stars as he saw them on January evenings in 1610. Galileo's work is not simply an extension of the preceding developments. Galileo played a pivotal role in developing empirical science; shaping and judging mathematical formulas by the way they fit controlled observations. In spite of his early exposure to Aristotelian natural philosopher, he took Archimedes, the prototypical mathematical scientist, as his ideal. However, these advances were only possible because of the developments we have been surveying. The mathematical treatment of motion and forces emerged from three centuries of muddling through quantities of qualities, verbal algebra based on proportions, and a gradual switch from a theological interpretative perspective to an observer-centered viewpoint.

A recent study, Brading and Jalobeanu (2002) demonstrates how Descartes gradually transformed the metaphysical problem of individuating bodies into a physical problem. In the scholastic tradition, the individuation of a body was explained through the composition of matter and form. Its perdurance through time as the same individual was explained through divine concurrence. Descartes transformed a principle of individuation into a model of individuation. Measurement, especially of spatio-temporal location, specifies an individual body. Conservation laws, not yet clearly formulated, conserve its state over time. The old tradition of treating quantities of qualities as a philosophical problem continued (See Solère 2001) but was not influential in the further development of physics.

[19] This brief summary is based on two articles, Frye (1981) and Chevalley (1993). Drake and Drabkin claim that there is no evidence to support Duhem's claim that da Vinci's notoriously difficult writings had an influence on the development of mechanics.

2.4 A New Physics Emerges

The new mathematical treatment of quantities culminated in the calculus of Newton and Leibniz and Newton's *Principia* (Newton 1952a). Since we are concerned with the developing relation between physical concepts and mathematical formulations, it is helpful to consider how this relation evolved. Three factors played a primary role. The first was the **co-evolution** of physics and mathematics. Co-evolution is featured in biological accounts of prey and predators, of hosts and parasites, but not in the physical sciences. Descartes, Newton, Leibniz, Euler, the Bernoullis, d'Alembert, Lagrange, Laplace, Fourier, Poisson, Gauss, and many more through Poincaré and Witten, made contributions to both physics and mathematics. The original formulations of the calculus were given a very physical interpretation that will be considered later. The result was a gradual accommodation of nested physical concepts to particular mathematical forms. Developing theories are interpreted through experimental results as reported in language. Here the structuring effect of concepts and conceptual networks is basic. We get at this by studying co-evolution. This shaping of physical concepts is totally lost in a more formal presentation of theories.

The second factor is a gradual change from a broadly theological perspective to an observer-centered perspective. In spite of the changes just noted all the leaders of the scientific revolution considered theology to be the fundamental arbiter of truth. Galileo deliberately set in motion the process that led to his first trial before the Inquisition. He thought of himself as a loyal Catholic and, in his widely circulated letter to the Grand Duchess Christina, tried to show how the interpretation of Scripture could be modified to accommodate a heliocentric cosmology. Kepler was officially denied communion in the Lutheran church primarily because of his belief that Lutherans, Calvinists, and even Catholics should respect each other's positions (See Connor 2004, pp. 243–251). One might suppose that the British Virtuosi (aka scientists) were more secular, because of the more tolerant regimes of the Restoration after the Puritan reign (1649–1660). This supposition hardly fits the titles of some of the basic scientific works produced.

- John Ray, *The Wisdom of God Manifested in the Works of Creation* (1601)
- Walter Charleton, *The Darknes of Athiesm Dispelled by the Light of Nature* (1652)
- Robert Boyle,
  - *The Excellence of Theology* (1674)
  - *Of the High Veneration Man's Intellect Owes to God* (1674)
  - *The Christian Virtuoso* (1690)
- Nehemiah Grew, *Cosmologia Sacra* (1701)

Boyle declared that the first requirement for being a virtuoso is that one be a proper Christian Gentleman. The advance of science requires trust, a belief that a virtuoso really performed the experiments and achieved the results reported. What kind of a person can really be trusted? The list of scientific books has one notable omission, Newton's *Principia* (1686), the exemplar of secular science. What requires an explanation is the question of how Newton came to develop a physics not dependent on a theological perspective or metaphysical assumptions. This will be treated later.

The third factor is problem solving. Here any general theory of the role of problem solving would induce a systematic distortion. Instead we will consider a few particular cases in passing, chiefly cases of creative thinkers struggling to make sense of breakthroughs. Kepler is a particularly interesting example, because his work precedes the development of calculus. The details of Kepler's development of his first and second laws remain matters of historical debate. I will offer a plausible reconstruction primarily to illustrate the developing interconnection of physical inferences and mathematical inferences. Ptolemy's geostatic system did not assume that the center of the earth is the center of each planet's motion. The basic motion for each planet was a circular motion around its eccentric center. Each planet had a different eccentric center of motion. Copernicus's heliostatic system and Brahe's system, in which the inner planets revolve around the sun, with some modifications. One takes circular orbits as basic and then accommodates deviations in the shape of the orbit or the velocity of a planet by adding: an eccentricity, the distance between the planet's center and the geometric center of a planet's orbit; a deferent, effectively the mean radius of a planet's orbit; epicycles, circular motions around the deferent point; and an equant point, with respect to which the radial velocity was uniform. Combinations of these factors supply a basis for accommodating any planetary motion and are still used in planetariums for projecting past and future appearances of the heavens. Tycho Brahe, with an assist from Kepler, made astronomical observations of unprecedented accuracy. His large cross-shaft sighting rods, precisely graded scales, and other pre-telescopic equipment led to a catalog of planets and fixed stars with an accuracy, in the best cases, of less than a minute of arc. Previous accounts were accurate only within ten minutes of arc. After Brahe's death, Kepler acquired (or absconded with) the data and extended the systematic analysis of the orbit of Mars that he had worked on under Brahe's direction. Now, however, he worked on the assumption that the Copernican, rather than the Brahian, system is correct.

What Kepler analyzed was not the raw observational data, but carefully corrected data. Mars was observed from earth, but its orbit had to be calculated with respect to the sun. Kepler needed a precise calculation of the earth's orbit to accommodate the moving platform of the observational basis. There were fairly good estimates of the earth-sun distance. This, however, did not determine the earth's orbit. It should be a circle around an eccentric point displaced at some distance from the center of the sun. This was a problem that did not exist in the Ptolomaic and Brahian systems and was not treated in the Copernican system.

Here Kepler resorted to an ingenious trick. An observer on earth can calculate the orbit of Venus on the assumption that when the angle between the observed position of Venus and the observed position of the sun is a maximum, then the line of sight from the earth to Venus is tangent to Venus's orbit. This gives a right triangle allowing a calculation of Venus's distance from the sun as a fraction of the earth's distance. An observer on Mars could calculate the ratio of earth and Mars's orbit by the same method. Kepler, in effect, made himself the Martian observer.

He chose three points on Mars's orbit known with accuracy. The position of the earth in its orbit could be determined for these times. From three points on the earth's calculated orbit, assumed to be circular, he could calculate the center of its eccentric circle. Comparing this radius with the calculated earth-sun distance yielded the earth's eccentricity. The results fit quite well. Then he picked four calculated positions of Mars and calculated its distance from the sun and its eccentricity. This overall argument may seem circular in a logical sense. Kepler used calculations of Martian positions to calculate earth's position, and then used the calculated earth positions and orbit to recalculate Mars's position. Today this type of problem would be treated as an iterative procedure, which could easily be handled by a computer program. Kepler claimed that he redid the calculations seventy times. In spite of all these repeated calculations he could not obtain consistent results. The basis of the inconsistency requires consideration.

Kepler could have used the Ptolomaic mathematical machinery of eccentrics, deferents, epicycles, and an equant point, to accommodate his corrected data on Mars's orbit. He believed, however, that the sun causes the motions of the planets, and that the attractive force exerted by the sun diminishes with distance. This was a novel concept. Both Ptolomy and Copernicus accepted the Aristotelian idea that the natural motion for heavenly bodies is circular. Even Galileo, who was developing an inertial physics, distinguished two types of inertial motion, linear and circular. It seems that Kepler inferred the idea of the sun's attractive force both from his quasi-divinization of the sun and from the decrease in orbital velocity with increasing orbital distance of the planets. This was the first anticipation of universal gravitation. Therefore, he inferred that the sun had to be the actual center of motion. This supplied the basis for his calculations. The assumption that Mars, like the earth, has uniform circular motion around the sun did not fit the corrected data.

Kepler's next attempt involved a method for calculating Mars's position in orbit based on his corrected data for selected positions. From the fact that the earth's velocity at perihelion and aphelion is inversely proportion to its distance from the sun and by an adaptation of Archimedes mathematics, he inferred that equal areas of arc are swept out in equal times. The orbit calculated on this basis worked fairly well, but had errors of eight minutes of arc in excess at the furthest Mars-sun distance and of defect at the shortest. Here, it seems, Kepler tried various stratagems, such as modifying the equal areas in equal times assumption. Finally, as a desperate expedient, he abandoned the assumption of circular orbits. This not only had the weight of tradition behind it. It seemed evident on intuitive grounds. Tie a stone to a string and swing it around. One does not have perfect circular motion because of arm and wrist motion. Yet, the stone's motion must be understood in terms of corrected circles, because the swinger causes it.

Kepler assumed an oviform circle, like a wooden hoop under pressure. This could accommodate both the excess and defect in Mars's orbit. There was no available mathematical formulation for such shapes. Kepler tried a variety of stratagems. Finally, he used an ellipse as a calculation device. On intuitive grounds, this could not be the correct shape. An ellipse has two foci. Since, in Kepler's view, the sun is the cause of the planet's motion, there can only be one center. An ellipse, however, had a distinct mathematical advantage. Kepler could adapt Apollonius's treatise on conic sections to calculate an entire elliptical orbit from three selected points. Then he made two sets of adjustments, one to accommodate the differences between the ellipse, which has two foci, and the oviform shape, which has one center. The second set of corrections adjusted the oviform shape to the calculated observation points. Here again, Kepler needed repeated recalculations to accommodate plausible orbital assumptions. It gradually became clear that the two sets of corrections cancelled. Mars has an elliptical orbit. This conclusion did not fit Kepler's ideal of proper heavenly motions. However, it did support a very significant physical inference. If Kepler had treated his repeated calculations merely as calculations, then there was no reason to suppose that the Martian calculations would apply to other planets. If these calculations are coupled to the assumption that the sun causes planetary motion, then he could infer that all planets travel in elliptical orbits. As Laplace noted, Kepler performed his calculations at exactly the right time. Prior to Brahe's data, there were no sufficiently accurate observations. Subsequent telescopic observations brought out departures from ellipticity. C. S. Pierce, who pioneered the study of physical inferences, evaluated Kepler's work, as "This is the greatest piece of Retroductive reasoning ever performed."[20] Kepler himself described it as a cartload of dung. It did not fit his account of orbits in terms of the shapes of regular solids.

Two aspects of this development illustrate the co-evolution of physical and mathematical concepts. The first is the idea that ellipses, rather than circles, should supply the basis for describing planetary motion. Now this seems trivial. Kepler required native ability, extreme stubbornness, and protracted effort to overcome the received wisdom of his predecessors. The second is the emergence of the concept 'gravity'. It played a basic role in Kepler's extension of his Martian studies to other planets. It did not yet have an adequate mathematical formulation. Conceptual, like biological, evolution often relies on punctuated equilibrium.

2.4.1 Newtonian Dynamics

On the assumption that the developments leading to Newtonian physics are widely known, I will omit one more summary and focus on the role of quantitative concepts in Newtonian physics. Newton's protracted efforts to clarify the concepts he took to be foundational were complicated both by the novel way he interrelated physics and mathematics and by the ontological significance he attached to foundational concepts. What I wish to consider here is the development of Newton's method of interrelating physical, mathematical, and philosophical concepts, and to clarify

[20] His brief analysis is in his Collected Papers (Peirce et al. 1931, Vol. I, pp. 28–31). Hanson (1961, pp. 72–86) extended Pierce's analysis and in Hanson et al. (1972, pp. 249–273) developed an account of cross-type inferences. The historical details of Kepler's development are summarized in Casper (1962, pp. 128–147).

the way in which Newton gradually and unintentionally made it possible for later Newtonians to ignore philosophical considerations which Newton himself thought indispensable. Accordingly, I will concentrate, not on Newton's well known successes, but on his conceptual struggles.

Newton's method of relating physical concepts and their mathematical expressions was conditioned by his failures as well as by his successes. Alan Shapiro, who edited Newton's early optical lectures, has clarified Newton's early efforts to develop a mathematical theory of light (Shapiro 1984). Newton began with phenomena, which could both be described and also be represented mathematically. The most important such phenomena was the decomposition of white light by a prism. To treat this mathematically Newton calculated the chromatic aberration of a planoconvex lens. He also developed mathematical formulas for refraction and dispersion in order to calculate the refractions of different colors at the interfaces of different substances, such as air and glass.

Newton's ideas on the proper way to develop a mathematical physics were closely related to his ideas on the methodological role of analysis and synthesis. Through analysis one resolves phenomena into their ultimate causes or constituents. Ideally this analysis leads to quantities that can be related by fundamental mathematical laws, from which a large variety of phenomena may be deduced. On a phenomenal level orange light looks the same whether it is pure orange or a mixture of red and yellow. The real difference between the two oranges is revealed by passing each through a prism. One remains pure orange, while the other is broken down into its components. This suggested to Newton that refrangibility, which could be directly correlated with all different colors, could serve as the key concept in developing a mathematical theory that could account for the phenomena. In his 1672 paper, "A new theory about light and colors" he proclaimed:

> A naturalist would scearce expect to see ye science [of colours] become mathematicall, & yet I dare affirm that there is as much certainty in it as in any other part of Opticks. for what I shall tell concerning them is not an Hypothesis but most rigid consequence, not conjectured by barely infering tis thus because not otherwise or because it satisfies all phenomena (the Philosophers universal Topick), but evinced by ye mediation of experiments concluding directly & w^{the} any suspicion of doubt. (Shapiro 1984, p. 34)

Newton's mathematical theory was to be based on three laws of dispersion.

1. Snell's sine law of refraction was assumed to be valid for each color separately;
2. Newton developed a dispersion law by adapting Descartes' model of the impulse corpuscles receive when they cross a refractive surface. Newton suppressed the mechanical aspects of the model, as well as any reference to Descartes, and presented a mathematical formula representing the dispersion, $\Delta\, n$, of light passing from a medium with an index of refraction, n, to one with an index of refraction, n', and a dispersion, $\Delta\, n'$,

$$\frac{\Delta n}{\Delta n'} = \frac{(1/n)(n^2 - 1)}{(1/n')(n'^2 - 1)}$$

This formula assumes (incorrectly) that the index of refraction is a property of light alone, not of the refracting substance.
3. The third law was one which enabled Newton to determine relative indices of refraction of any two media provided that their refraction is known with respect to some common medium, such as air.

Newton thought that the truth and certainty of these laws would obviate the need for any further dependence on experiments. These theoretical laws, however, led to false consequences. Newton's *experimental* analysis shared only a principle and a half with his mathematical theory. These are the principles that:

1. sunlight consists of unequally refrangible rays; and
2. there is a correspondence between refrangibility and color.

Shapiro interprets this as a principle and a half on the grounds that the second principle was only developed qualitatively.

The publication of the "New theory" led to the acrimonious debates and criticisms concerning Newton's reliance on hypotheses. Four months after sending off the "New theory" Newton suppressed his Optical lectures with their mathematical theory of light. When he published his *Opticks*, 34 years later, he repudiated his earlier Optical Lectures and insisted that : "My Design in this Book is not to explain the Properties of Light by Hypotheses, but to propose and prove them by Reason and Experiments" (Newton 1952b, p. 1).

Newton's developing position on the interrelation between mathematics, physics, and philosophy was complicated by various factors.[21] First, Newton gradually developed what I. Bernard Cohen calls 'the Newtonian style' (Cohen 1980 chap. 3). The mathematics proper to Books I and II was neither Descartes' *mathesis universalis*, nor pure mathematics in the contemporary sense. Newton, building on Descartes *Principles of Philosophy*, constructed imaginative idealizations of physical bodies, idealizations that admitted of precise mathematical expression, and then worked out the mathematical consequences of these hypotheses. This was generally an iterative process, beginning with the assumption of a central field force and its consequences, then a two-body interaction, then extended bodies, then many bodies. When this mathematical idealization and its consequences gave a sufficient approximation to the real world, then, in Book Three of the *Principia*, Newton compared his mathematical system with the observed phenomena. This was the second step, the physics.

The third step should be philosophy, an explanation of the phenomena through their true causes. This was the goal that Newton, as a philosopher of nature, set for himself. It was not, however, the goal that was realized in the *Principia*. The novel way in which Newton came to separate his mathematical physics from an explanation of phenomena through causes hinged on a symbiotic relationship between

[21] I am omitting the more familiar aspects of Newton's thought and its immediate background. A detailed treatment of this may be found in Cohen (1971) or Westfall (1980).

his concept of 'force' and his method of attaching mathematical values to physical concepts. Alan Gabbey (1980) has traced in detail the way in which Newton's conceptualization of 'force' emerged from the ongoing dialog on force and motion, and especially from Newton's adaptation of Cartesian notions. Descartes and his immediate predecessors thought of force in terms of a contest between the motive force of one body and the resisting force of another, with the excess determining the degree of acceleration or retardation. Newton eventually transformed this into his own idea of a '*vis inertiae*'. In the *Principia*, this term has two different, though related senses. First, it signifies a resisting force equal and opposite to the '*vis impressa*'. Second, it signifies a persevering or maintaining force equal to the body's total quantity of motion. The second sense implies that for Newton, as for the Aristotelians, unhindered motion in a straight line is an effect requiring a cause, and ultimately requiring God as first mover. This type of effect, however, was not an object of dynamical investigation, but a part of philosophy proper.

When Newton rejected the idea that a change of state is the result of a contest between unequal forces, he substituted a balance between a *vis impressa* acting on a body and the *vis inertiae* with which the body resisted. He gave this balance its canonical formulation in his Third Law of Motion. The *vis inertiae* could be identified with the *vis insita* or the moving force preserving a body in its state of rest or motion. This supplied an ontological presupposition of Newton's own thought, but one only indirectly reflected in his mathematical method. Thus, in deriving Kepler's second law, Newton thought in terms of a balance of forces, a *vis insita* preserving the state of motion and an impulsive centripetal force (See Gabbey 1980, p. 280). However, the only force that directly entered the mathematical account was the force responsible for a change of state. Since this could be measured by its effects, Newton could treat it mathematically while postponing an account in terms of true causes to philosophy proper (Book I, Prop. LXIX, Scholium).

The way Newton had come to relate mathematics, physics, and philosophy explains the secular quality of the *Principia*, its freedom from metaphysical and theological presuppositions. As the biographies by Westfall and Gleick show, before writing the *Principia* Newton had made an intensive study of the origins of Christian doctrine and concluded that the doctrines of the divinity of Christ and of the Trinity were not part of the original Christian traditions. He also analyzed the Book of Revelation (The Apocalypse) to determine the time of the second coming of Christ and the purging of the heresies, aka orthodox Christianity, that had corrupted the Christian tradition. Like Aristotle, he thought of God as the first cause of all motion. The philosophical explanation that should complete the work begun in the *Principia* should be based on God as first cause. In his Third Rule of Reasoning (*Principia*, p. 399) he took an explanation of gross matter in terms of the basic properties of atoms to be the foundation of all philosophy. Yet, neither type of cause could play a basic role while following the methodology of the *Principia.* Motion was not explained in terms of causes, but through a mathematical analysis of the effect of forces. The account of God as first cause and ruler of the universe was only treated in a pious patch, the General Scholium added to the *Principia* (pp. 542–547) As the developer of differential calculus, he effectively treated fluids as continuous, though he believed fluids were composed of atoms. The distinction between mathematics, physics, and philosophy excludes atomic explanations:

> But whether elastic fluids do really consist of particles so repelling each other is a physical question. We have here demonstrated mathematically the properties of fluids consisting of particles of this kind, that hence philosophers may take occasion to discuss this question. (*Principia*, p. 302)

The great innovation was the law of universal gravitation. Newton had developed and rejected two attempts to give a causal account of gravity. The General Scholium contains the most cited of Newton's claims:

> But hitherto I have not been able to discover the cause of these properties of gravity from phenomena, and I frame no hypothesis; for whatever is not deduced from the phenomena is to be called an hypothesis; and hypotheses, whether metaphysical or physical, whether of occult qualities or mechanical, have no place in experimental philosophy.

It is easy in retrospect to use this citation as a basis for imposing a positivistic interpretation on the *Principia* and regarding it as a mathematical systematization of measurable properties. Measurement, however, plays no role in the first two books and only a minor role in the third. What was emerging was a distinctive methodology. The competing traditions of natural philosophy, Aristotelianism, Cartesian, and the Leibnizian approach which Christian Wolff later systematized, all taught that natural philosophy should be based on intelligible principles. The Newtonian account of gravity did not qualify. In Newton's new methodology natural philosophy begins with careful observation of phenomena leading to inductive generalizations, which could be given a mathematical systematization.

The implementation of this methodology led to a conceptualization of bodies and their properties. This emerged from and originally functioned as part of the natural philosophy tradition, as the book's full title indicates. In this regard, the most salient feature is not Newton's success, but his failure. As indicated earlier, he envisaged a three-phase project. The first two phases fit into a causal account in a rather loose fashion. The operative assumption was that, though force considered as a cause requires a philosophical analysis, force considered in its effects may be studied through a mathematical analysis of the consequences of different force laws. Newton's third phase, philosophy, was notoriously unsuccessful. He never worked out an adequate philosophical account of forces or an explanation of the force of gravity. Accordingly, we concentrate on the concepts that came to play a functional role in the *Principia*, rather than the philosophical account that Newton intended, but never adequately developed.

In this perspective, we can consider the quantitative concepts that had a foundational role in the mathematical physics of the *Principia*, though not necessarily in Newton's philosophy of nature. These concepts are: 'quantity of motion', 'quantity of matter', 'inertia', 'acceleration', 'force', 'gravity', and 'essential properties of bodies'. Since this is deliberately streamlined, I will refer to other sources for the actual historical developments.

The medieval term 'quantity of motion' expressed a kinematic concept. The Latin term, *motus*, applied to change in general. Motion was regarded as a process requiring a cause. Rest was a privative notion, the absence of motion. Newton, following Descartes, treated motion as the state of a system. Rest was no longer an opposed concept. It was simply the ascription of a zero value to a state. Newton's definition, "The quantity of motion is the measure of the same, arising from the velocity and quantity of matter conjointly" (*Principia*, p. 1) makes quantity of motion a dynamic notion (momentum) by the introduction of 'quantity of matter'. This was a transformation of the kinematic notion that Galileo used in his law of falling bodies. It is intimately related to the notion, 'inertia'.

'Quantity of matter' was a more confused notion. The medieval discussions had intermixed the Aristotelian doctrine of substance and accident, the problem of dimensions, and the theological doctrine of transubstantiation, the substance of the bread and wine are replaced by the substance of the body and blood of Christ, though appearances remain unchanged. Newton's mechanistic predecessors had tried to explain quantity of matter either in terms of extension (Descartes), or as a function of the number and kind of atoms comprising a body. Newton's definition: "The quantity of matter is the measure of the same, arising from its density and bulk conjointly" (*Principia*, p. 1) was later attacked by Mach as involving a vicious circle. I think that McMullin's (1978) account clarifies this Newtonian concept. One should distinguish the dynamic concept of 'mass' as it functioned in Newton's mathematical physics from Newton's speculations about matter and forces. The dynamical concept 'quantity of matter' was essentially a more precise version of the more common sense concept, 'amount of stuff'. Its clarification comes, not merely from Definition 1, but from the eight definitions and three axioms. Then Definition 1 is not circular; it serves to relate an unfamiliar notion to notions presumed more familiar.

'Quantity of mass' and 'quantity of motion' are both *extensive* concepts. Though the term 'extensive' is not Newton's, he was quite clear about the concept. The mass of a body is equal to the sum of the masses of its parts. 'Quantity of motion' is also extensive; the overall quantity of motion is determined by calculating sums and differences (Corollary III). This extensive property is the basis for relating these new concepts to the mathematical system.

As his opponents insisted, and as even Newton himself acknowledged, 'force' was a more problematic concept. In spite of the complications involved in treating '*vis insita*', '*vis inertiae*', 'impressed force', and 'centripetal force', the relation of 'force' to the mathematical system is clear. Though force, considered as a cause, requires a philosophical explanation, force, considered through its effect, is an extensive property. This was all Newton required "For I here design only to give a mathematical notion of those forces, without considering their physical causes and seats" (*Principia*, p. 8). Earlier, we indicated how the novel aspects of this concept emerged out of an ongoing dialog within the tradition of philosophy of nature. It remains to show the special sense in which this became a quantitative concept. Cohen (1980, pp. 171–182) has traced the way in which Newton mathematicized this concept. The reigning mechanical philosophy treated forces as contact forces. This involved no appeal to occult qualities. Newton approached continuous forces, especially the force of gravity, by considering impulsive forces and then the limit of a sequence of impulsive forces. This enabled him to get at continuously acting forces mathematically without postulating action at a distance or introducing hypotheses concerning the cause of gravity.

'Inertia' is another concept which Newton accepted from his predecessors, especially Descartes and later Kepler (Cohen 1980, pp. 182–193), and transformed by making it a dynamic rather than a purely kinematic notion. Descartes is generally credited with being the first to give a clear correct statement of the law of inertia. On the point, however, I think that Gabbey (1980, pp. 286–297) is quite correct. 'Inertia', like other basic concepts cannot be understood through any sort of referential semantics. It relates to such other concepts as 'quantity of motion' and '*vis inertia*'. Newton transformed these concepts in the way sketched above. So, Newton should be credited with being the first to give a correct statement of the law of inertia. This culminated the process leading from the Aristotelian physics, in which motion required a continuous cause, to the Newtonian, where change of motion was explained in terms of impressed force. Newton thought of 'force' in causal terms. But, it was the quantitative relation between force, mass, and acceleration, not a theory of causes, that was basic in the functioning of the *Principia*. Though Newton speaks of the 'force of inertia', he never accords this inertial force the type of mathematical treatment, vectorial addition, accorded impressed force.

The difficulties presented by 'acceleration' are more mathematical than conceptual. The quantities that play a functional role in the *Principia* are those that can be treated by the method of the first and last ratio of quantities (*Principia*, pp. 29–39). 'Change of motion' is the limit of the ratio of two such quantities. The technique for determining limits is clear in spite of the difficulties in giving the mathematical operations a physical interpretation: "And in like manner, by the ultimate ratio of evanescent quantities is to be understood the ratio of the quantities not before they vanish, nor afterwards, but with which they vanish" (*Principia*, p. 39).

'Gravity' is the most crucial and problematic concept in the Principia. For our purposes we should make a sharp distinction between the problems Newton and his successors had in attempting to explain the cause of gravity and the way the concept functioned in the Newtonian system. The outline presented here is simply a summary of ideas developed by Westfall (1971, chap. 10) and Cohen (1980, chap. 5).

In November 1679, Hooke wrote Newton a letter in which he explained planetary motion as compounded of a direct motion by the tangent and an attractive motion towards the central body. Newton investigated this suggestion and found that whatever the law of force was he could derive Kepler's second and third laws from this assumption (though he did not credit Kepler with the second law).

Westfall sees this as the decisive turning point in Newton's treatment of forces, going beyond the mechanistic concept of contact forces and introducing forces of attraction and repulsion. Newton speculated on the nature of these forces and elaborated their mathematical consequences. His methodology prescribed the order of treatment. First work out the mathematical consequences of an assumed force law; then compare the mathematical consequences of these assumptions to physical reality. If the comparison establishes the truth of the assumption, then the philosopher should seek a causal explanation of these forces. As Cohen summarizes it: "But I believe that the key to Newton's creative thought in celestial dynamics was not that he considered forces to be real, primary, or essential properties of bodies or of gross matter, but rather that he could explore the conditions and properties of such forces as if they were real without needing to be able to find a satisfactory answer (or any answer at all) to questions about the independent realty of such forces" (Cohen 1980, p. 253).

This leads into Newton's discussion of the essential properties of bodies in the Third Rule of Reasoning (*Principia*, p. 399, MacKinnon 1982, pp. 32–37). There Newton argued that extension, hardness, impenetrability, mobility and inertia are the essential properties of all bodies. A few comments are in order. First, this was developed as an introduction to *Book III, The System of the World*. The mathematical physics of Books I and II did not need or use these essential properties. Second, Newton's claim that the ascription of these properties to all bodies is the foundation of all philosophy should be taken seriously, rather than dismissed as a throwaway line. Newton was consciously and deliberately trying to revise the mechanical philosophy: by attributing to matter a categorization of properties different from that of the Aristotelians (substance, quantity, ...), the Cartesians (extension the only essential property) or Boyle's mechanism (where extension is supplemented by dispositions which are to be explained mechanically); and by the transformation of the concept of force already considered. Yet, he was also clear on where philosophy, so conceived, fits into his account. First, one investigates the consequences of forces, regardless of how they are produced, on the assumption that forces are additive; next one compares the consequences of this mathematical analysis with the phenomena of nature: "And this preparation being made, we argue more safely concerning the physical species, causes and proportions of these forces" (*Principia*, p. 192). Thus, the properties thought basic to physical reality *as explained in a philosophical account* need not be the properties that play a basic role in an account of the phenomena, or physical reality as described.

The basic link between the concepts that play a foundational role in the *Principia* and those that should play a foundational role in philosophy is the requirement of extensiveness. 'Quantity of matter', 'quantity of motion', and 'force' are additive. If the philosopher is to explain the basic properties of observed bodies in terms of the basic properties of ultimate corpuscles he must assume properties that are extensive in that the properties of the whole can be explained in terms of the sum of the properties of the parts. For observed bodies, however, quantity of motion can be measured, especially for planetary motion and motions under laboratory conditions. Force can be measured by its effect, a change in motion. Quantity of matter can be measured by measuring weight. Newton assumes (*Principia*, p. 304) that the quantity of matter in a body is proportional to the weight and claims (p. 411) that his experiments demonstrate the adequacy of this assumption to one part in a thousand.

Thus, the conceptualization of reality that played a foundational role in Newton's *Principia* was different from and simpler than the philosophical account Newton aspired to. For the purposes of the mechanics, one treats the observable world as a collection of bodies endowed with the properties of extension, mass, and inertia. They all coexist in a space time framework that is absolute, i.e., has a reality independent of the bodies existing in this framework. The motions of these bodies are explained in terms of inertia, forces, and gravity. This can readily be interpreted as a simplified extension of the Strawsonian core considered earlier. Newton did not so interpret it. He thought that the system could be considered philosophically acceptable only if these properties of bodies could be explained in terms of the properties of atoms. The properties of atoms were not determined empirically, but by arguments based on conceptual necessity. The ultimate corpuscles must have all and only the properties common to all matter. These philosophical requirements never became a part of functioning physics. The later development of Newtonian mechanics will be considered in the next chapter.

2.5 Philosophical Reflections

Our primary concern is with the role of the language of physics. We do not yet have sufficient data for an analysis. The historical developments traced here place us at something of a conceptual point midway between the beginnings of science, where no one knew how to represent the quantity of a quality, and modern physics with its elaborate representations of quantities. To relate these developments to changes in linguistic usages we return to Davidson's perspective for the interpretation of language. Relating language to reality requires a triangulation of the individual speaker, the society of language users, and the reality known.

Individuals speak particular languages. We have been treating language abstractly, prescinding from what language the speaker is using. Here again we are at a conceptual midway point. The original language of the physics considered was Greek as spoken in Athens. Eventually, the vehicle of communication became scholastic Latin, as spoken in universities. This was succeeded by the French of the Enlightenment era and the current standard, English as spoken by a German professor, someone who learns and follows the rules. The language of physics is rooted in living languages, but has always been somewhat detached from the constraints imposed by the lived world.

The society of European language users was undergoing drastic changes: the rise of the venacular the Renaissance, Reformation, Counter-reformation, the age of discovery, the rise of nationalism, the beginning of capitalism, all affected it. We will bypass these presumably familiar developments and mention only two aspects pertinent to the development considered. The first is the establishment of vernacular languages as vehicles for expressing science. Here again, art preceded science. The troubadours in France, Dante and Petrach in Italy, Chaucer and Shakespeare in England, Luther in Germany, Cervantes in Spain, and many others, shaped the vernacular languages they spoke into vigorous, often earthy, vehicles for literary forms. The rise of nationalism led to the dominance of the language spoken in the court. French gradually replaced the Occitan of Southern France and the Celtic language of Britanny. Castilian came to be the language of Spain, in spite of the resistance of those speaking Catalan and Basque. My remote ancestors in the isles of Scotland fought in vain to preserve Gaelic over the encroachments of anglophiles. Copernicus, Galileo, Kepler, Descartes, Newton, Huygens, and Leibniz used Latin for official publications and vernaculars for popularizations. Increasingly, the Latin treatises were translated and read in vernacular form. This entailed adapting vernacular usage to accommodate terms and structures developed, for scientific purposes, in some other language, usually scholastic Latin.

The second societal change meriting mention is the rise of scientific societies. In Italy, Germany and elsewhere there were private groups seeking the advancement of knowledge of nature, like the Accademia dei Lincei, to which Galileo belonged. Most of these were short-lived. The Royal Society in England and the Paris Academy had the benefit of royal patronage and adequate funding. (See Mason 1962, chap. 22, for a survey.) They, and the journals they stimulated, set the initial style for international cooperation. This was another factor in decreasing a dependency on the resources and limits of individual languages.

The third factor in the triangulation is reality. It is enticing to think of this as reality, as it exists objectively, being progressively revealed through scientific discoveries. Our concern is with conceptions of reality implicit in and transmitted through language. Previously we focused on the features of ordinary language required for unambiguous reference and communication of information. What we are considering now is the partial breakaway from an ordinary language framework manifested in the early development of physics. Almost all the philosophers and early scientists who were the agents of change shared a basic theological perspective. In spite of the differences separating Judaism, Islam, Catholicism, Orthodoxy, and Protestantism, all accepted the idea of a single god who created the world in accord with a divine plan. Rather than the familiar focus on religion as a repressive influence, I wish to consider the explanatory significance of this theological perspective.

When Jocelyn Bell discovered the first pulsar in 1967 she entertained the LGM (little green men) hypothesis. In spite of popular support, this hypothesis was abandoned when a second pulsar was discovered and when the immense distance and energy, of pulsars was recognized. Consider the difference in the type of questions suggested by the LGM hypothesis and the rotating neutron star hypothesis. If one interprets the 50 millisecond pulses received every 1.3373 seconds as signals sent by intelligent aliens, then the urgent question is how to decode the message being sent. If one interprets these pulses as phenomenological manifestations of cosmic processes, then the pertinent questions concern the type of objects and processes causally responsible for such phenomena.

When one regards the universe as a product created by an infinitely wise being and designed with human beings as central, then this shapes the way the universe is understood and the type of questions that seem pertinent. In the sixteenth century opposed interpretations of God's goals for society contributed to the wars of religion and to the persecution of dissidents. This orientation carries over even to areas, like the philosophy of nature, that should be developed prior to and independent of any theological commitments. Aristotle introduced purpose in nature as a brute fact; something far removed form the first mover's self-contemplation. Medieval Aristotelians in the Islamic, Jewish, and Christian traditions also recognized purpose in nature, but attributed it to the plan of creation. This theological orientation remained intact through the scientific revolution. Galileo and Kepler spoke of God as the divine geometer. Descartes methodic doubt drew the line at divine deception. Newton's General Scholion at the end of the *Principia* and Leibniz's assumption that this is the best of all possible worlds, both put physics in an essentially theological context. In Quine's terms, physics did not function as the basic science of reality, because theology was the court of last appeal.

This intellectual orientation began to change, not because it was abandoned, but because its explanatory efficacy was eroding. The distinction between intellectualist and voluntarist theologies now seems like an archaic relic. It was as vital an issue to Newton and Leibniz as it had been to Aquinas and Ockham, and played a significant role in the methodology thought proper to investigations in physics. The intellectualist position, developed in different ways by Aquinas and Leibniz, stressed the idea that the plan of creation is something intrinsic to the universe. God is a creator, not an architect. The universe, accordingly, must be intelligible in terms of intrinsic principles. In explaining the universe, God should be invoked only as creator, final cause, and for events transcending the natural order. In principle, this made the study of nature an independent discipline. In practice, philosophers in this tradition tended to put too much reliance on a priori deductions. The voluntarist tradition, stemming from medieval nominalism and Lutheran theology, insisted that God is free to do whatever he wants without regard to the limitations of human understanding. Boyle, Newton, and perhaps Descartes[22] were in the voluntarist tradition. If one could not deduce basic features of reality from assumptions concerning the order of creation, then the only reasonable alternative is to investigate and speculate.

The wars of religion spawned by the Reformation and Counter Reformation led to edicts of toleration and a widespread rejection of religious zealotry. This spirit carried over to Enlightenment propaganda and popularizations of science. Bernard de Fontenelle (1657–1757), the first great popularizer of Cartesian and Newtonian physics, insisted on separating both the physical and biological sciences from any dependence on God as an explanatory principle (See Marsak 1959, sect. IV). Voltaire, du Châtelet, and the French Encyclopedists were, for the most part, neither materialists nor atheists. Yet, they opposed religious intrusions, whether social, political, or intellectual, in science or in their attempt to reform society in accord with ideals shaped by popularizations of Newtonian physics. The constraining influences on physics and its systematic expression moved from theology and metaphysics to systematic observations and mathematical formulations. Theology gradually shifted from an explanatory mode, relating the limited truths of science to the ultimate truths of theology, to a defensive mode, contending that advances in science do not really contradict properly interpreted revealed truth.

---

[22] Harry Frankfurt (1977) interprets Descartes as holding that the truths we take as intuitively evident, such as the laws of logic, are those God wills us to hold, not those that are necessarily true.

References

Anscombe, G. E. M., and P. T. Geach. (ed.) 1961. *Three Philosophers: Aristotle, Aquinas, Frege*. Ithaca, NY: Cornell University Press.

Baumgardt, Carola. (ed.) 1951. *Johannes Kepler: Life and Letters*. New York, NY: Philosophical Library.

Bochenski, I. M. 1961. *A History of Formal Logic*, trans. Ivo Thomas. South Bend, IN: University of Notre Dame Press.

Bochner, Salomon. 1966. *The Role of Mathematics in the Rise of Science*. Princeton, NJ: Princeton University Press.

Boyer, Carl B. 1968. *A History of Mathematics*. New York, NY: Wiley.

Brading, Katherine, and Dana Jalobeanu. 2002. All Alone in the Universe: Individuals in Descartes and Newton. *PhilSciArchives*, **# 671**.

Burtt, A. E. 1954. *The Metaphysical Foundations of Modern Science*. Garden City, NY: Doubleday.

Campbell, Joseph. 1959–1968. *The Masks of God*. New York, NY: Viking Press.

Casper, Max. 1962. *Kepler, trans.C. Doris Hellman*. New York, NY: Collier Books.

Cassirer, Ernst. 1967. Mathematical Mysticism and Mathematical Science. In Ernan McMullin (ed.), *Galileo: Man of Science*. New York, NY: Basic Books.

Chevalley, Catherine. 1993. Physics as an Art. In A. I. Tauber (ed.), *Aesthetics in Science* (pp. 1–2). Baltimore, MD: Johns Hopkins University Press.

Clagett, Marshall. 1959. *The Science of Mechanics in the Middle Ages*. Madison, WI: University of Wisconsin Press.

Clagett, Marshall. 1966. *Nicole Oresme and the Medieval Geometry of Qualities and Motions*. Madison, WI: University of Wisconsin Press.

Cohen, I. Bernard. 1971. *Introduction to Newton's Principia*. Cambridge, MA: Harvard University Press.

Cohen, I. Bernard. 1980. *The Newtonian Revolution, with illustrations of the transformation of scientific ideas*. Cambridge: Cambridge University Press.

Conacher, D. J. 1967. *Euripidean Drama: Myth, Theme and Structure*. Toronto, ON: University of Toronto Press.

Connor, James A. 2004. *Kepler's Witch: An Astronomer's Discovery of Cosmic Order amid Religious War, Political Intrigue, and the Heresy Trial of His Mother*. San Francisco, CA: Harper.

Crombie, A. E. 1959. *Medieval and Early Modern Science*. Garden City, NY: Doubleday.

Dijksterhuis, E. J. 1961. *The Mechanization of the World Picture*. Oxford: Clarendon Press.

Drake, Stillman, and I. E. Drabkin. 1969. *Mechanics in Sixteenth-Century Italy: Selections from Tartaglia*. Madison, WI: University of Wisconsin Press.

Frankfurt, Harry. 1977. Descartes on the Creation of Eternal Truths. *The Philosophical Review, 86*, 36.

Frye, Roland Mushat. 1981. Ways of Seeing, Or Epistemology in the Arts: Unities and Disunities in. In Jonathan E. Rhodes (ed.), *Papers Read at a Meeting June 5, 1980* (pp. 41–73). Philadelphia PA: The American Philosophical Society.

Gabbey, Alan. 1980. Force and Inertia in the 17th Century: Descartes and Newton. In Stephen Gaukroger (ed.), *Descartes: Philosophy, Mathematics and Physics* (pp. 240–280). Sussex: Harvester Press.

Galilei, Galileo. 1953. *Dialogue on the Great World Systems, trans. Salisbury, rev.de Santillana*. Chicago IL: University of Chicago Press.

Grene, Marjorie Glicksman. 1963. *A Portrait of Aristotle*. Chicago IL: University of Chicago Press.

Hacking, Ian. 2001. Aristotelian Categories and Cognitive Domains. *Synthese, 126*, 473–515.

Hanson, Norwood Russell. 1961. *Patterns of Discovery: An Inquiry into the Conceptual Foundations of Science*. Cambridge, UK: Cambridge University Press.

Hanson, Norwood Russell, Toulmin Stephen Edelston, and Woolf, Harry. 1972. *What I Do Not Believe and Other Essays*. Dordrecht: Reidel.

Harré, Rom. 1961. *Theories and Things*. London; New York, NY: Sheed and Ward.

Jaeger, Werner Wilhelm, and Richard Robinson. 1934. *Aristotle: Fundamentals of the History of His Development*. Oxford: The Clarendon Press.

Kirk, G. S. 1970. *Myth, Its Meaning and Function*. Cambridge: Cambridge University Press.

Kirk, G. S., and J. E. Raven. 1962. *The Pre-socratic Philosophers*. Cambridge, UK: Cambridge University Press.

Kline, Morris. 1972. *Mathematical Thought from Ancient to Modern Times*. New York, NY: Oxford University Press.

Koestler, Arthur. 1960. *The Watershed: A Biography of Johannes Kepler*. Garden City, NY: Doubleday Anchor.

Leach, Edmond. 1974. *Claude Lévi-Strauss*. New York, NY: Viking Press.

Levi-Strauss, Claude. 1962. *Totemism*. Boston, MA: Beacon Press.

Levi-Strauss, Claude. 1966. *The Savage Mind, The Nature of Human Society Series*. Chicago, IL: University of Chicago Press.

Levi-Strauss, Claude (ed.) 1969. *The raw and the cooked. His Introduction to a science of mythology, 1*. New York: Harper & Row.

Lindberg, David. 1987. Science as Handmaiden: Roger Bacon and the Patristic Tradition. *Isis, 87*, 518–536.

Lindberg, David C. 1992. *The Beginnings of Western Science*. Chicago, IL: University of Chicago Press.

MacKinnon, Edward. 1982. *Scientific Explanation and Atomic Physics*. Chicago, IL: University of Chicago Press.

Mahoney, Michael. 1987. Mathematics. In Joseph Strayer (ed.), *Dictionary of the Middle Ages* (pp. 205–222). New York, NY: Charles Scribner's Sons.

Maimonides, Moses. 1963. *The Guide of the Perplexed*. Chicago, IL: University of Chicago Press.

Mann, Wolfgang Rainer. 2000. *The Discovery of Things: Aristotle's Categories and Their Context*. Princeton, NJ: Princeton University Press.

Marsak, Leonard. 1959. *Bernard de Fontenelle: The Idea of Science in the French Enlightenment*. Philadelphia, PA: Transactions of the American Philosophical Society.

Mason, Stephen. 1962. *A History of the Sciences*. New York, NY: Collier Books.

McKeon, Richard ed. 1941. *The Basic Works of Aristotle*. New York, NY: Random House.

McMullin, E. 1978. Philosophy of science and its rational reconstructions. In G. Radnitzky, and G. Anderson (eds.), *Progress and Rationality in Science* (pp. 221–252). Dordrecht: Reidel.

Newton, Isaac. 1952a. *Mathematical Principles of Natural Philosophy*. Chicago, IL: Encyclopedia Britannica, Inc.

Newton, Isaac. 1952b. *Opticks*. New York, NY: Dover Reprints.

Peirce, Charles S., Charles Hartshorne, and Paul Weiss. 1931. *Collected Papers of Charles Sanders Peirce*. Cambridge: Harvard University Press.

Pritchard, James B. 1973. *The Ancient Near East. Vol. I*. Princeton, NJ: Princeton University Press.

Randall, John Herman. 1960. *Aristotle*. New York, NY: Columbia University Press.

Schmidt, Robert W. 1966. *The Domain of Logic According to Saint Thomas Aquinas*. The Hague: Martinus Nijhoff.

Segal, Charles. 1986. *Interpreting Greek Tragedy: Myth, Poetry, Text*. Ithaca, NY: Cornell University Press.

Sellars, Wilfrid. 1967. *Philosophical Perspectives*. Springfield, IL: C.C. Thomas.

Shapiro, Alan E. 1984. Experiment and Mathematics in Newton's Theory of Color. *Physics Today, 37*, 34.

Smith, A. Mark. 1981. Getting the Big Picture in Perspectivist Optics. *Isis, 72*, 568–589.

Snell, Bruno. (ed.) 1960. *The Discovery of the Mind, Harper Torchbooks, TB1018. Academy Library*. New York, NY: Harper.

Solère, Jean-Luc. 2001. The Question of Intensive Magnitudes According to Some Jesuits in the Seventeenth Century. *The Monist, 84*, 582.

Solmsen, Friedrich. 1960. *Aristotle's System of the Physical World: A Comparison with his Predecessors*. Ithaca, NY: Cornell University Press.

Strawson, Peter. 1959. *Individuals: An Essay in Descriptive Metaphysics*. London: Methuen.
Wallace, William A. 1981. *Prelude to Galileo*. Boston, MA: Reidel.
Westfall, R. S. 1971. *Force in Newton's Physics: The Science of Dynamics in the Seventeenth Century*. London: Macdonald.
Westfall, Richard S. 1980. *Never at Rest*. Cambridge: Cambridge University Press.

Chapter 3
The Unification of Classical Physics

Postulate I. Grant that two quantities, whose difference is an infinitely small quantity, may be indifferently used for each other: or (which is the same thing) that a quantity which is increased or decreased only by an infinitely small quantity, may be considered as remaining the same.
Marquise d l'Hospital, Analyse (1696)

Historians and philosophers of science have extensively treated the development of classical physics. Why one more survey? There is an important philosophical, or interpretative, problem that is rarely recognized and, to the best of my knowledge, has never received even a minimally adequate treatment. The problem is easily stated: What is classical physics? This admits of various answers. For its creators the issue was not *classical* physics, but simply physics. The shared goal was one of explaining physical reality through an articulation of its basic ingredients and fundamental laws. Even those like Fourier and Mach who argued for a phenomenological physics did so because they considered the goal unreachable. This can no longer be considered a goal for classical physics. Quantum mechanics has replaced classical physics as the foundational science.

In attempting to answer this question it helps to see why simple answers don't work. One might treat the classical–quantum transition in terms of paradigm replacement or a degenerating research program. But classical physics has not been abandoned. It is still indispensable in treating most aspects of macroscopic reality. Classical formulations still supply a springboard for setting up quantum formulations. The experimental programs that test and constrain quantum theories rely on the validity of classical laws.

Another facile solution is the claim that classical and quantum physics represent successive approximations in the description of physical reality. There are many examples of descriptions through successive approximations from computer programs that represent circles through polygons with a large number of sides to zooming in on a Google map to get more details. Here, however, the approximations are of the same nature. Classical physics describes reality through the application of large-scale deterministic laws. Quantum mechanics does so through small-scale probabilistic laws. These are incompatible foundations.

E. MacKinnon, *Interpreting Physics*, Boston Studies in the Philosophy of Science 289,
DOI 10.1007/978-94-007-2369-6_3, © Springer Science+Business Media B.V. 2012

This contrast suggests a simple dichotomy, If quantum physics is accepted as the true foundation, then classical physics must be false. This facile solution rests on treating classical physics as if it were a deductive theory. If the foundation is wrong then the theory is false. However, classical physics cannot be interpreted as a deductive theory. It has a different structure. In this chapter we will attempt to clarify the type of informal unification classical physics achieved. In the next we will consider how classical physics should be interpreted.

To get at the type of unification classical physics achieved we begin with a contrast between the normal functioning of classical physics and philosophical accounts of theory interpretation. In both the axiomatic and semantic modes of theory interpretation the mathematical formulation of a theory is considered the foundation for an interpretation. This requires a formulation that has a validity independent of any physical interpretation impose on it. Rigorous mathematics is required. This relates to the situation mathematicians faced earlier.

Histories of mathematics regularly present the early nineteenth century as a turning point in the development of calculus. Cauchy, Dirichlet, Weierstrass, Cantor, and many more, succeeded in putting the calculus on an arithmetical, and eventually a set-theoretical, rather than a physicalistic or geometric, foundation. To a man, classical physicists continued to rely on the Euler-Lagrange formulation of calculus, even after the reformation of the calculus was well established. The only woman prominently involved, Emmy Noether, was too much a disciple of David Hilbert to accept such sloppy math. The reasons for this are clear. Creative physicists were trying to develop physical concepts that were adequate to the phenomena and consistent with established physical concepts. Mathematical formulations were introduced chiefly as tools for expressing physical concepts. This effectively reversed the direction of philosophical interpretation. They began with physical concepts and then sought an appropriate mathematical formulation. The clearest example comes from the most creative physicist in the period being treated. Maxwell argued that since space, time, fields, and motion are continuous, they should be represented by continuous functions. (See, Niven 1965, Vol. II, pp. 215–229, Maxwell 1954 [1873], pp. 6–8.) Classical physicists also idealized matter and its basic properties as continuous. In dealing with continuous functions it was considered OK to speak of infinitesimals, to treat $x + dx$ as a legitimate addition, and even to treat dy/dx as the ratio of two infinitesimals. This remained routine practice among physicists and even in textbooks[1] long after 'infinitesimal' had become an *unconcept* for mainstream mathematicians.

To clarify the role of language in classical physics we will consider three interrelated topics. The first is the development of the basic concepts of physics as parts of an inference-supporting system. This cannot be brought out by the methods of theory interpretation just noted, which consider only formal inferences. We will consider the historical developments and the trend of proceeding from concepts to their

[1] See the highly popular calculus texts by Granville, later Granville and Smith, and finally Granville, Smith and Longley published between 1929 and 1962 (Granville et al. 1962).

mathematical expression, even though this involves sloppy mathematics. Next we will consider the spasmodic nineteenth century attempts to unify physics on a foundation of atomistic mechanism. Though this failed it left a significant residue, a linguistic framework for classical physics in which mechanical concepts play a foundational role. These supply a background for the third topic, an interpretation of the idealized classical physics that complements and contrasts with quantum physics. This will be treated in the next chapter. Thomas Kuhn (1977, chap. 3) introduced a useful distinction between the classical physical sciences; astronomy, statics, optics, and mechanics as analysis; and the Baconian sciences. In the classical sciences, experiments were performed chiefly to test the consequences of a hypothesis or to answer a question posed within the existing theory. With Bacon as nominal leader and chief publicist, the Baconian scientists performed experiments with the goal of learning something new, unanticipated from nature. Phenomena that did not lend themselves to familiar classification or fit standard theories were objects of qualitative studies and sources of aesthetic pleasure. Even the pages of the Philosophical Transactions of the Royal Society were filled with reports of marvelous phenomena and strange experiences, with the hope that an explanatory account would eventually emerge.[2] The practitioners of the new Baconian sciences were often people lacking scientific training and mathematical competence. Experimentation as a source of new knowledge, achieved a preeminent status in England, partially due to the precedent of Newton and the Royal Society and partially to the peculiar status of the religious dissidents. The non-Conformists, excluded from Oxford and Cambridge, started their own schools and societies. Since their schools did not teach much Latin, Greek, or classical literature, they featured mathematics and science. Two groups typify the new societies. The Birmingham Lunar Society (They met when the moon was full) was founded after James Watt moved to Birmingham in 1744 and became a partner with Matthew Boulton in building steam engines. The membership included Josiah Wedgwood, Erasmus Darwin, and Joseph Priestley. Slightly later, in 1781 people associated with the Unitarian Chapel founded the Manchester Literary and Philosophical Society. They built a library, a college, a mechanics institute and supported the experimental researches of Priestley, Dalton, and Joule.[3]

We will focus on the treatment of electrical and thermal phenomena and the struggle to achieve a mathematical account of quantities. Following Heilbron (1979), the development of the study of electricity prior to the nineteenth century may be divided into three stages. The first stage, lasting until 1700, is in the tradition of natural philosophy. Overtly, William Gilbert was opposed to Aristotle and Aristotelians, with the rather surprising exception of Thomas Aquinas. In spite of this overt opposition, he relied on Aristotelian methods. When he distinguished electrical (his term) from magnetic effects, he attributed electrical effects to a material

[2] A general survey of eighteenth century physical experiments is given in Hall (1954, chap. 12). Rueger (1997) illustrates the striking differences between the goals of Baconian and more quantitative experiments.

[3] A survey of this development is given in Dyson (1988, chap. 3).

cause and magnetic effects to a formal cause. The explanations that soon came to dominate this new science were those elaborated by Jesuit professors, like Athanasius Kircher and Francesco Lana, trained in an Aristotelian philosophy of nature. Rather than focus on a particular individual, I will simply indicate the type of argument they developed.

Magnetism, proper only to the lodestone and not requiring any activity on the part of the experimenter, might be explained in terms of an innate sympathy. The electrical effect, which required rubbing a rod, and which could attract diverse types of substances, could not be so explained. The effect was clear, the motion of chaff to the rubbed rod. Whatever is moved, Aristotle had insisted, is moved by another. Since motion is a continuing effect the moving agent must be in continuous contact with the object moved. Since the agent cannot be seen, it must be some sort of invisible effluvium.

The Aristotelian philosophy of nature was soon challenged and eventually replaced by competing philosophies of nature, stemming from Descartes, Boyle, Newton, and Christian Wolff. In spite of differences, they shared enough of a common core so that they could repeat the same experiments, adapt each other's explanations, and carry on a meaningful dialog. This necessitates a sharing of concepts. They also shared a reliance on some version of philosophical essentialism. The basic idea is that the causes on which the phenomena depend must have the properties common to all matter. Though the accounts of what these properties are differed, the accounts were all determined by a conceptual, rather than an empirical analysis. This sharing, in turn, tends to decrease the a priori aspects of philosophical systematization. The meaning of shared terms comes to depend on their common use, rather than on the scholastic type definitions of meaning found in natural philosophy textbooks.

By the mid 1730s common assumptions shared by the natural philosophy tradition in electricity were being undercut. Stephen Gray showed that the 'electrick Vertue' of a glass tube could be transferred to other bodies, suggesting that electricity was more like a fluid than an effluvium inseparably associated with the rubbed body. In 1733 Charles du Fay showed that a thin gold leaf brought near an electrified glass was attracted and then, after contact, repelled. This was not readily accommodated by the causal powers attributed to ether or effluvia. This led to an increasing reliance on experimentation and the introduction of various low-level hypotheses. Nollet's two-fluid theory, Franklin's one-fluid theory, the Leyden jar, Franklin's kite experiment.[4]

By the end of the eighteenth century, many physicists were abandoning such theories and natural philosophy explanations and concentrating on quantitative concepts (See Heilbron, chap. XIX). Thus, Aepenius stated the rule that similarly charged bodies repel while dissimilarly charged bodies attract, and advocated ignoring accounts of the mechanism involved. Three quantitative concepts came to play

---

[4] An older, but still valuable, summary of these developments is given in Whittaker (1960, Vol. I, chap. 2).

a basic role. The first was the concept of charge localized in a particular body. The experiments of Coulomb, Cavendish, and others led to Coulomb's law. The second idea was capacitance, which admitted of rough measurements. A typical spark-producing apparatus, by then a feature of sideshows, had plates of glass on a pulley rub against silk and then transfer the surface charge to a hollow metallic ball. A correlation was established between the number of rubbings and the amount of surface rubbed with the length of the sparks emitted by the ball. The third concept was 'tension', which later became 'voltage'. Alessandro Volta suggested that these are related by the formula, $Q = CT$. He also discovered current electricity, drastically changing the whole field. Subsequent developments will be considered later.

The development of quantitative concepts concerning heat was slower and later than the other developments considered.[5] There was confusion concerning what heat is, how it relates to temperature, to mass, to specific properties of bodies, to changes of state, to combustion, and to motion. Early attempts to perform measurements were complicated by uncertainty concerning: what was being measured, temperature, the heat in a body, the heat transferred to another body, or the matter of heat; how it was to be measured; and what sort of scales were appropriate. There was also a formidable philosophical difficulty, first enunciated by Galileo.

> But first I must consider what it is that we call heat, as I suspect that people in general have a concept of this which is very remote from the truth. For they believe that heat is a real phenomenon, or property, or quality, which actually resides in the material by which we feel ourselves warmed.[6]

Galileo held that heat, like other secondary qualities, resides properly in the consciousness of the observer as an effect of motions of imperceptible particles within the source. Enlightened philosophers, in both the rationalist and the empiricist traditions, shared the view that secondary qualities reside properly in the consciousness of the observer. Only naive amateurs, or people who deliberately separated their work from prevailing philosophical speculations, would study heat as a property of bodies. By the end of the eighteenth century a few basic concepts had emerged with some degree of clarity. This, in turn, led to a quantitative theory of heat phenomenon.

Tentative attempts to attach numbers to degrees of heat led to the first quantitative law concerning the temperature of mixtures that won any degree of acceptance, the law formulated by G. W. Richmann around 1744. If one mixes together two quantities, the first with mass, m_1 and degree of heat, t_1 and a second with mass, m_2 and degree of heat, t_2, then, 0, the temperature of the mixture is given by

$$O = \frac{m_1 t_1 + m_2 t_2}{m_1 + m_2} \tag{3.1}$$

---

[5] The early development of theories of heat is treated in McKie and Heathcote (1935), and in Tisza (1966, pp. 3–52).

[6] This is from *The Assayer* and cited from Drake's translation (Drake 1957, p. 274).

One might think that this formula could easily be refuted by simple experiments involving mixing and measuring. However, it was difficult to perform such experiments. There was not yet any clear distinction between heat and temperature, only quantitative measures of temperature, and no accepted methods of insulation.

The establishment and standardization of temperature scales was a gradual process. Galileo introduced the first thermoscope, a closed glass tube with a bulb at one end, which could be immersed in water. The oldest record of a thermometer, a liquid (brandy) in a glass bulb with a closed glass stem, was given in *A Fountain of Gardens*, published in 1629 by the physician and Rabbi, Joseph Solomon Delmedigo (*Discovery*, 18 Oct. 1997, 18), who was a student at Padua when Galileo taught there. Fahrenheit first developed alcohol based thermometers. Around 1714 he switched to mercury and in 1724 developed the scale where $o°$ corresponds to the temperature of a mixture of water saturated with salt and ice, while 32° corresponds to the temperature of a mixture of ordinary water and ice. In 1742 Anders Celsius, professor of astronomy at Upsala, proposed a scale in which the melting point of ice is 100° and the boiling point of water at standard pressure is 0°. Stromer, also of Upsala, inverted this to get the standard Swedish scale.

As McKie and Heathcote point out (1935, chap. 1), there is some uncertainty whether the distinction between specific and latent heats and related points should be attributed to Black or Wilcke. These were treated in Joseph Black's *Lectures on the Elements of Chemistry* (1803), which was not published until four years after his death. Much of the work it summarized, however, had been done in the 1760s. The students who attended his lectures in Glasgow and Edinburgh circulated some of his results. Johan Carl Wilcke, working independently of Black developed similar distinctions in the 1770s. He probably has some knowledge of Black's work. Both men were very much in the tradition of Baconian science, the non-philosophical, non-mathematical, hands-on approach to nature previously considered. Black systematically discounted both the Baconian motion theory of heat and the theory of heat as a subtle, highly elastic, penetrating fluid. (The term 'caloric' was not introduced until 1787.) Both men performed extensive experiments and relied much more on the results of experiments than on any philosophy of nature or any sort of philosophical theory.

The quantitative concepts that emerged from this work may easily be summarized. First, there was a clear distinction between temperature and heat. Where Black was more concerned with the heat transferred from one body to another, Wilcke was concerned with determining the amount of heat in a body. Secondly, there was a distinction between overt and latent heat. Latent heat is manifested in a change of state, from solid to liquid, or liquid to vapor, rather than with an increase in temperature. Finally, there was the term 'specific heat', introduced by Wilcke. Wilcke showed that he could measure the specific heats of iron, mercury, lead, and other substances by applying Richman's formula to mixing experiments.

These advances led to two consequences that concern us. The first was the rise of the quantitative science of heat, calorimetry. The heat transferred from one body to another can be determined by measuring masses, temperatures, the specific heats of the bodies involved, and including a consideration of latent heats when necessary.

Heat, with qualifications on specific and latent heats, became an additive quantity. As measuring instruments became more standardized and techniques improved researchers began to pay more careful attention to possible sources of error, such as loss of heat to air, to vessels, or to chemical reactions. Attempts to explain the heat in a body receded into metaphysical accounts of bodies and their essential properties. When attention was focused on heat transfer and the concepts needed to make this an additive quantity, then a quantitative science could be developed. This was a gradual transition. By its conclusion, a descriptive account of basic heat phenomena could be given.

1. Heat diffuses from hot bodies to cooler ones until an equilibrium condition is reached.
2. Heat is transmitted by conduction, convection, and radiation. This does not settle the issue of whether the same causes are involved in each case.
3. Different types of material vary greatly in their ability to conduct heat.
4. Bodies have different specific heats, i.e., the amount of heat required to raise one gram of the substance from 14° to 15°C.
5. Besides overt heat, manifested in temperature changes, there is also latent heat. This is measured by the amount of overt heat required to change a solid into a liquid at the same temperature, and a liquid into a gas.

3.1 Atomistic Mechanism

The architects of classical physics assumed that physics required a philosophical foundation. Descartes made his theory of knowledge foundational. Newton began his *Principia* by first clarifying the notions of absolute space and time. He developed accounts of impressed force and an atomic theory of matter. Leibniz advocated relative space and time, and living forces. Wolff's philosophy of nature drew heavily on Leibniz. It seemed that even an introductory account of the new physics had to take a stand on these disputed issues. Madam du Châtelet, who translated Newton's *Principia* into French and transposed his geometric arguments into the language of calculus, wrote *Institutiones de Physique* as an expanded version of the physics she taught her son.[7] The titles of the opening chapters indicate her fusion of Cartesian epistemology, Leibnizian metaphysics, and Newtonian physics: 1. Principles of Knowledge; 2. On the Existence of God; 3. Of Essence, Attributes, and Modes; 4. Of Hypotheses; 5. Of Space; 6. Of Time, 7. Of Elements of Matter; 8. Of the Nature of Bodies; 9. On the Divisibility and Subtlety of Matter. Euler, Lagrange and the French Newtonians effectively removed mechanics from its putative dependence on atomic assumptions. Lagrange dispensed with philosophical foundations by developing mechanics as a hypothetical-deductive system. The Lagrangian tradition of

[7] The text is on line at www.womeninscience.history.msu.edu. A summary account may be found in Zinsser (2006, chap. 4).

treating mechanics as a form of analysis, the newest member of the classical sciences, continued well into the nineteenth century with Gauss's Principle of Least Constraint and Hamilton's Principle of Least Action. Laplace initiated a new systematization, transforming Newton's mechanistic atomism in atomistic mechanism. He assigned mechanics, rather than a philosophical account of atoms, a foundational role. Then hypotheses concerning atoms and short-range forces not only revived atomism. They also supplied a potential for treating heat, chemical affinities, and other phenomena, within the framework of physics based on a mechanistic foundation. Laplace is usually interpreted as the last and greatest in a distinguished series of French Newtonians. What we wish to consider here, however, is not the continuity between Laplace and his predecessors or contemporaries, but the novel features in Laplace's treatment of mechanical concepts and methods.

Laplace initiated a new style that Poisson later dubbed 'physical mechanics'. He used mathematics as an instrument for physical calculations. Even by the lax standards of his time, his mathematics was not rigorous. He used approximations and power series in which he regularly dropped terms that were considered insignificant on physical grounds.[8] Laplace's work in physics began, not with the analytic mechanics of Lagrange, but with the grubbier observation-centered type of work that d'Alembert had labeled 'general and experimental physics'. In the 1770s Laplace collaborated with Lavoisier in an attempt to develop a chemical physics of heat. This involved models of matter and caloric. It also involved detailed experimental work, building an ice-calorimeter, and making measurements of specific properties of different types of bodies. This, I believe, conditioned Laplace's speculations on short-range forces. Where Boscovitch and Kant presented general force laws, which attempted to make action at a distance intelligible, Laplace introduced hypotheses about forces to explain specific properties of matter.

He thought of the normal state of a physical body as representing an equilibrium condition between heat, or the radiation of caloric, which tends to separate molecules, and the forces of chemical affinity, which draws molecules together. (Laplace 1912, Vol. IV, p. 1009) By 1784 he had introduced the notion (Green later introduced the term) of a potential from which force could be derived. From 1802 on considerations of short range forces were basic to Laplace's work. This is best illustrated by the treatment of heat of elastic fluids that he developed at the close of his career in the early 1820s (Ibid., Vol. 5, pp. 113–132).

After developing a mathematical formulation for spheres related by forces proportional to $r^{-2-\alpha}$ he shows that this reduces to Newtonian gravitational theory when $\alpha = 0$ and, when $\alpha = -3$, to the Newtonian model of an elastic fluid as a collection of molecules kept in static equilibrium by an inverse repulsive force. Since this does not fit observed results he makes the assumption that the intermolecular repulsive force operative at insensible distances is proportional to c^2, where c is the heat contained in each molecule of a gas. By assuming that the gas molecules

[8] A more detailed account of Laplace's mathematical methods may be found in Gillispie et al. (1978, Part IV).

are enclosed in a large sphere of arbitrary radius he is able to deduce Boyle's law, Gay-Lussac's law, and Dalton's law of partial pressures. With further assumptions about the heat of particles he deduced the correct results for the velocity of sound in air (Ibid., pp. 133–145). This illustrates the type of mixed inference proper to developing physics. In this way, mechanical concepts were extended to treatments of heat, light, and electricity by Laplace, his disciples, Poisson and Biot, and his associate, Ampère. This meant sacrificing the analytic rigor of Lagrange in favor of arguments based on the physical significance accorded mathematical expressions. At the same time he came to stress the importance of experimental verification on the grounds that the conclusion were very remote from the basic assumptions (Ibid., p. 469).

3.1.1 An Energetic Physics

The goal of unifying physics on the foundation of atomistic mechanism shaped the conceptual structure of much of nineteenth century physics (See Harman 1982a, p. 106). We will consider two stages of this. The attempt to fit thermodynamics into a mechanical picture and the later struggle to fit electrodynamics into this framework. The caloric theory of heat, in the form developed by Lavoisier and Laplace provided a theoretical explanation for the thermal phenomena previously listed.[9] The theory presuppose a distinction between the material atoms which constitute ordinary bodies and the much smaller, probably weightless, caloric atoms. The basic assumptions are:

1. Caloric atoms repel each other, but are differentially attracted to material atoms. The degree of attraction depends both on the type of substance and on its state.
2. Some caloric atoms are able to move freely from one atom to another, while other caloric atoms are bound. The diffusion of heat is a phenomenological manifestation of the flow of free caloric atoms.
3. Caloric atoms are neither created nor destroyed.
4. The amount of heat involved in changing the temperature of a body is directly proportional to the mass and the temperature change.
5. The amount of heat required for a change of state is proportional to the mass involved.

With these assumptions it is possible to give both a qualitative explanation of the phenomenological facts previously considered and also a quantitative account of some heat phenomena. If, following Lavoisier and Laplace, we assume that massive atoms exert mutual short range attractive forces, then it is relatively easy to account for the transition of a substance from a solid to a liquid to a gaseous state. An atom within a solid is surrounded by a layer of caloric atoms, as are its neighboring atoms. In the solid state the mutual repulsion of the caloric atoms is less than the mutual

[9] This is chiefly based on Lavoisier (1864, Vol. I, pp. 19–30), and on Morris (1972).

short-range atomic forces of attraction. This accounts for the stability of the solid state.

Heating a body means, according to caloric theory, transferring caloric atoms into the body. Since caloric atoms repel each other the newly introduced caloric atoms diffuse until an equilibrium condition is reached. Since the newly introduced caloric atoms are free caloric atoms two consequences follow. First, there is a temperature rise proportional to the number of caloric atoms introduced (or its phenomenological manifestation the amount of heat supplied). Secondly, the increased repulsive forces due to the free caloric cause the solid to expand. As more caloric atoms are introduced the repulsive forces grow until a new type of equilibrium is reached. The balance of forces still keeps an atom attracted to its near neighbors, but, due to the extra free caloric, not to distant atoms. This could account for a transition from the solid to the liquid state. The liquid state, like the solid state, would manifest a rise in temperature and an expansion as more free caloric is introduced until a new equilibrium condition is reached, one in which caloric repulsion is greater than atomic attraction. Because of caloric repulsion, the free caloric would differentially diffuse to the surface and preserve the liquid's cohesiveness through a net surface tension. The introduction of further caloric and the process of diffusion would lead to the expulsion of surface atoms. At this point heating produces a change of state, rather than a further temperature rise. For atoms in the gaseous state caloric repulsion is stronger than the short-range atomic attraction. This leads to a Newtonian model of a gas as stationary, rather than kinetic, molecules.

This network of physical concepts supports mathematical formulations. The key assumption that supplies a basis for quantitative calculations is (3): caloric atoms are neither created nor destroyed, together with the interpretative assumption that free caloric is manifested phenomenologically as heat. Consider a container, e.g., a glass jar enclosed in an insulating box. Suppose that the jar contains a fluid of mass m_f with a specific heat c_f and is at a temperature T_1. Then a metal with mass m_m and specific heat c_m and at a temperature T_2, higher than T_1, is put inside the jar. Since caloric atoms cannot cross the insulating walls all the caloric atoms that diffuse out of the metal are absorbed by the fluid, the container or the air, until they reach an equilibrium temperature, T_3. If we assume that the container and the air absorb a negligible amount of caloric, then the conservation of caloric atoms implies that the caloric lost by the metal equals the caloric absorbed by the fluid. Since this free caloric is manifested as heat we are led to the equation balancing heat loss and heat gained:

$$c_m m_m (T_2 - T_1) = c_f m_f (T_3 - T_1) \tag{3.2}$$

Black had developed a simple mathematics of mixing without a reliance on caloric theory. However, caloric theory gives an account of why this mixing happens and supplies a basis for extension. Equation (3.2) can easily be expanded to accommodate heat absorbed by the container, a mixture of more than two substances, and changes of state. It also manifests the ambivalent status of most of the laws we will be considering in this chapter. Its justification and its extension from observed to unobserved phenomena strongly depend on caloric theory. Yet all the terms in (3.2) refer to measurable properties of macroscopic bodies.[10]

The final contribution of caloric theory was given in Sadi Carnot's (1986 [1824]) treatise, "*Réflexions sûr la Puissance Motrice du Feu*." Though this treatise contained little mathematics, it had two novel concepts which strongly influenced the special correlation between thermodynamical concepts and mathematical expressions. The first, based on an idealization of the way steam engines work, is the idea of a *reversible cycle*. The second is the idea of the *state of a system*. These are closely interrelated. In a reversible cycle a system returns to its original state. In spite of their mechanical grounding, these concepts resisted any simple mechanical formulation. This has a bearing on the changing status of thermodynamics.

Count Rumford and Humphry Davy argued that caloric theory was unable to give a convincing account of frictional heating. Gillispie and Brush have defended the opinion that the decisive factor in the rejection of caloric theory was the acceptance of a wave theory of light. Since the sun radiates both heat and light, it is not plausible to consider one as particles and the other as waves.[11] The real trick here is to preserve the advances made by caloric theory, the first successful mathematical theory of heat, while changing the conceptual foundations. This fit in with a larger trend.

Around 1816 younger French physicists began to revolt against the Newtonian orthodoxy of Laplace and his disciples. The revolt occurred on two fronts: with the work of Fresnel and Arago on the wave theory of light; and with the work of Fourier, Petit, and Dulong on heat. Here we will only consider Fourier's work as illustrating the changing relation between a depth level and a phenomenological account. Laplace and his followers, especially Ampère and Poisson, professed a sort of split-level skepticism with regard to the ontological assumptions they made. When doing calculations they took them very seriously. When appraising, rather than doing, physics they regarded 'caloric' as a convenient name for the yet unknown cause of heat (Blondel 1985).

Fourier professed a more extreme skepticism: "Primary causes are unknown to us; but are subject to simple and constant laws, which may be discovered by observation, the study of them being the object of natural philosophy" (Fourier 1955 [1822], p. 1). On this ground, he concluded that heat could not be explained through the mechanical principles of motion and equilibrium. What this meant in practice was that Fourier could treat the distribution of heat in solids and its transmission across surfaces and through bodies by the now familiar mathematical methods without relying on depth assumptions about what heat is. Since Fourier's treatment of the difference between a phenomenological and a depth level influenced the rise of later phenomenology and positivism, it is helpful to see how he interpreted it:

---

[10] Truesdell (1980) refers to these developments as a tragicomical history because of the conceptual confusion and misleading data involved. I find his criticism of conceptual confusion too harsh and too dependent on later clarifications.

[11] Gillispie (1960, p. 406) and in the introductory survey, (pp. 3–103), and Brush (1976).

> It cannot be doubted that the mode of action of heat always consists, like that of light, in the reciprocal communication of rays, and this explanation is at the present time adopted by the majority of physicists; but it is not necessary to consider the phenomena under this aspect in order to establish the theory of heat. In the course of this work it will be seen how the laws of equilibrium and propagation of radiant heat, in solids or liquid mass, can be rigorously demonstrated, independently of any physical explanation, as the necessary consequence of common observations. (Fourier 1955 [1822], p. 40)

The common observations concerned the heat capacities of different bodies and their varying capacities to transmit heat. The necessity came from a conservation requirement. This led to Fourier's famous diffusion equation (in modern form)

$$\frac{\partial V}{\partial T} = \frac{K}{CD}\left(\frac{\partial V^2}{\partial x^2} + \frac{\partial V^2}{\partial y^2} + \frac{\partial V^2}{\partial z^2}\right) \tag{3.3}$$

Here, V stands for heat, K for conductivity, C for specific heat, and D for the weight of a unit volume. The notation for partial differentiation was not introduced until later in the century. In solving this equation for various geometrical configurations, Fourier developed the methods of trigonometric expansion, determination of what are now called Fourier coefficients, and integral solutions to partial differential equations. To a considerable degree, these mathematical formulations could be detached from particular physical interpretations. In the work of, and competition between, Fourier, Poisson, and Cauchy, equations were developed that could be applied to heat diffusion, wave motion, and vibrations of strings or elastic surfaces.

Fourier's work contributed to the development of positivism. Auguste Comte took it as a paradigm of scientific explanation and dedicated the six volumes of his *Cours de philosophie positive* to Fourier. The French physics that was transmitted to England and Germany in the 1830s was influenced by Fourier almost as much as Laplace.[12] However, the, British, and German physicists of the new generation were, for the most part, neither positivistic in their interpretation of physics, nor interested in cultivating a purely phenomenological physics. They consciously strove after a unified physics. After the 1847 publication of Helmholtz's paper, "*Uber die Erhaltung der Kraft*" (Kahl 1971, pp. 3–55) and the rise of thermodynamics as a distinct discipline, 'energy' rapidly emerged as the key concept unifying the conceptual structure of physics on a mechanical basis. The new energetic physics could easily take over the mathematical formulations of caloric theory. The terms occurring in equations like (3.2) referred to measurable values and relied on a conservation principle. Here energy conservation replaced caloric conservation.

When caloric theory was abandoned, physicists sought to retain the mathematics but reinterpret its significance. There are two ways in which this might have been done. One, advocated by Rankine, was to replace the old foundation by a new foundation based on molecular assumptions and the concept of energy. A second

---

[12] For the transmission of French physics to England see Crossland and Smith (1978); to Germany see Jungnickel and McCormmach (1986, Vol. II, pp. 3–45).

approach, initiated by Clausius,[13] was to develop thermodynamics as a science independent of such foundations. The latter method was adopted by Thomson, Maxwell, Gibbs, Boltzmann, and shaped the structure of classical thermodynamics. Thus classical thermodynamics was developed as a phenomenological science. There was a widely shared assumption that this phenomenological thermodynamics could be put on a mechanical foundation once an adequate theory at atoms and short-range forces was developed. Maxwell summarized the prevailing view:

> The first part of the growth of a physical science consists in the discovery of a system of quantities on which its phenomena may be conceived to depend. The next stage is the discovery of the mathematical form of the relations between these quantities. After this, the science may be treated as a mathematical science, and the verification of the laws effected by a theoretical investigation of the conditions under which certain quantities can be most accurately measured, followed by experimental realization of these conditions, and actual measurement of these quantities. (Niven 1965, Vol. II, p. 257)

In his *Theory of Heat* Maxwell applied this view to thermodynamics and explained the relation of thermodynamic to mechanical concepts. States of bodies are interpreted as varying continuously. Changes of state are described phenomenologically, in terms of gross properties, rather than through molecular motions. The solid state can sustain a longitudinal pressure, while a fluid cannot. A gas is distinguished from a liquid by its ability to expand until it fills the available boundaries. Without introducing any molecular hypotheses he expanded this phenomenological account to include the quantitative concepts characterizing thermal phenomena: 'temperature', 'heat', 'specific heat', 'latent heat', 'heat of fusion', 'heat of vaporization', and methods of measuring them (Maxwell 1872, pp. 16–31).

To develop *thermodynamics*, which for Maxwell is the explanation of thermal phenomena through mechanical principles, it is first necessary to introduce mechanical concepts. These, as Maxwell saw it, have a different status. In mechanics length, mass, and time are taken as fundamental. They supply the units in terms of which one can define other mechanical concepts. On this basis Maxwell defines: 'density', 'specific gravity', 'uniform velocity', 'momentum', 'force', 'work', 'kinetic energy', and 'potential energy' (Ibid., pp. 76–91). Thermal phenomena are related to mechanics on two distinct levels, that of empirical generalizations, and that of a causal account. On the first, or phenomenological level, the crucial law is energy conservation, which Maxwell treats as a mechanical, rather than a thermal, law for clear reasons. Energy conservation can be strictly proved only for dynamical systems meeting certain conditions. Kinetic energy is the most basic form of energy: "When we have acquired the notion of matter in motion, and know what is meant by the energy of that motion, we are unable to conceive that any possible addition to our knowledge could explain the energy of motion, or give us a more perfect knowledge of it than we have already" (Ibid., p. 281). Because of energy conversion, this principle can be extended to systems in which energy takes the form of heat, magnetization, or electrification.

---

[13] The pertinent text is given in (Brush 1965, p. 112).

We are concerned with classical reality, or reality as represented by the laws of classical physics. We will consider separate contributions before developing an integrated view. The laws of thermodynamics present an idealization of processes as continuous. There is a distinct conceptual gap between the idealized assumptions supporting this mathematics and a more realistic physical description. Thus the density of a gas is defined as the limit, $\rho = \lim_{\Delta \to 0}(\Delta m / \Delta V)$. An arbitrarily small volume of a gas may have either one or no molecules. Nevertheless, the density is treated as a strictly continuous function. As Kestin (1966, p. 33) explains it: "The assumption that a thermodynamic system can be treated as a continuum underlies all derivations employed in classical and continuum thermodynamics..."

3.1.2 Classical Electrodynamics

Maxwell's field theory derived from the experimental researches of Michael Faraday and the tension between Faraday's early polarization hypotheses and later reliance on lines of force. In his early work Maxwell followed Thomson's precedent of using the method of physical analogies as a tool for giving Faraday's geometric reasoning a mathematical form.[14] This requires a sharp distinction between a mathematical formulation, adapted from some other domain, and a physical interpretation (Papers I, p. 156). Maxwell adapted the mathematics of incompressible fluids, but did not regard electrical current as a fluid. Though he rejected the Continental action-at-a-distance theories, he did not yet have an alternative physical account.

In subsequent papers Maxwell introduced the hypothesis of magnetic vortices with idle wheels (Papers I, pp. 451–525, Wise 1979) and the hypothesis that light consists of transverse undulations of the ethereal medium. In his definitive paper (Papers I, pp. 526–597) the vortex model was replaced by the general assumption of an ethereal medium filling space, permeating bodies, and transmitting motion. The second novel feature is a reliance on dynamics, rather than mechanics. Since about 1838, British physicists had been using 'dynamical' for an explanation based on Lagrange's analytic mechanics, rather than on any particular mechanical model (Harman 1982b, pp. 25–27).

Our concern is with the representation of reality presented, or presupposed, in this paper. Maxwell accepts the existence of a medium as a datum established by the work of Faraday, Verdet, and the electromagnetic field are causally responsible for the phenomena of the transmission of light and heat (and probably of gravity, though he gave up on the attempt to include this). What this entails is that the

---

[14] Maxwell's Collected Papers will be cited as Papers with the volume and page; his Treatise by paragraph numbers. Whittaker's (1960) is still a basic source for the history of electromagnetism. The studies of Maxwell's development which have influenced the present appraisal are: Campbell and Garnett (1969 [1882]), Turner (1955), Hesse (1963), Kargon (1969), Heimann (1970), Bromberg (1968), Everitt (1975), Wise (1979, 1982), Nersessian (1984), and Siegel (1986). The role of models in Maxwell's development of electrodynamics is analyzed in Nersessian (2008, chap. 2).

basic explanatory concepts are mechanical, force and energy. This is presented in opposition to the distance theory assumption that charges are the causes of electrical phenomena. Maxwell assigned such distinctively electrical concepts as charge and current a phenomenological status. More particularly, electromotive force is called into play during the communication of motion from one part of a body to another. This electromagnetic force acting on a dielectric produces polarization. The assumption is that the positive and negative electricity in each molecule is displaced in opposite directions. However, all that Maxwell really needs is the theoretical notion, 'displacement'. This displacement can be thought of as the beginning of a current, one that quickly reaches equilibrium. The only phenomenological manifestation of this displacement is at the bounding surface of the dielectric, where it is manifest as a positive or negative charge. As Siegel points out, the displacement current has been promoted to the status of the only current in the dielectric medium and is now thought of as pointing in the same direction as the electric field. This, however, is not compatible with the way Maxwell had interrelated charge and displacement in his earlier mechanical model (See Siegel 1986, p. 143). Since electrical charge is a phenomenological manifestation of displacement, so too is current. Maxwell thought of current in terms of the electrical energy involved in displacement being absorbed and transformed into heat (Papers I, p. 586). Maxwell regarded the twenty equations based on these twenty variable quantities as equations derived from the definitions of mechanics and experimental research in electromagnetism, minimally supplemented by theoretical considerations. These considerations were 'displacement', Maxwell's key theoretical notion, and 'the electromotive force at a point', an extrapolation from measurements on large conductors. From these equations, Maxwell deduced his well-known results concerning electromagnetic vibrations and induction in circuits.

In 1865 Maxwell resigned his chair at King's College and retired to the family estate of Glenlair, where he spent much of his time preparing a comprehensive systematic account of electric theory. In 1871 he returned to Cambridge as professor of experimental physics and director of its new Cavendish laboratory. He published his *Treatise* two years later. This long confusing work is concerned with two basic difficulties. The first, and most perplexing, was that Maxwell did not understand electricity. He did not know what electricity is (p. 35), or the direction and velocity of current (p. 570). His earlier account involved stress in the medium, but he did not know how stress originates (p. 644), or what light really is (p. 821). He had a phenomenological, rather than a depth account (p. 574).

The second difficulty concerned the competition between his medium account and the competing distance accounts, which had been updated to include the derivation of an electromagnetic theory of light. Could the empirical data decide which is correct? The Treatise is divided into four parts: electrostatics, electrokinematics, magnetism, and electromagnetism. Each begins in the same fashion, summarizing the phenomena and the related empirical laws. In each case Maxwell finds both accounts empirically adequate. In his treatment of electromagnetism he cites three possible experiments that might support the medium account. However, the tests he performed did not supply such support (p. 577). Similarly, his electromagnetic theory of light led to three distinctive consequences that could be checked experimentally. With painstaking honesty Maxwell notes that there was no unambiguous empirical support for his predictions on the dielectrical capacity of a transparent medium (p. 788), for the transparency of conductors (p. 800), or the rotation of plane polarized light (p. 811). He reluctantly concluded that the two mutually incompatible theories were empirically equivalent, though he still preferred medium theory on intuitive grounds (pp. 846–866). Further developments came from a fusion of elements in both traditions.

The Maxwellians eventually dropped the primacy Maxwell had accorded displacement and accepted Helmholtz's idea of atoms of electricity, aka electrons, as sources (See Buchwald 1985). When Helmholtz returned to the study of electrodynamics he presented a general formula, which he thought might supply a basis for experimental tests of the competing theories. Franz Neumann had derived all the electrodynamic effects for a closed electric circuit from a formula for the potential. Helmholtz generalized this potential formula to express the potential which a current element *ds* carrying current *i* exerts on a current element ds' carrying current *j*, where the two are separated by a distance *r*:

$$P = \frac{1}{2} A_{ij}{}^{2} \left[ds \cdot ds' + (1 - k)(r \cdot ds) \frac{(r \cdot ds')}{r} \right] \tag{3.4}$$

A is a constant with the value $1/c$, where *c* is the velocity of light. The term, *k*, is a trial constant. If $k = -1$ this formula reduces to Weber's potential. If $k = +1$ the formula reproduces Franz Neumann's potential; while $k = 0$ reproduces Maxwell's potential. The parts of this expression multiplied by *k* become zero when integrated over a closed circuit. If experiments were to decide between the competing theories, Helmholtz concluded, it would be necessary to use open circuits as the testing ground, an extremely difficult process. The Maxwell limit, $k = 0$, is unique in two respects. First, it allowed only transverse vibrations. Second, only the Maxwell limit yielded electromagnetic vibrations with a velocity equal to the velocity of light. The experiments that Helmholtz and his aides conducted in the 1870s generally had negative results. Helmholtz himself gradually became convinced that the Maxwell potential formula was correct. However, interpreting Maxwell's theory as a limiting case of Helmholtz's potential formula, as most Continental electricians did, led to serious conceptual difficulties (Buchwald 1985, pp. 177–193).

Helmholtz's protégé, Heinrich Hertz, conducted a series of experiments supporting Maxwell's account of a dielectric medium supporting transverse vibrations. In 1885 he became a professor at Karlsruhe and in 1888 began his epochal researches. In experiments on the propagation of electromagnetic radiation in air he showed that these electromagnetic waves have a finite velocity and that they can be refracted, polarized, diffracted, reflected, and produce interference effects. This undercut distance theories and precipitated a consensus within the European physics community of the correctness of Maxwell's idea of a dielectric medium transmitting electromagnetic vibrations. However, this did not entail accepting Maxwell's theory. Hertz's diligent study of Maxwell's *Treatise* led him to conclude that he did not know what

Maxwell's theory was (Hertz 1962 [1892], p. 20). He concluded "Maxwell's theory is Maxwell's system of equations" (Ibid., p. 21). The transformation of Maxwell's mixture of a depth level account, largely unsuccessful, and a phenomenological account coupled to a mathematical formulation into a hypothetical-deductive system (Ibid., p. 138) undercut most of the proposed ontological props. The electrical charge, e, was determined by conservation laws and measurements, rather than particle assumptions. The displacement lost its mechanical significance. The specific dielectric capacity, ε, and the specific magnetic capacity, μ, were reinterpreted as characterizing properties of material bodies, rather than the medium. Lorentz focused on problem of re-establishing a relation between ponderable matter and the electromagnetic field (Hirosige 1969). He assumed that all ordinary matter contains a multitude of positively and negatively charged particles. In place of the previous assumption, proper to distance theories, that electrons interact with other electrons, Lorentz assumed that electrons interact with the medium, the ether. An electron moving in a magnetic field experiences a force, $\mathbf{F} = e\mathbf{E} + (e/c)(\mathbf{v}\, x\, \mathbf{H})$. This force can be transmitted to the ponderable body, with which the electron is associated, and can be dissipated in the form of radiation. This was the penultimate step in the emergence of the electromagnetic field as a distinct entity. After 1897, when Thomson measured the electron's charge to mass ratio and Lorentz used his simple harmonic oscillator model of electrons to explain the newly discovered Zeeman effect, this conception won widespread acceptance.

The final step in this transformation was special relativity and the abandonment of the ether. Though this cannot be considered as nineteenth century physics, it completed the process we have been considering. Subsequently, the study of electromagnetic phenomena involves a conceptual division into two parts, bodies and fields. In addition to such mechanical concepts as 'space', 'time', 'mass', and 'force', one adds a new distinctively electromagnetic concept, 'charge'. Classical electromagnetism includes the assumption that macroscopic bodies are composed of atoms and molecules, that atoms have electrons, that charged particles produce electrostatic fields, that charged particles in motion produce magnetic fields. Further properties depend on grouping bodies as insulators or conductors; dielectric, paramagnetic, ferromagnetic, etc., and measuring resistance, electrical polarizability, and related properties (See Cook 1975, chaps. 9–12).

The details are not our immediate concern. The pertinent point is the conceptual framework. It is still the familiar world of macroscopic spatio-temporal objects supplemented by the new properties required to explain and systematize electromagnetic phenomena. However, it also includes the assumption that macroscopic bodies are made up of atoms, and that atoms have parts, which are held together by electrical forces. It does not include any detailed models of the atom. The second component is the electromagnetic field. This was no longer based on any ontological interpretation of the electromagnetic field or any causal mechanism to explain how charged particles produce and interact with the electromagnetic field. The electromagnetic field is treated as a separate entity and assumed to be continuous. The field itself may be defined through the idealization of a dimensionless test particle. One puts a charge of strength q at rest relative to a source and them measures the force,

$F(r, t)$. If gravitational and any other forces acting on the particle are negligible or may be subtracted off then the field at a point may be defined by the limit,

$$E(r, t) = lim_{q \to 0} \frac{F(r, t)}{q}$$

The magnetic field may be defined through the Lorentz force, $\mathbf{F} = q\ (\mathbf{v}\, x\, \mathbf{B})/c$. If the **E** and **B** fields are taken as basic, then the auxiliary fields may be defined in terms of them: $D = \varepsilon\, E$; $H = B/\mu$. Thus the final concept of an electromagnetic field rests on the minimal ontological assumptions needed: charged particles produce fields; fields act on charged particles. Further extensions depend on measurements. This electrodynamics links with mechanics and thermodynamics as the chief components of classical reality. The question remains. How do they fit together conceptually?

References

Blondel, Christine. 1985. Propagation Mechanism in a Medium Within Ampere's Electrodynamics. *XVIIth International Congress of History of Science, 17.*

Bromberg, Joan. 1968. Maxwell's Displacement Current and His Theory of Light. *Archive for History of Exact Sciences, 4*, 218–234.

Brush, Stephen G. 1965. *Kinetic Theory*. Oxford: Pergamon Press.

Brush, Stephen G. 1976. *The Kind of Motion We Call Heat*. Amsterdam: North Holland.

Buchwald, Jed Z. 1985. *From Maxwell to Microphysics: Aspects of Electromagnetic Theory in the Last Quarter of the Nineteenth Century*. Chicago, IL: University of Chicago Press.

Campbell, L., and W. Garnett. 1969. *Life of James Clerk Maxwell, Reprint of 1882 Edition with Letters*. New York, NY: Johnson Reprint Corp.

Carnot, Sadi. 1986 [1824]. *Reflections on the Motive Power of Heat*. Manchester: Manchester University Press.

Cook, David M. 1975. *The Theory of the Electromagnetic Field*. Englewood Cliffs, NJ: Prentice-Hall.

Crossland, Maurice, and Crosbie Smith. 1978. The Transmission of Physics from France to Britain:1800–1840. *Historical Studies in the Physical Sciences, 9*, 1–61.

Drake, Stillman. 1957. *Discoveries and Opinions of Galileo*. Garden City, NY: Doubleday.

Dyson, Freeman. 1988. *Infinite in All Directions: The 1985 Gifford Lectures*. New York, NY: Harper & Row.

Everitt, C. W. F. 1975. *James Clerk Maxwell: Physicist and Natural Philosopher*. New York, NY: Scribner's.

Fourier, Joseph. 1955. *The Analytical Theory of Heat*. New York, NY: Dover.

Gillispie, Charles Coulston. 1960. *The Edge of Objectivity: An Essay in the History of Scientific Ideas*. Princeton, NJ: Princeton University Press.

Gillispie, Charles Coulston. et al. 1978. Laplace, Pierre-Simon, Marquise de. In Charles C. Gillispie (ed.), *Dictionary of Scientific Biography*. New York, NY: Charles Scribners' Sons.

Granville, William, Percy Smith, and William Longley. 1962. *Elements of the Differential and Integral Calculus*. New York, NY: Wiley.

Hall, A. R. 1954. *The Scientific Revolution 1500–1800: The Formation of the Modern Scientific Attitude*. Boston, MA: Beacon Press.

Harman, P. M. 1982a. *Metaphysics and Natural Philosophy: The Problem of Substance in Classical*. Sussex: The Harvester Press.

Harman, P. M. 1982b. *Energy, Force, and Matter: The Conceptual Development of Nineteenth-Century Physics*. Cambridge: Cambridge University Press.

Heilbron, John. 1979. *Electricity in the 17th and 18th Centuries: A Study of Early Modern*. Berkeley, CA: University of California Press.

Heimann, P. M. 1970. Maxwell and the Modes of Consistent Representation. *Archive for History of Exact Sciences, 6*, 170–213.

Hertz, Heinrich. 1962. *Electric Waves*. New York, NY: Dover Reprint (Original, 1892).

Hesse, Mary. 1963. *Models and Analogies in Science*. London: Sheed and Ward.

Hirosige, T. 1969. Origins of Lorentz' Theory of Electrons and the Concept of the Electromagnetic Field. *Historical Studies in the Physical Sciences, 1*, 151–209.

Jungnickel, Christa, and Russell McCormmach. 1986. *Intellectual Mastery of Nature: Theoretical Physics from Ohm to Einstein*. Chicago, IL: University of Chicago Press.

Kahl, Russell. 1971. *Selected Writings of Hermann von Helmholtz*. Middletion, CT: Wesleyan University Press.

Kargon, Robert. 1969. Model and Analogy in Victorian Science: Maxwell and the French Physicists. *Journal of the History of Ideas, 30*, 423–436.

Kestin, Joseph. 1966. *A Course in Thermodynamics*. Waltham, MA: Blaisdell Pub. Co.

Kuhn, Thomas S. 1977. *The Essential Tension: Selected Studies in Scientific Tradition and Change*. Chicago, IL: University of Chicage Press.

Laplace, Pierre S. 1912 [1878]. *Oeuvres Complètes*. Paris: Academie des sciences Gauthier-Villars.

Lavoisier, Antoine. 1864. Traité Elementaire de Chimie. In *Oeuvres de Lavoisier. Tome Premier* (pp. 1–2). Paris: Imprimerie Imperiale.

Maxwell, J. Clerk. 1872. *Theory of Heat*. London: Longmans, Green and Co.

Maxwell, James Clerk. 1954 [1873]. A *Treatise on Electricity and Magnetism*. New York, NY: Dover Reprint.

McKie, Douglas, and Niels de V. Heathcote. 1935. *The Discovery of Specific and Latent Heats*. London: Edward Arnold & Co.

Morris, Robert J. 1972. Lavoisier and the Caloric Theory. *British Journal for Philosophy of Science, 6*, 1–38.

Nersessian, Nancy. 1984. *Faraday to Einstein: Constructing Meaning in Scientific Theories*. Dordrecht: Martinus Nijhoff.

Nersessian, Nancy. 2008. *Creating Scientific Concepts*. Cambridge: The MIT Press.

Niven, W. D. 1965. *The Scientific Papers of James Clerk Maxwell*. New York, NY: Dover.

Rueger, Alexander. 1997. Experiments, Nature and Aesthetic Experience in the Eighteenth Century. *The British Journal of Aesthetics, 37*, 305–327.

Siegel, Daniel. 1986. The Origins of the Displacement Current. *Historical Studies in the Physical Sciences, 17*, 99–146.

Tisza, Laszlo. 1966. *Generalized Thermodynamics*. Cambridge, MA: MIT Press.

Truesdell, Clifford. 1980. *The Tragicomical History of Thermodynamics: 1822–1854*. New York, NY: Springer.

Turner, Joseph. 1955. Maxwell and the Method of Physical Analogy. *British Journal for Philosophy of Science, 6*, 226–238.

Whittaker, Edmund. 1960. *A History of the Theories of Aether & Electricity: Vol. I: The Classical*. New York, NY: Harper Reprints.

Wise, M. Norton. 1979. The Mutual Embrace of Electricity and Magnetism. *Science, 203*, 1310–1318.

Wise, M. N. 1982. The Maxwell Literature and British Dynamical Theory. *Historical Studies in the Physical Sciences, 13*, 175–201.

Zinsser, Judith P. 2006. *La Dame d'Esprit: A Biography of the Marquise Du Châtelet*. New York, NY: Viking.

Chapter 4
The Interpretation of Classical Physics

The limits of my language are the limits of my world
Ludwig Wittgenstein

By the end of the nineteenth century many physicists were beginning to realize that the project of unifying physics on a foundation of atomistic mechanism had failed.[1] Einstein and Bohr emerged as the leading developers of a non-classical physics. In doing so they, more clearly than anyone else, established the limits of validity of classical physics. To look at their early work from this perspective is a bit like looking through the wrong end of a telescope. Yet, this is what we need.

4.1 The Limits of Classical Physics

Einstein's distinctive way of doing physics has been analyzed elsewhere.[2] Here I will focus on the role 'state of a system' (SOS) played in Einstein's thought and how it helped to unify and limit classical physics. In his 'Brownian motion' paper[3] he interpreted the Boltzmann constant k (or R/N), whose value Planck had calculated, as characterizing the scale at which molecular fluctuation phenomena become significant, and, as a consequence, the limits of valid applicability of thermodynamics (Einstein Papers, Vol. 2, p. 224). In his light-quantum paper (Einstein 1905b trans. in Miller 1981). Einstein showed that the assumption that radiation confined to box with molecules and both free and bound electrons could be in an equilibrium state

[1] Kelvin concluded that his 50 years of work were characterized by one work, 'FAILURE', because he still knew nothing of atoms and forces (See Thompson 1910, Vol. 2, p. 984). Helmholtz modified the goal and sought only a consistency between empirical laws and mechanical principles (See Jurkowitz 2002). Hertz (1956 [1894]) developed mechanics on a simpler, non-ontological basis. Boltzmann (1974, p. 227) recognized the possibility that future developments of atomic physics might have a non-mechanical foundation.

[2] See MacKinnon (1982, chaps. 9 and 10), Pais (1982), and Fine (1986).

[3] Einstein (1905a), translated in Furth (1956, pp. 1–18) and in *Einstein Papers*, Vol. 2. It is analyzed in MacKinnon (2005). See Faye and Folse (1994).

E. MacKinnon, *Interpreting Physics*, Boston Studies in the Philosophy of Science 289,
DOI 10.1007/978-94-007-2369-6_4, © Springer Science+Business Media B.V. 2012

leads to a divergence, which Ehrenfest dubbed 'the Ultra-Violet catastrophe'. Most physicists, had they noted this divergence, would have attributed it to a reliance on the equipartition theorem in deriving the average energy. Einstein saw this as indicating the limited validity of electromagnetism and related wave theory. Instead of relying on Planck's law, whose derivation he rejected, Einstein assumed that in the limit where Wien's law is valid each frequency may be treated as a separate unit. The light-quantum hypothesis was the assumption that this treatment of light as individual light quanta could be extended beyond the Wein limit.

The basic contention of the relativity paper (Einstein 1905c, translated in Miller 1981) is now one of the best-known arguments in physics. Postulate the invariance of the speed of light and the covariance of the laws of physics. Then deduce the Lorentz transformation equations for lengths, time intervals, forces, and fields. I will simply consider the points which have a bearing on the interpretation and limitations of classical physics. First, there is a global issue. Einstein's thermodynamic style of reasoning gives physics a type of coherence that it had not previously possessed. In place of Lorentz's successive order of magnitude corrections Einstein relied on the more general notion of the state of a system: "The laws by which the states of physical systems undergo changes are independent of whether these changes are referred to one or the other of two coordinate systems moving relatively to each other in uniform translational motion" (Miller, p. 395). The special theory of relativity does not give further laws of physics; it gives *constraints* binding on all the laws of physics, giving these laws a functional unification. These constraints effectively specified classical reality as the model of these laws.

Bohr's general point of view and his way of doing physics were different from most of his contemporaries. Bohr believed that a physical account must precede and ground a mathematical formulation. This personal orientation was reinforced by Bohr's professional role. As the presiding figure in the atomic physics community, he encouraged a close collaboration between experimenters and theoreticians and repeatedly modified his theoretical formulations to accommodate experimental results. In this context of an intimate ongoing dialog between theoreticians and experimenters, the issue that emerged as epistemologically problematic was the use of mechanics and electrodynamics in small-scale descriptions. This contrasts sharply with Einstein's approach. Einstein's treatment of states relied on assumptions concerning equilibrium conditions, not mechanical specifications.

The experimental sources that contributed to the development of atomic physics can be roughly divided into two types: those that obtain information from the light emitted, absorbed, or modified by atomic processes; and those that obtain information by hitting a target with a projectile and examining the debris. In neither case did formal theories supply the basic interpretative perspective. In spectroscopy the fundamental high-precision instrument then, as now, was the diffraction grating[4] The underlying presupposition in diffraction analysis is the assumption that the

[4] A general survey of the theory and practice of diffraction gratings may be found in Stroke (1969) or in Hecht and Zajac (1974, chap. 10).

wavelength, λ, characterizes a real periodic disturbance in space and time. This can be spelled out more precisely. The diffraction grating, the interferometer, and other high-precision optical instruments essentially register interference effects. The interference occurs on a very small scale, of the order of less than one tenth of a wavelength. The sharp reinforcements and cancellations seem utterly inexplicable unless the wave description of electromagnetic vibrations is accurate to at least this order of magnitude.

The second basic source of information concerning atomic states and systems was collision experiments. In the pre-cyclotron era this generally meant using as projectiles either electrostatically accelerated electrons or alpha particles from radioactive decay. Collision analysis never approximated the precision of spectroscopic analysis and usually involved something of a black-box approach. The controlled input and measured output were related by assumptions concerning single or multiple collisions, short-range forces, screening effects, radiation, excitation, ionization, energy loss and electron capture.[5] The unifying assumption here was the trajectory of an electron. An incoming electron is characterized by sharply defined velocity. Its trajectory through the scattering material can be decomposed into discrete scattering incidents. Each incident is characterized by an *impact parameter*, the minimal distance that would obtain between the projectile and the scatterer if there were no scattering forces. The treatment of an individual collision depends on the impact parameter and on assumptions concerning bound electrons.

In the Bohr-Sommerfeld program, theory and experiment interacted in such a way that they modified each other. The acceptance of the Bohr model of the atom changed the interpretation of the formulas governing spectral lines (Robotti 1983). It also modified the interpretation of scattering experiments by the assumption that a projectile's energy loss came in discrete units correlated with atomic energy levels. Experiments, in turn, modified the interpretation of the theory. I will consider one spectroscopic and one collision example. Quantum numbers served a double function. In addition to characterizing aspects of electronic orbits; they were also used to classify spectral lines. Orbital considerations suggested transitions that did not correspond to any observed lines. So Bohr introduced the selection rule, $\Delta k = \pm 1$ 1to eliminate the unobserved lines. To relate spectra to observed doublets and triplets Sommerfeld introduced a new, *inner* quantum number, j, where j is restricted to the values $k, k-1$. The corresponding selection rule is $\Delta j = 0, \pm 1$. Originally, he used this as a bookkeeping device for classifying spectra, not as a means of characterizing atomic states or electronic orbits. Others, however, soon related this in different ways to angular momenta of electrons. Soon the states of atomic systems came to be characterized by further quantum numbers and selection rules introduced to accommodate experimental results on spectral lines.

Sommerfeld had introduced the idea of space quantization as a means of determining changes in energy levels due to the presence of electrical and magnetic fields. This was widely regarded as a mere calculational device, even by those who relied

[5] Bohr's papers on scattering are in Bohr, *Works*, Vol. 8.

on space quantization in calculating the Stark effect and the Zeeman effect. The Stern-Gerlach established the reality of space quantization.[6] So space quantization, characterized by the magnetic quantum number, m, was incorporated into the language used to describe atomic events. Since m is the projection of j along some direction in space, and the significance of j was not clear, neither was the significance of m. Here again the focus uniting mechanics, electrodynamics, and quantum hypotheses was the concept of a silver atom as an object whose trajectory could be precisely described.

Subsequent developments shaped Bohr's position on the limits of validity of classical physics and on the role of language in the interpretation of physics. This development, however, is widely misunderstood. The 'standard' history of the Bohr-Sommerfeld (B-S) theory (or program) is easily summarized. The method of quantizing elliptical orbits and the introduction of generalized coordinates in a Hamiltonian-Jacobi formulation led to the general conception of atomic structure as a conditionally periodic system. When more complex problems, such as the anomalous Zeeman effect, did not yield to this approach, physicists began introducing specialized models, such as the core model, the successive Landé models, and various ad hoc hypotheses. By 1923 there was a widespread recognition that the B-S program had reached a crisis stage. In a desperate attempt to control the crisis, 'Bohr made some uncharacteristically bad physical assumptions'. He continued to reject the light-quantum hypothesis in spite of the overwhelming support provided by the Compton effect. In defense of this rejection he advocated, in the Bohr-Kramers-Slater (BKS) paper the bizarre idea of stationary states emitting continuous radiation and the highly unphysical notion of virtual oscillators replacing orbiting electrons. This house of cards soon collapsed under internal criticism and experimental refutation. Then Heisenberg, de Broglie, and Schrödinger redirected physics along more productive lines. These should have supplied a new beginning. Unfortunately Bohr again intervened and 'imposed a muddled metaphysics on these new mathematical formulations'.[7]

As recent studies have demonstrated[8] Bohr's way of doing and interpreting physics during this period was quite different from the physics practiced in Munich, Göttingen, or Berlin, or taught in textbooks. Bohr focused on problems the B-S 'theory' could not handle. He was much more concerned with underlying concepts and the consistency of the frameworks in which they functioned than with closed theories. What I wish to do here is to attempt to make this perspective intelligible,

[6] A detailed historical account is given in Mehra and Rechenberg (1982, Vol. I, pp. 422–445). See Boorse and Motz (1966, pp. 939–939) for a translation of the original paper and Trigg (1971, pp. 88–96), for an analysis.

[7] The primary sources for this interpretation are Jammer (1966, chaps. 3 and 4), and Van der Waerden (1967, Introduction). The most authoritative historical support for this interpretation is given in Heilbron (1985) and Cushing (1994, chaps. 3, 7, 10).

[8] See Hendry (1984, chaps. 3–5), Röseberg (1984, chap. 3), Folse (1977, chaps. 2–3), MacKinnon (1985, 1994), Chevalley (1991, Introduction), Faye (1991), Darrigol (1992, Part B), and Petruccioli (1993).

and especially to show how problems of conceptual consistency came to focus on the limits of validity of classical physics. Bohr gradually came to organize his work in terms of three groups of principles he took as established. The first group comprises his distinctive quantum principles, stationary states, and discrete transitions between states with spectral lines proportional to the energy difference. The second and third group must be understood in tandem. He thought of realistic principles as principles embodying classical concepts that could be used to describe reality objectively. Thus one could describe the orbits of electrons and the propagation of radiation in free space, but not orbital transitions or the production and absorption of radiation. Formal principles, in contrast, have significance only through their functioning in a network of concepts. In his development of the periodic table, Bohr relied on three formal principles, The Correspondence principle, the adiabatic principle, and the *Aufbauprinzip*. Bohr used the Correspondence Principle in a forward sense, employing classical formulations as a guide in guessing the proper quantum formulation. The adiabatic principle states that in a slow change of a periodic system the ratio of the kinetic energy to the frequency remains constant. The *Aufbauprinzip* guided the buildup of the periodic table by adding one electron to the previous atom, with the assumptions that the valence electron produces the spectral lines and that the core is in the lowest possible energy state.

The problems of the era will be grouped in families: problems of atomic structure; radiation problems; and related developments.[9] The first family of problems begins with the specification of states in mechanical terms and ends by abandoning mechanical descriptions in favor of state specification through the assignment of quantum numbers and energy values. In building the periodic table Bohr assumed that the quantum number n specifies an orbit; k specifies its degree of ellipticity; and j the orientation of the orbit of the outer electron relative to the core. Only the descriptive significance of the k quantum number seemed secure. Subsequent developments gradually eroded the descriptive significance attached to these quantum numbers. As Darrigol (1992, pp. 150–165) has shown, the pivotal building block in the construction of the periodic table was the parahelium configuration of helium. This not only supplied the inner core of all higher atoms; it also functioned as the paradigm for applying data, principles, and inferences, because it was the only case that could be treated with mathematical precision. A perturbation method, applying action-angle variables to the atom considered as a conditionally periodic system, was eventually generalized by Born and Heisenberg to consider all types of electron orbits compatible with the quantum conditions. This still left an irreducible gap between theoretical calculations of ionization energies and experimental results. When this gap was finally accepted as something that could not be covered over by better experiments or further calculations then there seemed to be only two choices: (a) there are no stationary states; and (b) it is not possible to describe electronic motions in quantitative terms. Bohr reluctantly accepted (b) and concluded that the

[9] See Jammer (1966, chaps. 3 and 4), MacKinnon (1982, chap. 5), Mehra and Rechenberg (1982, Vol. I, Part 2).

framework of spatio-temporal descriptions breaks down at atomic dimensions. All that remained from his three descriptive props was the requirement of the existence and permanence of quantum numbers. In abandoning descriptive accounts, he was following Pauli's suggestion.

In addition to his hypercritical positivistic orientation, two problem areas induced Pauli to stress quantum numbers over descriptive accounts. The first was the anomalous Zeeman effect, the complex splitting of spectral lines in weak magnetic fields. (See Forman 1968, Mehra, Vol. I, pp. 445–510). Pauli concluded that the half-integral numbers attributed to the interaction of the valence electron and the core must be due to the valence electron alone. This, in turn, meant that the valence electron is characterized by two azimuthal quantum numbers, k_1 and k_2, whose projections on the magnetic field are m_1 and μ. Since the energy requires a factor of $m_1 + 2\mu$, the μ quantum number makes a double contribution to the energy. Pauli characterized this as a 'classically non-describable double-valuedness'. This was the first clear separation between the assignment of quantum numbers and the possibility of a classical description.

Pauli's second contribution, his exclusion principle[10] stemmed from his new interpretation of quantum numbers and his acceptance of E. C. Stoner's systematization of orbits. Stoner gave a **formal** account of level occupation in Bohr's sense of 'formal'. He did not rely on the descriptive significance attached to the quantum numbers. This led to the same number of electrons per shell as in the Bohr account, but to different numbers of electrons for some sub-shells. Pauli suggested the plausible rule: *It shall be forbidden for more than one electron with the same n to have the same values for the three quantum numbers, k, m_1, and μ. This principle should hold as the magnetic field is weakened to zero* (Das Pauli Verbot).

The second family of problems, the interaction between radiation and atoms supplied a more rigorous testing ground for the B-S program than any theory had hitherto encountered. The theory was expected to yield precise values concerning frequencies, splittings, polarizations, relative intensities and selection rules, together with the effects of weak, intermediate, and strong electrical and magnetic fields. The intrinsic complexity of the situation was exacerbated by extrinsic factors. Frequencies were explained by differences in energy between stationary states. However, these were known with precision only for hydrogenic atoms. Bohr's abandonment of descriptive accounts of orbital motion occasioned the Bohr-Kramers-Slater (BKS) paper.[11]

If virtual oscillators (*Ersatzocillatoren*) were substituted for orbital transitions, then one had a mechanism that accorded with the Correspondence Principle and allowed calculations of polarizations and frequencies. This was never intended as a

---

[10] Massimi's (2005) provides an excellent account of Pauli's principle and its extended significance.

[11] The best account of the role of virtual oscillators is in Petruccioli (1993, chap. 4). A detailed treatment of the radiation problems is given in Mehra and Rechenberg, Vol. I, pp. 445–453. Less detailed accounts may be found in Jammer (1966, pp. 109–133) and MacKinnon (1982, pp. 185–190).

realistic model of atoms. The idea is that there is a statistical equivalence between two models of a collection of atoms interacting with radiation, one involving orbital transitions and one involving virtual oscillators. Neither admitted of a realistic interpretation. Bohr set up the strategy for using the Correspondence Principle in the treatment of radiation, but it was H. A. Kramers, Bohr's first disciple, who really developed it.[12] Kramers used the virtual oscillator model and the Correspondence Principle as guides in developing a dispersion formula that eventually incorporated earlier work of Ladenburg and Smekal. It had one peculiar feature. It required the inclusion of both positive and negative oscillators. The full mathematical treatment, developed by Kramers and Heisenberg, led to a dispersion formula, which looked so right that it was assumed (correctly) to be a formula of the to-be-developed quantum theory. As I showed in some detail elsewhere (MacKinnon 1977), the virtual oscillator model played a crucial role in Heisenberg's development of quantum mechanics. Keeping the order of virtual transitions straight necessitated non-commutative multiplication rules, which led to the introduction of matrices and matrix mechanics.

The virtual oscillator model was never intended as a literal model of atoms and radiation processes. It is a classical model that should be statistically equivalent to the to-be-discovered quantum theory of radiation, at least in the limits of high quantum numbers. Nevertheless, if the model is to be used, it must be developed with an overall consistency. Hence, Bohr and Kramers (but not Slater) postulated virtual fields with frequencies proper to transitions, $\nu_{12} = (E_2 - E_1)/h$. The fields surrounding a particular atom account for the spontaneous transitions Einstein had introduced, those surrounding other atoms account for induced transitions. For this to happen the fields must communicate with each other. The connection was statistical, rather than causal. Turning on a virtual field in one atom is independent of whether or not the reciprocal processes occurs in a neighboring atom. Einstein had argued from momentum conservation to light-quanta with momentum $h\nu/c$. A denial of light-quanta, something Bohr insisted on, reversed Einstein's argument and led to the non-conservation of energy and momentum. The new usage abandoned any reliance on spatio-temporal descriptions of electronic orbits and relied on damped simple harmonic oscillators providing **all** the allowed frequencies. Only one descriptive classical prop remained, the spatio-temporal description of radiation in free space.

In spite of its utility in dispersion theory and some related problems the BKS theory soon succumbed to internal criticism and experimental refutation. Bohr interpreted this refutation as a conceptual advance. It conclusively showed that no spatio-temporal description of radiation was possible for distances small compared to the wavelengths involved. It induced abandoning the hopeless project of using descriptive accounts of state transitions as a means of explaining the production of continuous radiation. Finally, it strongly supported a theme Bohr was increasingly

[12] Kramers's dispersion papers may be found in Van der Waerden (1967). Kramers's contributions to the development of quantum theory are analyzed in Radder (1982) and Konno (1993). Kramers (1957, pp. 92–95) shows how quantum mechanics yields his dispersion formula.

stressing, the necessity of a radical departure from the spatio-temporal descriptions on which the foundations of physical theory have hitherto been built (See, e.g., Bohr, *Works*, III, p. 375).

By the mid 1920s the practice of atomic and particle physics was generating many contradictions. Most of the participants did not see this as a crisis in classical physics, but as difficulties concerning the systems treated. Electrons and photons appeared to have contradictory properties. The development of quantum mechanics and the interpretative difficulties it generated make it extremely difficult to retrieve the crisis of 1924 as a crisis in **classical physics**. I will indicate the type of contradictions physicists encountered.

1a Electromagnetic radiation is continuously distributed in space. The high precision optical instruments used in measurements depend on interference. This in turn depends on the physical reality of wavelengths.

1b Electromagnetic radiation is not continuously distributed in space. This is most clearly shown in the analysis of X-rays as needle radiation and in Compton's interpretation of his eponymous effect as a localized collision between a photon and an electron.

2a Electromagnetic radiation propagates in wave fronts. This is an immediate consequence of Maxwell's equations. Experimentally this is manifested not only by interference, but also by the distinctively wave phenomena of superposition.

2b Electromagnetic radiation travels in trajectories. Again theory and observation support his. The theory is Einstein's account of directed radiation. The observations concern X-ray absorption. When X-rays are emitted at point A and totally absorbed at point B, then the only reasonable way to describe this is in terms of something traveling from A to B.

3a Photons function as discrete individual units. The key assumption used to explain the three effects treated in Einstein's 1905 paper is that an individual photon is either absorbed as a unit or not absorbed at all. Subsequent experiments of Millikan and others confirmed this.

3b Photons cannot be counted as discrete units. Physicists backed into this by fudging Boltzmann statistics. It became explicit in Bose-Einstein statistics.

4a Atomic Electrons travel in elliptical orbits. Here we should get away from the role of models or pictures and consider how this served as a semantic presupposition in three different types of inferences.

1. The basic rule for the B-S program was $\oint pdq = nh$, the action around a closed orbit is quantized.
2. A charged particle moving in an elliptical orbit produces a magnetic moment perpendicular to the plane of the orbit. This was a basic presupposition in all treatments of the Zeeman effect.
3. The explanation of atomic structure involved some tentative models and some features which did not depend on any particular model. A basic assumption was that the outer, or valence, electron explains an atom's chemical properties and spectral lines. In explaining X-ray production the basic

assumption was that short-wave length radiation penetrates the outer electrons and is absorbed by an inner electron leading to its ejection. Here 'inner', 'outer', 'orbit', 'distorted orbit' and related locutions presuppose orbital trajectories.

4b Atomic electrons do not travel in elliptical orbits. The treatment of dispersion and related phenomena led to this conclusion. The use of virtual oscillators in treating dispersion, resonant fluorescent polarization, and the anomalous Zeeman effect signaled an abandonment of the model of orbiting electrons.
5a Free particles travel in trajectories. This was routinely presupposed in setting up and interpreting such experiments as Franck-Hertz and Stern-Gerlach. It was also presupposed by the electron collectors and detectors used in The Compton, Davisson-Germer, Bothe-Geiger, and Compton-Wilson experiments. It was visibly manifested in the Wilson cloud chamber photographs.
5b Free particles do not travel in trajectories. This was an assumption that seemed unavoidable in attempts to explain the scattering of electrons from noble gases and nickel crystals.

In all these cases, we are not dealing with the falsification of theories by experiments. We are treating contradictions that emerged in the normal practice of physics. They are all rooted in the material inferences proper to experimental physics and the dialog between experimenters and theoreticians. In this context, these are not isolated contradictions, but one more indication that the spatio-temporal description of objects and events that physicists routinely rely upon could not be extended to submicroscopic objects and events in a holistic fashion. For most physicists these difficulties signaled a need for revised concepts and better theories. In Bohr's case, this linguistic crisis precipitated something akin to a Gestalt shift, a shift in focal attention from the objects studied to the conceptual system used to study them. This effect was somewhat similar to the epoché advocated by Husserl. He stressed the need for philosophers to bracket the natural standpoint and consider it as a representational system, rather than the reality represented. Similarly, Bohr's realization of the essential failure of the pictures in space and time on which the description of natural phenomena have hitherto been based shifted the focus of attention from the phenomena represented to the system that served as a vehicle of representation. He could not, however, emulate Husserl's detachment. He was soon engulfed in the problem of interpreting the new quantum mechanics. We, who lack this sense of urgency, can take a more detached view.

4.2 The Interpretation of Classical Physics

We can consider the interpretation of classical physics in two stages: pre-critical and critical. Historically, pre-critical interpretations were concerned with interpretation as a clarification of what physics reveals about objective reality. A critical interpretation is concerned with the status of classical physics as a representation.

The pre-critical interpretation can be regarded as an extension of the standard way of interpreting classical physics.

Before the twentieth century the goal of atomistic mechanism was to supply a basis for a functional unification of classical physics. To illustrate how this worked we turn to a text that functioned as the bible of atomistic mechanism, before higher critics questioned the biblical foundations. This is the *Treatise on Natural Philosophy* by William Thomson and Peter Guthrie Tait (Thomson and Tait 1867), (Henceforth T-T').[13] After developing mechanics they state what a perfect science should be:

> Where, as in the case of planetary motions and disturbances, the forces concerned are thoroughly known, the mathematical theory is absolutely true, and requires only analysis to work out its remotest details. It is thus, in general, far ahead of observations, and is competent to predict effects not yet even observed as, for instance, Lunar Inequalities due to the action of Venus upon the Earth, etc., etc. (Ibid., Vol. I, p. 405)

This ideal could not be implemented in any treatment of material bodies. Thus, a complete account of the problem of lifting a mass with a crowbar would involve a simultaneous treatment of every part of the crowbar, fulcrum and mass. This, however, is impossible: "... and from our almost compete ignorance of the nature of matter and molecular forces, it is clear that such a treatment of the problem is impossible" (Ibid., Vol. I, p. 337). In place of this idealized treatment, based on a mechanics of atoms, they outline a practical approach based on macroscopic representations and successive approximations. The first approximation represents the crowbar by a perfectly rigid rod and the fulcrum by an immovable point. The second approximation includes corrections due to the bending of the rod. Here again, the ideal treatment would be based on a theory of molecular forces. Since this too is impossible, they substitute the assumption that the mass is homogeneous and that the forces consequent on a dilation, compression, or distortion are proportional in magnitude and opposite in direction to the deformations. A third approximation is based on the consideration that compression produces heat, while extension develops cold and that both can change the operative length of the rod. Further approximations could include corrections due to heat conduction, thermoelectric currents, and departures from homogeneity.

Two aspects of these citations should be noted. First, they assumed that the mechanistic foundations were secure. The two cases differed in that they knew the force laws for planetary attraction, but not for molecular attraction. They assumed that when molecular force laws became known, then one would no longer have to rely on successive approximations. After an adequate law was established, the Schrödinger equation, people still rely on successive approximations for such problems, rather than ab initio calculations. The second, and more pertinent point, is that the functional unification of physics is material, rather than formal. The nature of the object treated, whether a planet or a crowbar, and the level at which it is treated determines

---

[13] Maxwell dubbed this book, and its authors, T and T'. Tait extended the terminological game by developing an equation, dp/dt = JCM, and subsequently referring to Maxwell as dp/dt.

the appropriate bits of physics needed. In this pre-critical period, as in subsequent uses of classical physics in experimental contexts, physicists took objects as given. They did not inquire into thestatus of ‘object’ as a presupposition.

4.2.1 Critical Classical Physics

Earlier I argued that the acceptance of an ordinary language claim as true (This shirt is yellow) does not entail accepting a functional presupposition as an ontological claim (Color is a property of bodies). To consider the role of presuppositions in generating contradictions we turn to a different example.

John murdered Henry. (S1)
John is not guilty of murdering Henry. (S2)

Though these claims seem to be in conflict, they may easily be interpreted as not contradictory. If S2 is taken as a legal claim, as in ‘guilty of murder in the first degree’, then its proper applicability depends on the legal presuppositions concerning premeditation, malice aforethought, and reasonable doubt. When this legalistic extension of ordinary language is put in the same context as the ordinary language claim, S1, then it does not seem referentially transparent. Its reference to a particular act is filtered through a categorical framework of legal concepts, precedents, and implicit rules. S1 does seem referentially transparent because we do not normally consider or even recognize the presuppositions implicit in ordinary language use. We use S1 to refer to a murder, not to reflect on ‘murder’. For these reasons a juror could accept both S1 and S2 without being involved in a contradiction. Suppose she accepted

John is guilty of murder in the first degree. (S3)
John is not guilty of murder in the first degree. (S4)

No appeal to presuppositions would be of any avail in avoiding an explicit contradiction, since S3 and S4 both rest on the same presuppositions. Suppose the juror accepted both S1 and

John did not murder Henry. (S5)

Then the question of presuppositions would never arise. Both S1 and S5 are taken as referring to a murder not to ‘murder’. One would fall back on Aristotle’s contention that there is no point in discussing anything with a person who fails to recognize the principle of non-contradiction, and hope that such a person is not selected for jury duty.

In the preceding case, an apparent contradiction was avoided by a clear-cut separation of two systems, ordinary language and legalese, resting on different presuppositions. There are other methods of avoiding apparent contradictions by functional separations. The civil right movement brought to light the radical inconsistency of preaching the universal brotherhood of man in strictly segregated churches. When this was made explicit, it was seen as an embarrassing contradiction, rather than a custom hallowed by tradition. Many scientists avoid apparent contradictions between their religious and scientific beliefs by consigning each to logic-tight compartments, faith and reason. Physicists teach their students that colors are not real properties of bodies and yet guide their experiments by directions such as: Put the red wire in the red hole and the green wire in the green hole. All of these stratagems have a similar effect. Contradictions are avoided by not putting the conflicting claims in the same framework of discourse, but in frameworks that rely on different presuppositions. This may seem irrational in the context of logical rules. It often supplies a viable resolution to clashes in the context of personal or collective behavior.

These considerations change the status of the problem of interpretation. In the syntactic interpretation of a theory, if the axioms are true, then any consequences of the axioms must be true. The truth status of axioms and conclusions is on the same level. In the semantic conception a theory is true of a model. Here again, a mathematical structure and the conclusions function on the same truth level. In the legal example claims and presuppositions do not function on the same truth level. Classical physics is not a theory. It relies on 'object' as a presupposition, but the systems considered represent idiontology, rather than ontology. To see how interpretation functions in contexts involving different levels we must consider how interpretation functions in other contexts.

The myth of Sisyphus relates the punishment inflicted on a legendary king of Corinth for trying to avoid death by a trick. Camus's famous essay interprets this myth as symbolizing the ultimate absurdity of death. The Emperor Constantine's conversion initiated a decisive change in the public status of Christianity within the Roman Empire. Historians dispute whether this should be interpreted as a religious conversion or a political ploy. After swearing revenge, Hamlet passes up a clear opportunity to kill the new king and avenge his murdered father. Freud interpreted this indecision as a consequence of Hamlet's subconscious identification with the man who killed his father and married his mother.

In each case we seem to be dealing with two distinct levels, a fact of some sort and a superimposed interpretation. In a more critical scrutiny, the fact/interpretation distinction looks a bit fuzzy. It rests on a myth of uninterpreted facts. What is the myth of Sisyphus, the account that functioned in early Greek culture, or the later literary exposition? What was Constantine's conversion: a psychological change after the legendary vision at the battle of the Milvian bridge; his convocation and control of the Council of Nicea to preserve orthodox Christianity rather than the Arian heresy; or his deathbed baptism by an Arian bishop? The answer depends on how one reads ancient texts and on how one interprets becoming a Christian. What aspect of Hamlet's indecision requires interpretation, the failure to avenge his father's murder, his recognition of kingship as divinely sanctioned, or his soliloquy justifying his procrastination? Does it make sense to psychoanalyze a literary character?

No interpretation builds on a bedrock of uninterpreted facts. The factual level relies on an implicit functional interpretation. There can be no reading of a text, accommodation to a practice, or following of a tradition without an interpretation. Yet, the basic distinction of levels retains some significance. Anyone carrying out a literary, historical, or scientific interpretation must first accept, at least provisionally, something as the phenomena (data, facts, events) to be interpreted. In the interpretation of classical physics, as in the simple examples just cited, we encounter two levels: a depth level, where core concepts play an integrating role, and a phenomenological level, where classical physics separates into distinct theories. Our initial concern is with the depth level and the two constraints it must meet. It must supply a basis for consistent informal inferences. It must also supply a basis for mathematical representations of quantities and structures. The dual-inference model was introduced to meet these constraints. Now we should consider the problem of the presuppositions generating contradictions.

The contradictions (1–5) previously listed all have the same general form. The presuppositions, e.g. an electron is a localized particle, an electron is a wave, that support experimental reports in one context cannot be extended to further contexts without generating contradictions. Most physicists saw this as an ontological conflict. How can an electron have both wave and particle properties. Bohr's Gestalt shift changed this to an epistemological problem. We are not concerned with the truth of the claims: "The electron is a particle" of "The electron is a wave". The question is: in which contexts can these serve as presuppositions? To answer this we will develop *Bohrian semantics*. This is a systematized extrapolation from Bohr's scattered and often opaque pronouncements. We will consider three points, the basic framework, the problematic extensions, and rules for avoiding contradictions. When we view classical physics as a conceptual system, then its conceptual core is a streamlined version of the core of ordinary language. We take as the core of ordinary language a subject/object distinction and a basic conceptualization of reality as interrelated spatio-temporal objects with properties. Classical physics extends the object side of the basic dichotomy and implicitly replaces the subject by a detached observer who can view reality without changing it. To understand the conceptual framework for experimental reports it is imperative to regard classical physics as an extension of an ordinary language framework. Typical reports include claims about the states of the system studied, states of the apparatus, decisions of agents, and the implementation of these decisions through computer programs, calibrations of the recording instruments, and preparations of the equipment. If there are fundamental inconsistencies in the system, then no inferences are reliable. This includes the submerged inferences that are regarded as a factual basis. The streamlining of this framework is effected by the systematization of space, time, and properties through representations that support mathematical formulations.

On a phenomenological level the assignment of numbers to properties depends on the measurement of quantities. Since measurement situations will also play a crucial role in the interpretation of quantum mechanics a brief review of Measurement theory is appropriate. In the second half of the nineteenth century, when the practice of assigning numbers to quantities was well established, there were systematic discussion of the interrelated problems of classifying quantities, measuring their values, and developing a general theory of measurement. Two somewhat different approaches to the general problem of measurement emerged. The first, stemming from Maxwell (Niven 1965, Vol. II, pp. 257–266), Helmholtz (1977), and Campbell (1920), focused on systematizing and generalizing the methods of measurement employed in physics. The second tradition, beginning with Hölder's mathematical analysis of measurement and the discussions of measurement in psychology by Stevins (1946) and of utility measures by Von Neumann and Morgenstern (1947), focused more on the empirical significance of assigning numerical values to non-mathematical objects. Before considering the differences, we can summarize some more or less standard terminology. Measurements may be classified as: *fundamental (or basic)*, not depending on prior measurements; *derivative*; depending on prior measurements; or *associative*, measuring some property like temperature or current by a direct measurement of some correlated property like expansion of mercury in a tube or deflection of a needle. We will begin with a standard treatment of fundamental measurements. The question that concern us is whether or not this should be accorded a *conceptually* fundamental role in classical physics.[14]

Consider a set of objects, x_i, in a domain, A, each with a property, P, that admits of an empirical ordering relation, $x_i \geq x_j$. If this relation is: asymmetric, $((x_i \geq x_j) \supset \neg(x_j \geq x_i))$; transitive $(((x_i \geq x_j) \,\&\, (x_j \geq x_k)) \supset (x_i \geq x_k))$; and connected, $((x)(y)((x \in A) \,\&(y \in A)) \supset ((x \geq y) \, v \, (y \geq x)))$ then it groups the objects of the domain into equivalence classes that may be homomorphicly embedded into the ordinal numbers. Measurements involving the imposition of cardinal numbers require a unit, a process of concatenation ($\circ$), that is associative and commutative, and a scale. The standard example is a collinear concatenation of m copies of the unit to measure the length of an object.

Generalization from such stock examples leads to the idea that measurement involves the homomorphic embedding of an empirical relational structure, $\mathcal{E} = \langle A, \zeta, \circ \rangle$ into a mathematical relational structure, $\mathcal{N} = \langle \Re^{+}, \geq, + \rangle$, where A is a non-empty set, ζ is an empirical ordering of the members of A, $\circ$ is a binary operation on members of A producing a new member, $\Re^{+}$ are the positive real numbers, and $\geq$ and $+$ have their standard meanings. There is no a priori reason why raw data should have an empirical structure isomorphic to some theoretical structure. However, as Giere (2006) shows, testing in science is best understood in terms of agents comparing theoretical models with models of the data. The data modeling is shaped within a perspective determined by the questions imposed on nature and the particular instruments used. We will consider particular examples testing predictions in particle physics in more detail in subsequent chapters.

The representative theory of measurement was developed to clarify the empirical significance of measurements. When one uses questionnaires, statistical surveys,

---

[14] The basic concepts of measurement theory are treated in Ellis (1968). The definitive treatment of the representative theory of measurement is Krantz et al. (1971). See also Adams (1979), Narens (1985), and for a critical reaction, Kyburg (1984).

and economic indicators to 'measure' I.Q., degree of sibling rivalry, economic utility, or proneness to delinquent behavior, one can not but wonder whether the resulting numbers really characterize objective properties of the things measured. The representative theory of measurement seeks to answer such questions by establishing representation theorems, guaranteeing the homomorphic embedding, and uniqueness theorems, showing that any other scale is equivalent within a linear transformation. Such considerations are rarely needed in physics.

In the practice of physics measurements rely on increasingly sophisticated instruments that presuppose a network of prior measurements and established standards. Measurements are conceptually structured and the conceptual structures function within a phenomenological level of classical physics. The empirical structures are idiontological, a systematization of properties. This presupposes a loose informal unification. Thus, one uses aspects of mechanics and electrodynamics to measure temperatures. Measurement, accordingly, presupposes, rather than supplies, a conceptual foundation. This conceptual foundation supports the empirical structures that make systematic measurement possible.

Space and time are not measured. Conceptually we measure lengths in terms of multiples of a unit and durations in terms of repetitions of a cyclic process. Classically, measurements of lengths and durations were taken as measurements of space and time. We postpone a consideration of relativity and, for analyzing discourse, accord the speaker's framework a privileged status. The proper mathematical representation of space, time and motion from an observer's perspective stem from Galileo's arguments for linear inertia. In modern terms, he laws of physics are invariant with respect to uniform local motion. This leads to the Galilei group, a continuous 10 parameter Lie group, essentially a non-relativistic version of the Poincaré group. Since the mathematical representation of this group is something we will need in discussing quantum field theory we will postpone the mathematics and give a qualitative summary. In a famous paper, Eugene Wigner (1939) showed that irreducible representations of the Poincaré group (or the inhomogeneous Lorentz group) correspond to the representation of particles of fixed mass and spin. The extension to the Galilei group (Levy-Leblond 1963) shows that irreducible representations of this group represent particles of fixed mass and angular momentum.

We can give a simplified representation of what this means for the mathematical representation of space, time, and motion from the observer's framework. The spatio-temporal location of a body (or its center) relative to a reference system is given by three position and one temporal coordinates. The operations of spatial displacement, temporal displacement, motion, and rotation can all be represented by operators. Putting these together, we have

$$\mathbf{x}' = \mathbf{x} + \mathbf{a} + \mathbf{v}t + R(\mathbf{x}) \qquad t' = t + b \tag{4.1}$$

This fits a model of space as continuous, Euclidean, and homogeneous and of time as continuous.

The properties measured are represented as properties of bodies. 'Body' plays a presuppositional role. To understand the dualistic (particle/field) divergence characterizing the last stages of classical physics we should consider the ordinary language bases, of which these are extensions. In *Word & Object* Quine stressed some basic differences between *mass* (now used in a generic sense) and *count* terms, a topic that has provoked extensive discussion.[15] Though the literature refers to these issues as metaphysical, I will interpret it exclusively as descriptive metaphysics. Count nouns and mass nouns exhibit syntactical and semantical differences. Syntactically count nouns admit of pluralization, can take numerical values (5 apples), while mass nouns do not admit of plural forms, do not occur with numerals, and are qualified by terms like 'much' and 'little', rather than 'few' and 'many'. These differences relate to what Quine dubbed 'divided reference'. Count nouns presuppose individuated perduring objects: "He ate the apple that I bought yesterday". The syntactic criteria for mass nouns puts terms like 'gold', 'water', 'mass', 'freedom', and 'happiness' in the same category. Physical mass terms differ from non-physical mass terms is some interesting respects concerning reference and numbers. As Helen Cartwright (1965) pointed out, an unstressed occurrence of 'some' plays the individuating role of an indefinite article for mass nouns. Compare "Heraclitus bathed in some water yesterday and bathed in it again today," with "Heraclitus experienced some happiness yesterday and experienced it again today". The first occurrence of 'it' plays a referential role, while the second is ambiguous.

The idea of concatenation is built into the grammar of mass terms. If I pour some water into Heraclitus's tub then it has more water. But, this concatenation does not normally admit of a set-theoretic reconstruction. Compare "The students who were in the classroom are now in the auditorium" with "The water that was in Heraclitus's tub is now in a puddle". In the first case, the students are counted as the same individuals regardless of the room that contains them. In the second case the water is individuated by its container, not by the set of elements it contains. These functional differences relate to the way these types of terms are represented mathematically. Count terms are represented by integers, mass terms by a continuum of values. Since this is ordinary, rather than regimented, language, both norms admit of exceptions. One can use 'apple' as a mass term, as in "Put more apple in the fruit salad". One can use water as a count term, as in "the waters of the seven seas". The distinctions can be arbitrary. We could describe the food on a plate as 12 noodles, but not as 12 spaghettis. The language of physics is an extension of normal usage, not the exceptions.

Objects are the exemplary count nouns. In classical physics, a particle is an object whose inner structure, if such there be, is contextually irrelevant. In appropriate contexts electrons, atoms, planets, stars, and galaxies function as particles. This functional concept of a particle is the simplified extension of the notion of a public object. It is not an ontological term. The term 'particle' is at the center of

[15] See Quine (1960, § 19 and also note 3, p. 91) for the background to the distinction. Helen Cartwright's (1965) played a pivotal role in further discussions. See Zimmerman (1995), for a recent treatment.

a conceptual cluster. To call anything a particle entails that it is individuated by its spatio-temporal location, that it travels in a space-time trajectory, that it can collide with a target, that it collides at a definite point, that it can impinge on it, penetrate it or recoil from it. 'Particle' as the term is used in EOL, has the properties of count terms plus such quantitative refinements as mass, rest energy, and charge. Charged particles are attracted by electrical fields. The orbits of moving charged particles are altered by magnetic fields. We will use 'particle$_c$' to designate this classical conception of particles.

The term 'wave' is at the center of a different conceptual cluster. Waves do not have sharp locations. They do not strike, impinge on, penetrate, or recoil from targets. Rather, waves can interfere with each other, be diffracted by a medium or refracted by an object in their path. They have their own distinctive properties, such as frequencies, wavelengths, superposition. This emerged from the OL basis of mass terms, with the medium supporting the waves functioning as a transformed mass term. Eventually electromagnetic waves were conceptually separated from a dependence on a supporting medium and functioned as part of a descriptive account on a phenomenological level.

The term 'classical physics' can be used in different ways. We will consider two. The first is classical physics as a tool. The problematic area concerns a very restricted context, measurement situations. Many philosophers who take theories as units of interpretation still follow the positivistic tradition of regarding experiments as suppliers of facts that check theoretical predictions. Any analysis of an experiment involves informal inferences. If the inferential basis is inconsistent then no inferences are reliable. Yet most of these informal inferences become so submerged that they are no longer recognized as inferences. In the experiment cited in Chapter 1 Franck and Hertz assumed that electrons are units of negative energy, that they travel in trajectories, that they can be accelerated by electrical fields, that in a collision with a mercury atom an electron either suffers no energy loss or a discrete loss of $h\nu$, that a mercury atom has an outer valence electron, and that a collision can excite this valence electron rather than the atom as a whole. At earlier times all of these 'facts' had the status of hypotheses. High energy particle accelerators like the Tevatron, SLAC, and the Large Hardron Collider (LHC) produce billions of collisions per second. The collider is supplemented by a collection of enormous, very complex detectors. To record only the potentially interesting collisions the detector must be able to infer almost instantaneously that a particular collision may have involved the production of an exotic particle. A particular detector and some of its components will be considered later. The goal is to assemble the information from these various subsystems and infer the location, charge, and energy of the originating particle. The organizing principle in many of these inferences remains particle trajectories. These inferential system are submerged in the accepted facts and now in computer programs that make the inferences and trigger the production of photographic records. To see the consistency problem suppose that at some stage in either the Franck-Hertz experiment or in the functioning of a high-energy particle detector. If the incoming particle was reflected from the face of a crystal, then one could no longer rely on trajectories as an inferential basis and could draw no reliable conclusions. This is the key point in Bohrian semantics. In setting up experiments and reporting results one must rely on the systematic extension of classical concepts and either the particle cluster of concepts or the wave cluster. Either cluster supplies a basis for valid informal inferences within a particular context. Neither supplies a basis for inferences across complementary contexts. This principle is based on an analysis of the conditions for the unambiguous communication of experimental results. It precedes and is independent from any analysis of quantum mechanics as a theory. The second notion we wish to consider, 'classical reality', involves a change of perspective. Instead of considering the contextual limitations in the extension of classical concepts we consider an idealized model of reality embodying a consistent interrelation of the laws of mechanics, thermodynamics, and electrodynamics. All of these laws rely on idealized models of continuous systems. Two conclusions follow. First, classical reality is characterized by large-scale deterministic laws. Second, classical reality is not a realistic model of even the objects, properties, and processes treated in classical physics. We will return to the problematic status of classical reality at the end the final chapter.

4.3 A Dual Inference Model

The treatment of a dual inference model of scientific explanation has been postponed because it builds on the material we have covered and sets the stage for the special perspective for viewing quantum mechanics. The basic idea is simple. The normal process of solving problems in physics usually involves two distinct inference systems, which we will refer to as formal and physical. By 'formal' in this context we simply mean a system like formal logic or mathematical deduction where the inferential rules are independent of the material to which they are applied. This has been thoroughly explored by philosophers in developing axiomatic and semantic models of physical theories. For both models syntax precedes and grounds semantics. One sets up a mathematical formalism and then imposes an interpretation. By 'physical' inference we mean an extension of the material inferences that Sellars treats and which we summarized in Chapter 1. Material inferences are content-dependent and are based on a common sense understanding of the normal properties, activities, and interrelations of familiar objects. Similar inferences that presuppose and utilize established physics will be called 'physical inferences'. We begin with a simple model that illustrates what is meant by a dual inference system and then focus on the special difficulties involved in clarifying physical inferences and relating them to formal inferences.

A game of bridge begins when the dealer distributes fifty-two cards from a normal deck to four players in two teams, and then commences bidding. As an aid to bidding, various partial representational systems have been developed: the Goren system, the Culbertson system, the Italian system, and specialized systems proper to tournament players. To use this as an analogy, we begin with the categorial system of 52 objects divided into 4 suits each having 13 types with a rank ordering of suits and types. The Goren system imposes a mathematical order that corresponds to selected aspects of the categorial system and its rank ordering. There are values of 4, 3, 2, 1 for A, K, Q, J, plus 3 for a void, 2 for a singleton, 1 for a doubleton.

This combined system supports three different types of inferences. The first, which could be called 'card reasoning' to parallel 'physical reasoning', is based on the cards and their players. The simplest example, which plays a presuppositional role, is categorial inference. If I have the K ◇, I can infer that no one else has it, an inference that would not be true in pinochle or four-deck blackjack. Intuitive, non-formalizable reasoning plays a distinct role, in evaluating opponent's skill level, facial expressions, and even subconscious signals. The formal system is trivial, based on the additive properties of small integers. However, it supplies a helpful guide to bidding my hand and evaluating the strength of the other hands on the basis of their bidding. These two inference systems are independent. There is a partial isomorphism between the rank ordering of the cards and the values of the points. Each has surplus structure not represented by a partial isomorphism. Thus, cards from two to ten are not represented in the Goren system. Large numbers, and multiplication and division of numbers do not match anything in the card system. The two systems are linked materially, not formally. Both are applied to the same objects.

There is also mixed reasoning, based on the cards, the numerical system and the correspondence rules connecting them. Thus, from the bidding I, as South, might infer that West has 10–13 points. From the cards played and those I see in my hand and my partner's exposed hand I infer that West probably has the Q ♣. On these grounds, I assume that a finesse would not work and try to set up an endplay. Here the dual-inference system is needed because each component by itself is inadequate to the task of making this simple inference.

The normal practice of physics exhibits physical reasoning, formal reasoning, and mixed reasoning. We use 'reasoning' as a general term to cover both inferring and simply muddling through problems. The physical reasoning is most evident in the early stages of scientific development. The development of the Baconian sciences was chiefly a matter of physical reasoning without the benefit of formal theories. It is also evident in both theoretical and experimental work. The clearest instance of this in theoretical analysis is in thought experiments. Experimental analysis exhibits both physicalistic and mixed inferences. All of these points are best developed by analyzing historical examples, rather than by imposing a theory. I am attempting to clarify the way physics actually functions and am not imposing a theory of how it should function.

The developments considered fit this schema. Newton's treatment of forces was preceded by a long process of controversial reasoning concerning impetus, inertia, impressed forces, living forces, impulsive forces, and impressed forces. The critical breakthrough was Newton's presentation of impressed force as an additive concept. The isomorphism between concatenating forces and adding (or integrating) numbers established the link between physical and mathematical reasoning. This simple isomorphism carried over to other mechanical notions that could be represented by extensive concepts, quantity of motion (or momentum), velocity, acceleration.

An extensive concept of heat required a clarification of related concepts: 'temperature', 'overt heat', 'latent heat', 'specific heat', and methods of measurement. Humphrey Davy's refutation of caloric theory supplies a clear example of what we mean by a physicalistic inference. He arranged a way of rubbing two ice cubes together in a sealed container until they melted. If heat is a manifestation of overt caloric then rubbing should not cause melting. No caloric is lost from the container. When heat is regarded as a manifestation of kinetic energy then the phenomenon is explained. Early electricians muddled through Descartes' account of electrical and magnetic forces in terms of lines of particles with hooks, one and two-fluid theories, the distinction between positive and negative electricity before developing extensive quantitative concepts: 'charge', 'capacitance', and 'potential'.

These examples feature physicalistic reasoning about the properties and activities of real and hypothetical entities. It is an extension of the type of reasoning that Sellars labeled 'material inferences'. In recent years this has drawn increasing attention from two different groups. The first group includes philosophers and other who teach and write books about clear thinking, critical thinking, or basic reasoning, to cite my entry into the field (MacKinnon 1985). These studies have led to strategy considerations, helpful paradigms, and a multitude of examples. They have not led to a logic of informal inferences. The second group studies artificial intelligence (AI), a highly disputed field. My own rather superficial evaluation is that AI is succeeding well in specialized studies such as medical diagnoses, mining data, and engineering problems. Deep Blue defeated world chess champion Garry Kasparov while IBM's Watson beat the reigning champions in Jeopardy. Yet, AI has yet to match normal humans at general common-sense reasoning. This presupposes a broad knowledge of many different types of things and uses fast intuitive judgments. AI programs must have some kind of representation of all the facts presupposed and rely on explicit rules for all inferences. In a common-sense situation this leads to a combinatorial explosion. The most dramatic example of this comes from the test: What's wrong with this picture? (Koch and Tononi 2011). A person and an AI program can both be presented with a picture in which something is obviously wrong: a person is standing two feet off the floor; a computer monitor is plugged into a plant, an elephant is sitting atop the Eiffel tower. The normal person instantly recognizes the anomaly. The AI program does not unless it has explicit rules about inferring normal relations between objects. This illustrates two points. The first is the enormous amount of information of different forms that goes into common sense knowledge. The second is the way in which material inferences and the common sense knowledge they presuppose resists systematization. The intuitive grasp of the fact that a man does not hover two feet above the floor is not recognized, much less analyzed, as an inference. It is simply a fact about people and places.

A training in physics supplies the basis for an extension of the common sense reasoning relying on EOL, the extended language of physics. To bring out the significance of the shift from common sense to physicalistic reasoning we begin with a couple of simple examples. A painter wants to use a long ladder to paint the walls on the upper floors of a house. If the angle between the ladder and the ground is too large then he may fall backwards. If the angle is too small then the ladder may slide when he is near the top rung. What should be the angle between the ladder and the ground? A beginning physicist learns to translate this problem into EOL. The painter could fall backwards if a line from the center of mass of the painter and ladder to the center of the earth (or perpendicular to the supporting surface) is behind the base of the ladder. This is a problem when he is on the lower rungs and can be solved by leaning against the ladder. The problem of slippage is most acute when he is on the top rung. Here one must consider the torques due to the weight of the ladder acting at the ladder's center of mass and of the weight of the painter on the top rung. The force these torques exerts on the bottom of the ladder can be split into two components, a vertical component proportional to the sine of the ladder's angle and a horizontal component proportional to the cosine of this angle. If this horizontal component is greater than the horizontal component of the force of friction between the base of the ladder and the ground then the ladder slides. When the problem is cast in these terms then one can plug in numbers to calculate the minimal allowed angle, a simple example of a mixed inference. A baseball player wants to run at the right speed and in the right direction to catch a fly ball. How can he determine the speed and direction? The coach's answer is "Practice, Practice, Practice." A physicist's analysis would begin by splitting the problem into components. First, assume that the ball is coming straight towards the outfielder. Then a bit of calculation shows that he must run at such a speed that the rate of change of the tangent of the ball's angle of elevation remains constant. Suppose that the ball is traveling in such a direction that there is an angle, α between the horizontal component of the ball's trajectory and a line between the player and the ball. Then he must run in a direction that keeps α constant. This ensures that he gets to the right spot. To catch the ball he must adjust the speed and direction of his gloved hand, proprioceptivly perceived, so that it intersects the damped parabolic arc of the baseball. The normal training of a physicist involves solving many textbook problems of increasing complexity that recasts problematic situations in the terminology of physics and then plugs in in numbers.

As with common-sense reasoning routine physicalistic inferences are rarely recognized as inferences. They are submerged through retroactive realism. As noted earlier, Franck and Hertz treated as established facts claims that had the status of hypotheses a generation earlier. This physicalistic reasoning also resists systematization. Russ Hanson, to whom this book is dedicated, kept trying to develop a logic of discovery. He never succeeded. There is no such logic. However, the concern that inspired the search remains a vital issue. The great breakthroughs in physics can not be explained by a systematic analysis of the theories they produce. Yet, they surely deserve a philosophical analysis. Most of them, especially in the early stage in scientific development, are chiefly, sometimes exclusively, based on physicalistic reasoning. I will illustrate this by sketching the developments that led up to atomic and particle physics.

In 1895 Wilhelm Roentgen discovered that the cathode-ray discharge from a Lenard or a Crookes tube could produce fluorescent radiation in crystals on a paper screen. To prove this was an effect caused by the cathode rays he had evacuated the tube linking the source to the target, installed appropriate insulation, and conducted the experiment in the dark. The novel conclusion was that the agent inducing the fluorescence could pass through cardboard, human flesh, and other types of material. The degree of transparency depended only on the density of the material involved. He inferred the existence of a new type of radiation, X-rays, and addressed the question of the nature of this radiation. It is like light in casting shadows, but differs from ultra-violet radiation in not showing refraction and reflection. His tentative conclusion was that this radiation should be ascribed to longitudinal vibrations of the ether.

Henri Becquerel extended these investigations by covering photographic plates with thick black paper and then exposing them to sunlight. Even a two-day exposure produced no effect. However, when different types of salts were placed on the paper then sunlight induced a phosphorescence that clouded the plates. Since uranium salts worked best he repeated this experiment using uranium salts for three consecutive days at the end of February, 1896. The weather refused to cooperate. The sun was covered by clouds. Yet, to his surprise, the plates showed the same result. From this he inferred that uranium salt was the source of this radiation, not the sun's power to induce fluorescence. He tried different types of uranium salts and concluded that uranium alone was the source of the radiation. It must be produced by some sort of discharge from uranium atoms. This was a remarkable inference. Though there were no good theories of the atom, one assumption had seemed incontestable. Atoms, as the name implies, are indestructible. The Curies extended these investigations. After testing different substances, they came to recognize pitchblende as a potent source. Their heroic efforts to purify the pitchblende led to the identification of two new much more powerful radiation sources, polonium and radium. What is the nature of the radiation these substances emit? Rutherford's famous tabletop experiments dominated the efforts to answer such questions. By various combinations of insulating materials between sources and targets, different distances, and different arrangements of electrical and magnetic fields he eventually distinguished α, β, and γ rays. Further testing, especially of the ratio of weights to charge and techniques for collecting α-rays in sealed vacuum tubes led to the identification of α-rays as helium nuclei and β-rays as electrons. From his famous scattering of α particles from nuclei he inferred that an atom contains a small positively-charged nucleus.

Few would deny that such physicalistic reasoning plays a basic role in the fact-gathering stage of a science's development and even in the creation of theories. However, the implicit contention is that it may be neglected in theory interpretation, which is essentially a matter of relating mathematical formalisms to the reality of which they are theories. This neglect is defensible only when two conditions are met. First, what is interpreted is a theory that is rationally reconstructed to fit the interpretative schema. Second, experiments enter only as supplying facts that confirm or falsify theories. I contend that informal physicalistic inferences relate to theories in at least three significant ways. First, the normal application of a theory to a problematic situation requires a prior physical analysis of a problematic situation, as illustrated in the ladder and baseball examples. Second, informal inferences play a role in developing and advancing theories through thought experiments. Finally, experimentation and the dialog between theoreticians and experimentalists requires informal inferences and mixed inferences. We will consider each of the last two points. Here again, Newton set the precedent. In Book One of the *Principia* (p. 10) he argued for absolute space and time by considering the relative motions of a rotating bucket and the water it contained. The key point is that the concave shape of the rotating water's surface is due to the absolute motion of the water, not its motion relative to its container. In Book Three (p. 551) he argues that the circular motion of the moon's orbit should be explained by the same combination of inertial motion and gravitational force that explains projectile motion. The argument relies on a thought experiment of considering projectiles shot off high mountain peaks. By increasing the height of the peaks and the velocity of the projectiles until the projectiles are put into orbits.

The concept of entropy stemmed from Sadi Carnot's thought experiment. He considered an ideal an ideal heat engine going through a reversible cycle. When run in one direction it required work, in the other direction it did work. The work cycle starts with a working substance t a temperature T_2. It is adiabatically compressed until its temperature reaches T_1. The working substance is allowed to expand isothermally at temperature T_1 taking in Q_1 units of heat. Then it expands adiabatically until the temperature is back to T_2. Finally the working substance is compressed isothermally at temperature T_2 giving off Q_2 units of heat. The efficiency of the engine is defined as $(Q_1 - Q_2)/Q_1$. This argument demonstrated that no engine can be more efficient than a reversible engine and that no ideal engine can have an efficiency of 1. Late Maxwell explored the limits of the entropy concept by imagining a demon opening and closing a shutter to sort out molecules in a gas.

I suspect that no one in the history of physics made a more fertile use of thought experiments than Einstein. Some of the thought-experiments that figured in the Bohr-Einstein debates will be considered later. Here we will consider the argument he advanced to establish the equivalence of gravitational force and acceleration, a basic postulate of general relativity. Consider a man standing in an isolated elevator who experiences a force pulling him downward. This could be explained either through the gravitational attraction of a massive body beneath the elevator or through the acceleration produced by a rope pulling the elevator up. No activities the man can perform inside his isolated elevator can settle this issue. So gravitational force and acceleration are equivalent.

These disparate arguments have a common feature. They are not based on mathematical inferences. They rely on inferences about the properties, activities, and interrelation of bodies. The concepts developed through these arguments, absolute space, reversible cycle, entropy, and the equivalence of gravitation and acceleration assume a foundational role in setting up a mathematical formalism. After a formalism is established, informal inferences still play a role in relating theory to experiment. In classical physics the inferential duality is submerged because the same quantitative concepts function in theories and experimental discourse. In quantum physics the language of experimentation does not carry over to theories without explicit restrictions. The concept of an object that plays a foundational role in ordinary language and in classical physics does not play a foundational role in quantum physics. To illustrate this we will consider two famous experiments, one classical and one quantum, and then two thought experimenters introduced by the two founders of quantum mechanics.

James Joule, who had a bit of training in chemistry but none in mathematical physics, spent the years 1840–1850 conducting experiments concerned with relating electrical and mechanical energy to heat. Energy conservation was still regarded as a speculative hypothesis of German scientists. Joule used 'vis viva' and did not refer to a conservation law. The question that concerned him was how to relate vis viva, or mechanical energy to heat. This required physicalistic reasoning, arranging an experiment that transformed measurable mechanical energy into measurable heat. His simple, but ingenious, paddle-wheel experiment is now routinely performed in high-school physics classes. It supplied a basis for determining the heat equivalent of the mechanical power expended when a massive body drops a certain distance. It amazed Thomson, the leading theoretician of heat phenomena. Here the breakthrough came from devising a way of using mechanical force to produce thermal effects. Then the mathematics was trivial. Measure the weight of the body and the distance it descended. Measure the quantity of water and the rise in temperature. Also measure the weight and temperature change of the copper vessel and the brass paddle wheel. This led to the conversion factor that the quantity of heat capable of raising the temperature of a pound of water by 1°F requires the expenditure of a mechanical force represented by a fall of 772 lbs through a space of one foot. (Shamos 1959, p. 182)

For a pivotal quantum experiment we will consider the process culminating in Compton's 1923 experiment. Einstein's introduction of the light-quantum hypothesis exploited the idea of one light-quantum interacting with one electron to explain the photoelectric effect. It failed to convince the majority of physicists, including Planck and Bohr. The account of light as electromagnetic vibrations was too successful to be abandoned for a tentative counterintuitive hypothesis. A series of experiments by Millikan, Ellis, Maurice de Broglie, and Compton lent increasing credence to the light quantum hypothesis, but did not supply convincing proof. Compton reasoned that the best way to test this hypothesis was to scatter hard X-rays, whose wavelengths are comparable to the still unknown size of the electron. However, this presented many difficulties. Both J. J. Thompson's theory of X-ray scattering and experiments supported the claim that scattered X-rays were unchanged in frequency and coherent with the primary radiation. However, a few experiments indicated that a component of some scattered radiation had a longer wavelength than the primary radiation. This could be explained: by assigning a larger size to the electron (Compton's early hypothesis); or by fluorescence, his later hypothesis; or by a tertiary radiation of the bremsstrahlung type, a hypothesis advanced by William Duane, director of the Harvard X-ray research lab where Compton was conducting his experiments.

Compton attacked this problem by a combination of experimental and theoretical reasoning. Experimentally the problem was to develop an experimental arrangement where his hypothesis would lead to results different from the competing hypotheses. Theoretically, he compared two explanations of the scattering of radiation off electrons. In the classical (Thompson) account the scattering of X-rays off electrons induces vibrations. The vibrating electron produces secondary radiation of the same wave length as the primary wave length. This Compton called the unmodified ray. For a quantum account Compton considered the interaction between an individual X-ray quantum and a free electron. The recoil of the electron absorbs some of the momentum of the incident X-ray quantum ('photon' was introduced later), which then is deflected in a different direction and with a lesser frequency (or a longer wavelength). Compton called this the modified ray. This quantum treatment poses new difficulties. One had to use bound, rather than free, electrons. According to the reigning B-S model a bound electron jumping to a higher orbit could only absorb the energy equivalent to the energy difference between the orbits. Compton's decisive experiment is schematized in the following diagram.

As illustrated in Fig. 4.1, a beam of X-rays of known frequency emanating from a molybdenum anticathode, T, strikes electrons weakly bound in carbon atoms in a graphite target, T. The X-rays leaving the atom in a given direction pass through slits, omitted in the diagram, which filter out extraneous secondary radiation, are deflected by grazing a crystal face, and then enter an ionization chamber. For the modified ray momentum and energy conservation yield

$$\lambda_\theta = \lambda_0 + (2h/mc)\, sin^2(\theta/2) = \lambda_0 + 0.0484\, sin^2(\theta/2), \tag{4.2}$$

where λ_0 is the primary wave length, 0.711 A, and θ is the angle between the ray emitted from T and the X-ray emitted from the carbon target, R and we use

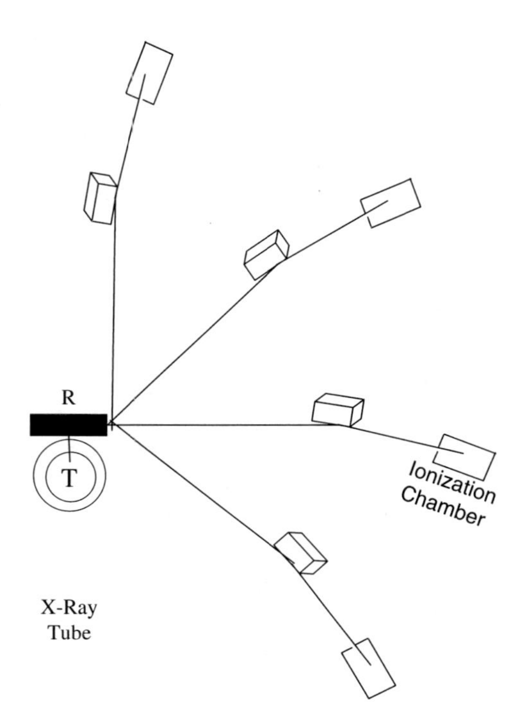

Fig. 4.1 Compton scattering experiment

Compton's values for h, m, and c. The diagram schematically shows angles of $0°, 45°, 90°$, and $135°$. According to (4.2), the modified radiation for these angles should have the values 0.711 A, 0.718 A, 0.735 A, 0.752 A. These fit the observed value within 0.07 A. The alternative hypotheses did not yield these angular dependent values.

This was widely, and reasonably, accepted as establishing the reality of light quanta. Yet it left residual ambiguities in terms of the way reality is described. In the classical view a wavelength is the distance between crests (or troughs) in an indefinitely extended wave, ideally a plane wave. The crystal spectrometer effectively measures a wave-like property of the secondary X-rays, wave length. It does so by exploiting the most basic of wave features, interference. The lattice of atoms on the crystal face functions like the slits in the diffraction grating discussed earlier. Compton was treating X-rays as photons in scattering off the carbon target and as waves in deflection by the Calcite crystal. Compton concluded that the manner in which interference occurs is not yet clear, and expressed the hope that further studies might shed some light on the difficult question of the relation between interference and the quantum theory. This ambiguity carried over to the most famous exploitation of Compton's reasoning. In developing his uncertainty principle Heisenberg introduced a thought experiment of an electron moving in the positive x-direction and striking a photon which enters the objective lens of a γ-ray microscope (heisenberg 1927). If λ is the wave length of the radiation then, according to the laws governing the optics of microscopes, the uncertainty in the measurement of the x-coordinate is $\Delta x = \lambda / sin\, \varepsilon$. If one treats the radiation in terms of a single photon scattering off an electron then the electron recoil is not known precisely, because the direction of the scattered photon is only known within the cone entering the objective lens. This gives an uncertainty of $\Delta p_x = (h/\lambda) sin\, \varepsilon$. The product of the two uncertainties is h. Here the radiation is treated as a photon scattered off an electron, while the γ-ray microscope is treated by the laws of wave optics. In 1935 Schrödinger published a survey paper on quantum mechanics in which he included his notorious thought experiment (Schrödinger 1935). A cat is penned up in a steel chamber with a diabolical device. A Geiger counter contains such a small bit of radioactive material that there is only a probability of 1/2 that a single atom will decay within an hour. If it does decay the click of the Geiger counter initiates a reaction that exposes the cat to poison gas. If there are no observations then the probabilistic laws of quantum mechanics lead to the conclusion that the after one hour the radioactive material is in a superposition of two states, decayed and not decayed. The mechanism leads to the conclusion that the cat is in a superposition of states, alive and dead, or 1/2 |cat dead> +1/2|cat alive>. The physics involved will be treated later. The point here is that the use of classical concepts in a quantum context engenders a paradox. In quantum mechanics a superposition of states is allowed. However, terms like |cat alive> and |cat dead> do not represent quantum states. They are quasiclassical place holders indicating that there should be a quantum state, or a large equivalence class of states, corresponding to the condition described in classical terms by saying "The cat is alive" or "The cat is dead". Since no one knows how such state functions could be formulated, state specification must rely on classical place holders. The classical terms are contradictory. A cat is either alive or dead. It cannot be both at the same time. Quantum physics requires the use of classical terminology to specify observable states, whether the condition of cats or the results of experiments. Yet, the presuppositions governing the inferential relations of classical terms, if alive then not dead, do not carry over to quantum physics. Hence the need to analyze the distinctive role of classical concepts in the quantum realm.

Philosophers of physics do not, to the best of my knowledge, analyze the role of informal inferences in the functioning and the interpretation of physics. Three reasons for this neglect seem clear. Informal inference and networks of informal inferences do not admit of a systematic analysis or a logical reconstruction. When philosophers of science analyze theories they only invoke experiments as supplying data that might confirm or falsify theoretical predictions. There have been some good case studies of experimental physics and attempts to analyze experimentation through Bayesian analysis, or some other form of inductive inference. Such analyses treat the relation between evidence and hypotheses. They do not treat the role of informal analysis. The second reason is the conviction that informal analysis, though operative in the development of theories, drops out of theory interpretation after theories have been developed. The third reason is occupational. The methods and tools philosophers of science cultivate do not apply to informal inferences.

In the practice of classical physics there was no clear separation of the two inferential systems. Kline (1972, p. 616) summarized the situation:

> Far more than in any other century, the mathematical work of the eighteenth was directly inspired by physical problems. In fact, one can say that the goal of the work was not mathematics, but rather the solution of physical problems; mathematics was a means to a physical end.

The following table illustrates the branches of mathematics that emerged from branches of physics.

Physics	Mathematics
Mechanics	Differential and integral calculus Differential equations Calculus of variations
Acoustics Thermodynamics	Fourier series
Acoustics Hydrodynamics Electrodynamics	Complex analysis
Elasticity Hydrodynamics Electrodynamics	Partial differential equations

In the nineteenth century mathematics was developed as a discipline with its own foundation leading to a divergence. Almost all classical physicists, and many contemporary physicists, preferred the more physical treatment of differentials and functions stemming from Euler and Lagrange. Thus, a function is a relation between two variables. Scalars, vectors, tensors, and pseudoscalars are defined through physical examples, rather than by their properties under transformation. Furthermore, prior to the twentieth century there was no occupational division between experimenters and theoreticians. Most physicists did both. Since then only very few, like Bridgman, Fermi, and Lamb, have made both theoretical and experimental contributions. Most philosophers of science are heirs of the mathematical reformation. The extended language of physics could accommodate both experimental practice and classical theories. The separation of theoreticians and experimentalists was occupational, not a reflection of an underlying tension.

The crisis physicists experienced in 1924–1927 centered on the application of classical concepts in quantum situations. The contradictions itemized earlier exhibit a common feature. Assumptions concerning the motion of electrons or electromagnetic radiation that supplied a basis for valid inferences in one context had to be replaced by contrary assumptions in a different context. Most thought of this as a problem with the objects treated. How can an electron behave like both a particle and a wave? Bohr instituted a Gestalt shift from objects to 'objects'. This made the framework of inferences a problematic situation. The core of both ordinary language and EOL involves objects with properties. In quantum experiments the concept of an object may not carry over from one experimental situation to another. The concept of an object does not play a foundational role in quantum mechanics. Yet the basic ling between material and formal inferences comes from their application to the same objects. How can this conceptual situation be handled?

In the discussion of rights in Chapter 1 we showed how reasoning about problems could shift from an ontological to an epistemological mode. The ontological mode made for simpler and clearer inferences based on the accepted properties and activities of familiar objects. When, however, this direct reasoning was perceived as relying on dubious presuppositions one could switch to an epistemological mode. This involved the difficulty of analyzing the use of basic concepts in different situations and the related network of concepts. As we will see later, this switch from ontological to epistemological analysis is sometimes done in analyzing theories. Any attempt to apply this to experimental physics would involve endless complexities. A point that One may call into question any particular presupposition. However an attempt to call all presuppositions into question paralyzes the process of inquiry.

The alternative is an adaption of Quine's strategy, considered in Chapter 1. When contradictions arise he advocates the strategy of moving them as far from the core as possible, resolving them by any feasible means, and then carrying on. The contradictions physicists were experiencing in the mid 1920s could not be removed from the core of EOL. They concerned the applicability of the concept of an object. A revision of Quinean strategy would involve developing semantic rules to accommodate, or at least tame, the problem and then carrying on the work of physics with the new semantics. Bohrian semantics supplies the initial resolution. Can the mathematical formulation of quantum mechanics be developed in a way that fits this semantics? This, of course, reverses the normal methodology of theory interpretation. Philosophers begin with a mathematically formulated theory and then seek an interpretation. This inevitably slights the role of language in interpreting quantum mechanics. Accordingly I will pursue a reversal of this methodology by making language considerations basic and them attempting to analyze how well the mathematical formulations fit this basis.

References

Adams, Ernest. 1979. Measurement Theory. In P. Asquith, and H. Kyburg (eds.), *Current Research in the Philosophy of Science* (pp. 207–227). East Lansing, MI: Philosophy of Science Association.

Boltzmann, Ludwig. 1974. *Theoretical Physics and Philosophical Problems*. Dordrecht: D. Reidel.

Boorse, Henry A., and Lloyd Motz. 1966. *The World of the Atom*. New York, NY: Basic Books.

Campbell, Norman. 1920. *Physics: The Elements*. Cambridge: Cambridge University Press.

Cartwright, Helen. 1965. Heraclitus and the Bath Water. *The Philosophical Review, 74*, 466.

Chevalley, Catherine. 1991. *Niels Bohr: Physique atomique et connaissance humaine*. Paris: Gallimard.

Cushing, James T. 1994. *Quantum Mechanics: Historical Contingency and the Copenhagen Hegemony*. Chicago, IL: University of Chicago Press.

Darrigol, Olivier. 1992. *From C-Numbers to Q-Numbers: The Classical Analogy in the History of Quantum Theory*. Berkeley, CA: University of California Press.

Einstein, Albert. 1905a. Über die von der molekularkinetischen Theorie der Wärme geforderte Bewegung von in ruhenden Flüssigkeiten suspendierten Teilchen. *Ann. der Phys., 17*, 549–560.

Einstein, Albert. 1905b. Zur Elektrodynamik bewegter Koerper. *Ann. der Phys., 17*, 891–921.

Einstein, Albert. 1905c. Über einen die Erzeugung und Verwandlung des Lichtes betreffended heuristischen Gesichtspunkt. *Ann. der Phys., 17*, 132–148.

Ellis, Brian. 1968. *Basic Concepts of Measurement*. Cambridge: Cambridge University Press.

Faye, Jan. 1991. *Niels Bohr: His Heritage and Legacy: An Anti-Realist View of Quantum*. Dordrecht: Kluwer Academic Publishers.

Faye, Jan., and Henry J. Folse. 1994. *Niels Bohr and Contemporary Philosophy*. Dordrecht; Holland: Kluwer Academic Publishers.

Fine, Arthur. 1986. *The Shaky Game*. Chicago, IL: University of Chicago Press.

Folse, Henry J. 1977. Complementarity and the Description of Experience. *International Philosophical Quarterly, 17*, 378.

Forman, Paul. 1968. The Doublet Riddle and Atomic Physics Circa 1924. *Isis, 59*, 156–174.

Furth, R. 1956. *Investigations on the Theory of Brownian Motion*. New York, NY: Dover.

Giere, Ronald N. 2006. *Scientific Perspectivism*. Chicago, IL: University of Chicago Press.

Hecht, Eugene, and Alfred Zajac. 1974. *Optics*. Reading, MA: Addison-Wesley.

Heilbron, John. 1985. The Earliest Missionaries of the Coppenhagen Spirit. *Revue d'histoire de sciences, 38*, 195–203.

Helmholtz, Hermann von. 1977. *Epistemological Writings: The Paul Hertz/Moritz Schlick Centenary Edition*. Dordrecht: D. Reidel.

Hendry, John. 1984. *The Creation of Quantum Mechanics and the Bohr-Pauli Dialogue*. Dordrecht: D. Reidel.

Hertz, Heinrich. 1956. *The Principles of Mechanics Presented in a New Form*. New York, NY: Dover.

Jammer, Max. 1966. *The Conceptual Development of Quantum Mechanics*. New York, NY: McGraw-Hill.

Jurkowitz, Edward. 2002. Helmholtz and the Liberal Unification of Science. *Historical Studies in the Physical Sciences, 32*, 291–317.

Kline, M. 1972. *Mathematical Thought from Ancient to Modern Times*. New York: Oxford University Press.

Koch, Christof, and Giulio Tononi. 2011. A Test for Consciousness. *Scientific American, 304*(6), 44–47.

Konno, Hiroyuki. 1993. Kramers' Negative Dispersion, the Virtual Oscillator Model, and the Correspondence Principle. *Centaurus, 36*, 117–166.

Kramers, H.A. 1957. *Quantum Mechanics*. Amsterdam: North Holland.

Krantz, D., P. Suppes, and A. Tversky. 1971. *Foundations of Measurement: Volume I*. New York, NY: Academic Press.

Kyburg, H. 1984. *Theory and Measurement*. Cambridge: Cambridge University Press.

Levy-Leblond, Jean Marc. 1963. Galilei Group and Nonrelativistic Quantum Mechanics. *Journal of mathematical Physics, 4*, 776–788.

MacKinnon, Edward. 1977. Heisenberg, Models, and the Rise of Matrix Mechanics. *Historical Studies in the Physical Sciences, 8*, 137–188.

MacKinnon, Edward. 1982. *Scientific Explanation and Atomic Physics*. Chicago, IL: University of Chicago Press.

MacKinnon, Edward. 1985. Bohr on the Foundations of Quantum Theory. In A. P. French, and P. J. Kennedy (eds.), *Niels Bohr: A Centenary Volume* (pp. 101–120). Cambridge: Harvard University Press.

MacKinnon, Edward. 1994. Bohr and the Realism Debates. In J. Faye, and H. Folse (eds.), *Niels Bohr and Contemporary Physics* (pp. 279–304). Dordrecht: Kluwer.

MacKinnon, Edward. 2005. Einstein's 1905 Brownian Motion Paper. *CSI Communications, 29*(6), 6–8.

Massimi, Michela. (ed.) 2005. *Pauli's Exclusion Principle: The Origin and Validation of a Scientific Principle*. New York, NY: Cambridge University Press.

Mehra, Jagdish, and Helmut Rechenberg (eds.) 1982. The Quantum Theory of Planck, Einstein, Bohr, and Sommerfeld: Its Foundation and the Rise of Its Difficulties, 1900–1925. In Jagdish Mehra and Helmut Rechenberg. (eds.), *The Historical Development of Quantum Theory* (vol. 1, part 1–2). New York, NY: Springer.

Miller, Arthur I. 1981. *Albert Einstein's Special Theory of Relativity: Emergence (1905) and Early*. Reading, MA: Addison-Wesley.

Narens, Louis. 1985. *Abstract Measurement Theory*. Cambridge, MA: MIT Press.

Niven, W. D. 1965. *The Scientific Papers of James Clerk Maxwell*. New York, NY: Dover.

Pais, Abraham. 1982. *Subtle Is the Lord: The Science and Life of Albert Einstein*. Oxford: Clarendon Press.

Petruccioli, Sandro. 1993. *Atoms, Metaphors and Paradoxes: Niels Bohr and the Construction of a New*. Cambridge: Cambridge University Press.

Quine, Willard Van Orman. 1960. *Word & Object*. Cambridge, MA: MIT Press.

Radder, Hans. 1982. Between Bohr's Atomic Theory and Heisenberg's Matrix Mechanics. A Study of the Role of the Dutch Physicist, H. A. Kramers. *Janus, 69*, 223–252.

Robotti, Nadia. 1983. The Spectrum of (zeta) Puppis and the Historical Evolution of Empirical Spectroscopy. *Historical Studies in the Physical Sciences, 14*, 123–145.

Röseberg, Ulrich. 1984. *Szenarium einer Revolution*. Berlin: Akademie-Verlag.

Rosenfeld, L. et al. 1972. *Niels Bohr: Collected Works*. Amsterdam: North Holland.

Schrödinger, Erwin. 1935. The Present Situation in Quantum Mechanics. In J. Wheeler, and W. Zurek (eds.), *Quantum Theory and Measurement* (pp. 152–167). Princeton, NJ: Princeton University Press, 1983.

Shamos, Morris H. 1959. *Great Experiments in Physics*. New York, NY: Henry Holt and Company.

Stachel, John et al. 1987. *The Collected Papers of Albert Einstein: Vol. I: The Early Years, 1879–1905*. Princeton, NJ: Princeton University Press.

Stevins, S. S. 1946. On the Theory of Scales of Measurement. *Science, 103*, 677–680.

Stroke, G. W. 1969. *An Introduction to Coherent Optics and Holography*. New York, NY: Academic Press.

Thompson, Silvanus P. 1910. *The Life of William Thomson, Baron Kelvin of Largs* (2 Vols.). London: MacMillan.

Thomson, William P., and P. G. Tait. 1867. *Treatise on Natural Philosophy*. Oxford: Clarendon Press.

Trigg, G. L. 1971. *Crucial Experiments in Modern Physics*. New York: Van Nostrand Reinhold.

Van der Waerden, Bartel. 1967. *Sources of Quantum Mechanics*. New York, NY: Dover.

Von Neumann, John, and Oscar Morgenstern. 1947. *Theory of Games and Economic Behavior* (2nd. edn.). Princeton, NJ: Princeton University Press.

Wigner, Eugene. 1939. On Unitary Representations of the Inhomogeneous Lorentz Group. *Annals of Mathematics, 40*, 149–157.

Zimmerman, Dean. 1995. Theories of Masses and Problems of Constitution. *The Philosophical Review, 104*, 53–110.

Part II
The Classical/Quantum Divide

Intermezzo

Part I was primarily concerned with the development of the language of physics and the ways in which it came to relate to the mathematical formulations used in physics. Part II has a different function. It develops a novel perspective for the interpretation of quantum physics. The present introduction is intended to serve a double function. First, since this is a very unfamiliar perspective for the interpretation of quantum physics, I will indicate why it is needed and how it will be developed. Second I presume that this book may be read by some philosophers and others who are not proficient in the mathematics of quantum physics. As an aid I will present a non-technical summary of the developments that follow.

The mid-1920s witnessed the birth of quantum mechanics, wave mechanics, and a linguistic crisis in attempts to give a coherent formulation of experimental results. Bohr effectively mitigated this crisis by introducing a semantic shift from ontological to linguistic considerations. Experimental physicists, like Clint Davisson were agonizing over the problem of how an electron could be both a particle and a wave. Bohr transformed this into an issue of determining the circumstances in which either particle or wave language supplies the appropriate vehicle for describing experimental situations and reporting results. After some version of the Bohrian semantic guidelines were assimilated and accepted, the linguistic crisis that precipitated the changes submerged beneath the conscious awareness of the physics community.

The interpretation of quantum mechanics as a system regularly reemerged as a problematic issue through the Bohr-Einstein debates, the introduction of hidden variable interpretations, through criticisms by philosophers like Karl Popper, and through paradigm shifts in the philosophy of science. These shifts led to attempts to give quantum mechanics and quantum field theory a rigorous mathematical formulation in accord with either an axiomatic or a semantic model of theories. Both methods shared a common feature. Language as such plays no role in the interpretation. Thus a realistic approach to interpreting a formulation of quantum mechanics that accords with the semantic method of interpretation asks: What must the world be like if this theory is true of it? If answers to this question are not convincing, then philosophers can switch from an ontological to an epistemological perspective, interpreting quantum mechanics or quantum field theory in terms of observables. There are variants. An anti-realist interpretation insists that a theory need not be true to be acceptable. A modal interpretation asks what might the world be like. In all these methods of interpretation language, considered as a separate unit, plays no role. Even a switch to an epistemological foundation relies on empiricists accounts of knowledge, rather than analysis or phenomenology. The most unfortunate consequence of this exclusion of linguistic analysis is an exclusion of experimental analysis. Experiments are cited only as supplying data that test theoretical predictions. This leads to a logical lacuna. In contemporary particle physics theoretical predictions are never tested by comparison with data as such. They are tested against inferences drawn from the data. These inferences are informal. As was indicated earlier we call inferences formal if they depend on syntactic rules whose validity does not depend an the matter to which the rules are applied. In an extension of this method, a formal interpretation of a theory imposes a semantics on an already formulated syntactic structure. Informal inferences rely on the meanings of the basic concepts employed and are generally context-dependent. Galison, Franklin, Pickering, and others have brought out the richness of the traditions in experimental research and their quasi-independence. Experimental inferences cannot be interpreted as phenomenological models of the theories being tested. To deal with these separated but coordinated inference systems we need a dual-inference model of the practice of physics. Earlier we discussed dual-inference systems and had some remarks on their role in classical physics. Formal and informal inference systems were coordinated by their relations to the same objects. However, we did not develop a general theory of dual-inference systems. I believe that an attempt to develop and impose a theory of dual-inference systems would be misleading, since informal inferences are context-dependent. Fortunately, there is a better approach. The development of atomic, and particularly of particle, physics has been characterized by a broad and detailed collaboration between theoreticians and experimenters. Accordingly, we will consider the discourse between theoreticians and experimenters. Our limited focus will be on the role of informal inferences and on how the framework of informal inferences relates to the mathematical formulations used in quantum physics. Two preliminary remarks give an initial orientation. First, in informal inferences there is no sharp distinction between syntax, semantics, and pragmatics. Material inferences often hinge on the meanings of the terms used. The practice depends on choices made by experimenters. Second, informal inferences rest on presuppositions, not axioms. In a formal system anything follows from axioms that allow contradictions. This also applies to informal inferences, as indicated by the dictum of medieval logicians: *Ex falso sequitur quodlibet*. Hence, the need to consider the presuppositions that are implicit in experimental inferences and methods of restricting their use to avoid contradictions. If the inferential system allows contradictions, then no inferences are reliable. Since theoretical predictions are tested against the conclusions of informal inferences, an unreliable inferential system does not supply an adequate basis for testing theoretical predictions.

With this background we can outline the inner logic implicit in the material that follows. We begin with the linguistic crisis of the mid 1920s and the semantic guidelines stemming from Bohr that avoided the contradictions physicists were encountering. These developments led to the famous, or notorious, Copenhagen interpretation of quantum mechanics. This is now widely rejected by philosophers as a seriously misleading guide to the interpretation of quantum mechanics as a theory and rejected by some of the physicists who pay attention to such issues as inadequate to advances in quantum field theory and particle physics. When quantum mechanics is interpreted *as a theory* and the Copenhagen interpretation is taken as an interpretation of that theory, then it is almost impossible to understand the original position of the Copenhagen patriarchs. Bohr, Pauli, and Heisenberg all defended the idea that quantum physics is best understood as a rational generalization of classical physics. Before rejecting this idea as preposterous we should indicate how this approach can be developed and whether it can make a contribution to interpreting quantum physics.

Bohr's basic program was very simple. The resolution of the linguistic crisis led to guidelines for extending basic classical concepts and for restricting their usage in quantum contexts. He regarded the mathematical formalism as an inferential tool, not as a theory to be interpreted. The development of the mathematical formalism hinged on representing basic classical terms, such as 'position', 'momentum', and 'energy' by mathematical operators and translating restrictions on the use of these terms into restrictions on the use of the corresponding operators. Then one has the basis for a mathematical formalism that relates in a coherent way to the proper usage of terms in experimental contexts and the inferences experimental analysis supports.

This is a polar opposite to formal interpretations. It also seems like a preposterous example of a tail wagging a dog. Quantum mechanics has clearly replaced classical mechanics as the basic science of reality. Nevertheless, this approach is worth exploring for at least two reasons. The first is an Occamist approach to interpreting physics. If the theory can function on this minimalist basis then it is not necessary to interpret the mathematical formalism as a theory. A functional interpretation suffices. Second, we are concerned with the interrelation of formal and informal inference systems. They cannot be related by any method that trivializes the role of experimentation. Hence we will attempt to relate them by beginning on the other end, with informal inferences. Does an analysis of the distinctive features of quantum experiments supply a basis for developing the mathematical formalism of quantum mechanics? My answer to this question relies on an exploitation of the work of two outstanding quantum theoreticians, Paul Dirac and Julian Schwinger. Dirac introduced and Schwinger developed the idea that the distinctive features of quantum measurements expressed in classical terms supply a basis for developing the mathematical formalism of quantum mechanics. I will call this the 'measurement interpretation'. It is essentially an austere version of the Copenhagen interpretation. The new label is needed to avoid identifying the measurement interpretation with the various versions and misinterpretations of Copenhagen found in the literature.

This background clarifies the argument threading through the tapestry of issues that unfolds. We begin with the formation of the orthodox, or Copenhagen, interpretation of quantum mechanics and focus on Bohr's clarification of the use and limits of classical concepts in quantum contexts. Then we consider his relatively unknown analysis of the roles of 'particle' and 'field' in nuclear physics and quantum field theory. Next we show how these semantic rules can guide the formation of the mathematical formalism. We use Schwinger's extension of the measurement interpretation to quantum electrodynamics and quantum field theory to assess the limits of validity of the measurement interpretation. There are two types of limitations. First, the measurement interpretation, or any version of the Copenhagen interpretation, does not supply a proper basis for evaluating quantum mechanics as a theory. This is because the measurement interpretation treats the mathematical formalism as a calculational tool, not as a theory. Second, and more pertinent to the present development, the measurement interpretation is inadequate to advances in quantum field theory and quantum cosmology. However, it is functionally adequate to non-relativistic quantum mechanics, quantum electrodynamics, and basic quantum field theory.

The measurement interpretation is a semi-classical, or phenomenological, interpretation of quantum mechanics. It uses properly restricted classical concepts as a semantic basis for interpretation. Since quantum mechanics is the fundamental science of physical reality, an interpretation of quantum mechanics should rest on a quantum, rather than a classical, foundation. Many interpretations attempt to accomplish this in an ontological way by relating a suitably reconstructed mathematical formulation to the reality it is a theory of. I follow the practice of physics in relating the mathematical formalism to its experimental basis, a framework of consistently reportable claims. Physicists go beyond this basis through experimental discoveries and theoretical hypotheses. I attempt a very limited philosophical advance. First I clarify the relative ontology of the measurement interpretation and then consider characteristic quantum properties and processes that this framework cannot accommodate and that a quantum interpretation must accommodate. A *relative* ontology is an explicitation of an account of reality implicit in or presupposed by a systematization of some branch of knowledge. As discussed earlier, one systematization of our ordinary language relies on a basic subject/object distinction and represents physical realty as an interrelated collection of spatio-temporal objects with characteristic properties and causal properties. We defer a consideration of the subject aspect to the final chapter. This is a minimal *lived-world* ontology, not an account of physical reality as it exists objectively. Particular sciences may have a relative ontology that plays a presuppositional role. Thus, much of chemistry relies on an account of atoms and molecules with definite sizes and shapes.

An ontology of objects with properties is inadequate to the quantum realm. Quantum mechanics treats *systems* with properties. Three distinctively non-classical properties characterize quantum systems: superposition of states, interference, and non-locality. Also, in a sharp break with classical methods, quantum physics treats *virtual* processes in the same way as observable processes. This claim is elaborated by examining the treatment of virtual processes in quantum electrodynamics, where they emerged into prominence. Before attempting an interpretation of this physics we consider a preliminary question. What function does any interpretation of a scientific theory, or a scientific practice, fulfill?

We distinguish an implicit functional interpretation from an explicit imposed interpretation. An explicit interpretation is useful either when the functional interpretation is perceived as inadequate to advances in physics, or when one is asking external questions about a theory or practice, or tradition. My method of interpretation focuses on the practice of physics, rather than on reconstructed theories, and looks for a revised interpretation that meets three requirements. It must incorporate the features of the measurement (or Copenhagen) interpretation that account for its unprecedented empirical success. To put quantum physics on a quantum, rather than a semi-classical, foundation, it it should assign a foundational role to the characteristic features that distinguish quantum from classical physics. Finally, it should be capable of accommodating advances in quantum field theory and quantum cosmology. The two leading replacement candidates are a *many-worlds* interpretation and a *consistent-histories* interpretation. I indicate why I consider the Gell-Mann–Hartle version of the consistent histories interpretation a reasonable choice.

The concluding chapter has a novel purpose. Contemporary philosophy manifest a sharp gap between analytic or phenomenological treatments of the lived world and analyses of scientific theories as relatively isolated units. The present work brings out the underlying conceptual continuity between an ordinary-language framework and the developing language of physics. This modifies both ends of the philosophical gap. It undercuts the presuppositions analysis and phenomenology often rely on to downgrade the role of science. It undercuts the methodology of interpreting scientific theories as isolated units insulated from lived-world ontology. I examine the bearing this change has on some basic philosophical problems. I am more concerned with analyzing how the *problematic*, or the implicit presuppositions, must be modified than in proposing solutions. The issues considered are: the continuity underlying scientific development; realism, reductionism versus emergence, and the interrelation of the human and scientific realms.

Chapter 5
Orthodox Quantum Mechanics

These things, therefore, having been expressed by us with the greatest accuracy and attention, the Holy Ecumenical Synod declares that no one shall be allowed to bring forward, nor to write, nor to put together, nor to frame a different faith, nor to teach others anything different.
Council of Chalcedon, 451 A. D.

Any consideration of the role of language in interpreting quantum mechanics (QM) must consider Bohr who expressed his distinctive perspective with the claim: We are suspended in language (See Petersen 1968). In spite of his leading role in forming the Copenhagen, or orthodox[1] interpretation of QM Bohr's position is widely misinterpreted. I will present a redevelopment of his position as a minimal interpretation of QM. To situate this I will indicate how Bohr's position developed and came to be misinterpreted. The reason for the redevelopment is to appraise the limits of valid applicability of orthodox QM. This, in turn. supplies a basis for evaluating attempts to go beyond a minimal basis.

In the mid 1920s the development of a coherent functional interpretation of QM was an urgent concern. Routine reporting of experimental results generated contradictions. The new theoretical breakthroughs were couched in different formulations: the matrix formulation that limited interpretation to observables; de Broglie's wave-particle and later pilot wave interpretation; Schrödinger's wave mechanics; and Dirac's transformation theory. On a functional level the basic interpretative problem was one of relating theoretical terms, like ψ or matrix components to aspects of actual and thought experiments. Even after the equivalence of wave and matrix mechanics was established there was still a conflict between Born's interpretation of $\int \psi \dagger \psi dx$ as a probability and Schrödinger's interpretation of it as charge density.

[1] The term 'orthodox' stems from the Council of Chalcedon, which set the standards of orthodoxy accepted by the Eastern Orthodox, Roman Catholic, and mainstream Protestant Churches. By a curious turn some theologians are now using Bohr's doctrine of complementarity to explain the Chalcedonian decrees. See Richardson and Wildman (1996), pp. 253–298.

E. MacKinnon, *Interpreting Physics*, Boston Studies in the Philosophy of Science 289,
DOI 10.1007/978-94-007-2369-6_5, © Springer Science+Business Media B.V. 2012

Bohr's underdeveloped and loosely assimilated ideas on complementarity helped experimenters avoid contradictions in reporting and extrapolating results.

Theoreticians assimilated the new QM by learning how to solve problems using properly formulated data. Much of the initial work involved treating problems whose solutions were already known through the Bohr-Sommerfeld (B-S)program and its various modifications. A more challenging test came from the problems the B-S program did not resolve and from previously unanticipated consequences of the new formalisms. Matrix mechanics had difficulty adapting the method of action-angle variables. Pauli found a way to calculate hydrogen energy levels and the Stark effect for hydrogen. The problem of the rotator was independently treated by: Lucie Mensing in Göttingen, Gregor Wentzel in Hamburg; Otto Halpern in Vienna; Igor Tamm and Lev Landau in Russia, and David Dennison, an American visiting Copenhagen.[2] After the development of wave mechanics the problems treated were: the hydrogen atom (Schrödinger, Dirac); the Stark effect (Schrödinger, Wentzel); the anomalous Zeeman effect (Heisenberg, Jordan); motion of a free particle (Ehrenfest); the Compton effect (Wentzel, Beck), the fine structure of hydrogen (Dirac); and the Kramers-Heisenberg radiation formula (Klein, Dirac).[3] These were old problems done in a new way.

There were also some new developments that went beyond the B-S program, notably: collision theory (Born); the helium atom (Heisenberg, Hylleras, Bethe); Fermi-Dirac statistics; treatment of electrons in metals as a degenerate Fermi-Dirac gas (Sommerfeld); an explanation of the extreme density of white dwarf stars (Fowler); spectra of complex atoms (von Neumann, Wigner, Slater); penetration of a potential barrier (Gamow, Condon and Gurney); an account of ferromagnetism (Heisenberg); paramagnetism (Pauli); the inclusion of spin (Pauli); the existence of exchange forces (Heitler and London); exchange interaction in scattering (Oppenheimer, Mott); molecules (Born and Oppenheimer); details of chemical bonding (Pauling); and various approximation techniques (Born, Fock, Hartree, Fermi, Thomas, Wentzel, Kramers, Brillouin). The Raman effect had been predicted by a heuristic argument in the old quantum theory, but only really fit the new theory.

These solutions articulated the way quantum mechanics (from now on used as a general term including matrix and wave mechanics) related to and went beyond classical physics. Classical terms, like 'mass', 'energy', 'momentum', and 'angular momentum' entered in the same basic formulas, such as, $\mathbf{p} = m\mathbf{v}$, $\mathbf{L} = \mathbf{rxp}$, and the conservation laws. Following the correspondence principle (CP) tradition, classical physics served as a starting point and guide for setting up quantum mechanics. The standard way of doing this was to analyze a problem in classical terms, set up the classical Hamiltonian, replace dynamical variables by quantum operators, and then attempt to solve the resulting differential equation. Most physicists,

---

[2] Surveys of the problems treated by matrix mechanics may be found in Mehra-Rechenberg (1982, Vol. 4, Part 2); and in Max Born's 1926 lectures (Born 1962, p. 68–129).

[3] For more details see: Mehra-Rechenberg (1982, Vol. 5, Part 2, pp. 838–854); Hund (1974), chaps. 12–14; Jammer (1966), 362–365; Pauli (1947 [1932]), 161–214; Bethe (1999) and Kuhn et al. (1962).

even those concerned with foundational issues, apparently felt that the practice of physics should not depend on settling issues about the meanings of concepts, the role of observability, or whether ultimate reality is continuous or discontinuous. Bohr's conceptual subtleties and Dirac's c-number/q-number distinction were largely ignored. Dirac's transformation theory was rarely used. However, its development was widely regarded as proof that wave and matrix mechanics were special cases of a more general system, quantum mechanics.

By 1929 non-relativistic quantum mechanics (NRQM) was no longer seen as a problematic field. Though much remained to be done, the foundations seemed secure. With this brief background we may list the basic features of the Copenhagen interpretation that became orthodox quantum mechanics: Heisenberg's uncertainty principle; the idea that photons, electrons, and other particles exhibit both wave and particle properties; the probabilistic interpretation of the wave function; the correspondence between eigenvalues derived from the mathematical formalism and values of quantities obtained from measurements; the idea that wave and matrix mechanics are special representations of a more general formalism; and some sort of complementary relation between classical and quantum physics. In 1927 most of these features seemed novel and more than a bit bizarre. By 1929, they were generally accepted as part of the normal practice of quantum physics. Bohr's underdeveloped doctrines that classical concepts stemming from ordinary language play a definitive role in measurement, and that quantum physics is a rational generalization of classical physics were widely regarded as speculative philosophical issues. After the development of quantum field theory (QFT), relativistic quantum mechanics (RQM), and especially after the discovery of the neutron, the leading European physicists concentrated on these new fields and on nuclear physics.

5.1 The Development of Bohr's Position

For most physicists the functional interpretation of QM no longer seemed problematic. There was one strong dissent. Bohr saw the developments just cited as a challenge to his way of handling problems in QM. His way of resolving these difficulties reflects and clarifies the unique aspects of his conceptual analyses. In developing his wave equation, Dirac expressed the hope that he could avoid the negative energy states allowed by the Klein-Gordon equation (Dirac 1928). He originally did this by simply ignoring the negative energy states. After Klein demonstrated the possibility of transitions to negative energy states, these could not be ignored. Bohr's evaluation of the situation was expressed in a letter to Dirac:

> In the difficulties of your old theory I still feel inclined to see a limit of the fundamental concepts on which atomic theory hitherto rests rather than a problem of interpreting the experimental evidence in a proper way by means of those concepts. Indeed according to my view the fatal transitions from positive to negative energy should not be regarded as an indication of what may happen under certain conditions, but rather as a limitation in the applicability of the energy concept. (Sources: Bohr Scientific Correspondence, sect. 4, letter of 5 December, 1929)

Bohr's previous analyses had used classical concepts to interpret experimental information. RQM seemed to show that this method could not be extended to relativistic phenomena. Other considerations seemed to show that it could not be extended to nuclear physics or quantum field theory either. Before 1932 nuclear physics had two outstanding and apparently related problems, electron confinement and nuclear statistics. Electrons, it was agreed, must be in the nucleus since they are emitted in β decay. Yet, electrons confined in such a small volume should have very high kinetic energies. These energies should not only allow escape. They should also require RQM. Whether a nucleus obeyed Bose-Einstein of Fermi-Dirac statistics should be determined by counting the number of protons and electrons in the nucleus. This gave the wrong results for nitrogen. Bohr had enthusiastically accepted Dirac's QFT as the only reasonable account of photons. Yet, as Oppenheimer showed, this theory encountered divergence difficulties when the interaction of an electron with a radiation field is treated in terms of the emission and absorption of virtual particles. For most physicists, these were separate problems on the frontiers of physics. For Bohr, the common feature these difficulties shared was the problematic extension of the classical concepts needed to give a descriptive account.

Bohr's resolution of these problems focused on the use and limitation of the concept, 'particle' and the informal inferences it supports. It provides a foundation for the application of other concepts such as 'space-time location', 'momentum', 'energy' and 'trajectory'. These quantitative concepts supply the correspondence principle basis for the introduction of the mathematical formalism of quantum mechanics. Nevertheless, 'particle' remains a classical concept either when used as the basis for a description of a particle's trajectory or as a peg for the CP. Bohr was concerned with showing how the concept 'particle' can be extended to quantum applications. We will summarize how he did this in nuclear physics and in scattering theory, two topics that are rarely considered in philosophical accounts of Bohr's position.

The applicability of 'particle' as applied to electrons broke down somewhere between the Compton wavelength of an electron, $\lambda = h/mc = 2.4 \times 10^{-10}$ cm. and the classical radius of the electron, $e^2/mc = 2.8 \times 10^{-13}$ cm. If the concept of an electron as a localized particle is inapplicable within the nucleus then so too is energy conservation for these electrons, though their charge is conserved. The statistical problem is dissolved, since electrons within a nucleus cannot be counted as particles. The Klein argument is moot. It requires extremely strong electrical fields over very small distances. Such fields must ultimately be due to the presence of charged particles. Yet, by Bohr's new argument, it is impossible to localize enough particles in a small enough region to produce such a strong field. This argument, in turn, set limits to the applicability of quantum mechanics since it had to be suspended from pegs of classical ideas.[4]

[4] This is a summary of ideas Bohr presented in October, 1931. A more detailed analysis is given in MacKinnon (1982a, chap. 8) and MacKinnon (1985).

Bohr's provisional solution was undercut by new developments. The discovery of the neutron and the Fermi theory of beta decay eliminated the problem of electrons within the nucleus. These advances obviated some of the difficulties Bohr had in extending the concept 'particle'. Since neutrons and protons have much greater masses than electrons, they could be treated as particles confined within the nucleus and having kinetic energies in the non-relativistic range. On this basis, Bohr went on to develop the two models of the nucleus that dominated research in the field. The collective model, stimulated by Fermi's experiments with slow neutron capture, assumed that an incoming slow neutron is absorbed by the nucleus leading to a compound state that can decay through any one of a number of competing processes. Later Bohr introduced the liquid drop model and after reports of fission, used this model both to explain fission and also to conclude that the fission of uranium was due to the relatively rare isotope, U^{235}. (Bohr, Works, Vol. 9, 365–389). Both models shared two assumptions: that visualizable models are useful in the limits within which one can use classical concepts to give descriptive accounts; and, an implication of his earlier conceptual analysis, one cannot model individual particles within the nucleus. Here again Bohr insisted on the limits of applicability of basic concepts. One could speak of protons and neutrons as particles and as confined within the nucleus. However, there was no meaningful basis for ascribing positions or trajectories to any particle within the nucleus. So, both models relied on continuous potentials, rather than discrete particles.

The second assumption was sharply challenged by the success of the individual particle (or shell) model of the nucleus developed independently by M. G. Mayer, Haxel, Jensen, and Suess. Bohr eventually found a way to interpret collective and individual particle models as complementary, rather than contradictory. Since protons and neutrons are both Fermi-Dirac particles, the Pauli's exclusion principle applied to nuclear particles, effectively gives each of them an infinite mean free path. Since it was meaningful to speak of particle trajectories, it was also meaningful to speak of individual particles having these trajectories. His suggestion led to the collective model developed by his son, Aage, and Ben Mottelson, and rewarded with a Nobel Prize.

Bohr's lifelong concern with scattering theory illustrates the way he related the particle concept to mathematical formulations treated as computational tools. (See Bohr, Works, Vol. 8 and MacKinnon 1994) To see the complications we can begin with the perspective that characterized Bohr's earliest work on scattering of electrons from atoms. At low energies classical approximations are valid. When the energy of the incident electron is high enough to induce orbital transitions, then quantum effects must be included. In the 1940s Bohr was concerned with particles incident upon nuclei and effectively reversed his earlier standards. At very high energies the incident particle can be thought of as striking an individual nucleon. At low energies it is absorbed by the nucleus and quantum levels must be considered. In 1940 he promised a general paper on collision theory, but did not complete it until 1948 (Bohr, Works, Vol. 8, 423–568). Here the quantum/classical division was determined by a parameter, ζ, the ratio of the collision diameter to the screening factor. When $\zeta \ll 1$, then one has a pure classical picture. When $\zeta \gg 1$ one has pure wave diffraction. Models are required for the intermediate cases. These are not models of the mathematical formalism. They are models of the reality treated that are introduced when the formalism seems inapplicable. The appropriate model depends on the problem. In a relatively low energy collision between an electron and an atom, one may use the orbital model of the atom. When the energy is high enough so that the incident electron effectively interacts with all the bound electrons the Fermi-Thomas model is appropriate. Similar considerations determine when it is appropriate to model the incident electron as a particle or as a wave packet (Born approximation). The intermediate cases must include many special effects: exchange phenomena in the collision of two identical particles; the Ramsauer effect for slow electrons interacting with noble gases, the capture and loss of electrons by fission fragments. The problematic feature in these cases was the development of a consistent descriptive account adequate to the phenomena treated. Different contexts required different accounts and an analysis of the valid applicability of the concepts used. When that was accomplished, the mathematical formulation was routine. When Bohr's epistemological comments are cited out of context, then they may seem pontifical and arbitrary. However, they are best understood as emerging from his abiding concern with making the practice of physics conceptually consistent.

QFT seemed to fail a Bohrian analysis. Landau and Peierls (1931) argued that quantum mechanics could not be applied in the range of relativistic energies. They interpreted the difficulties in RQM (negative energy states) and QFT (divergences) as indicating the failure of these two theories and sought to explain the reason for the failure along Bohrian lines. A necessary condition for the applicability of quantum mechanics is the existence of predictable measurements. By adapting the time-energy indeterminacy principle to measurements of electrons and photons they concluded that measurements precise enough to support predictions can be made only for systems that vary little in the time required to achieve this precision. On this basis, they inferred that quantum mechanics does not apply at all to photons and only to non-relativistic electrons. They visited Bohr's institute and were amazed at the strength of his rejection. (See Peierls's Introduction to BCW, Vol. 9 (1985)).

Bohr's response manifested a way of doing physics that was uniquely his. He concluded that the Landau-Peierls position was wrong on conceptual grounds, and then began to learn the mathematics of quantum field theory. Two years of intense work with Rosenfeld yielded a paper (Bohr and Rosenfeld 1933), which, as the authors noted, was more respected than read. This paper convinced Bohr that his manner of interpreting quantum physics was correct. Yet, it is rarely treated in any discussions of Bohr's position. I will try to bring out the point of the paper and refer to Darrigol (1991) for a more complete account. The paper is not concerned with quantum field theory as a theory, or with the fundamental difficulties concerning divergences. It is exclusively concerned with a consistency problem. The definition of the quantities that quantum field theory uses is set by the CP and the uncertainty principle. Testing means measuring field components individually, or in combinations. A necessary condition for the extension of quantum mechanics to the electromagnetic field is that definitions of field quantities must be used in a way that is consistent with the possibility of measurement. It was here that Landau and Peierls argued that the theory was inconsistent.

The argument given in the Bohr-Rosenfeld paper is essentially a peculiar form of double-entry bookkeeping. The credits come from the application of the correspondence principle and the uncertainty principle to the electromagnetic field. The debits come from measurement of field quantities. The details present a double problem. First, they are technical and difficult. It took Bohr and Rosenfeld two years of intensive work to get all the details straight. Secondly, the proposed measurements are so grossly unrealistic that it is difficult to see what the authors are getting at. We begin with the credits. The CP extends basic concepts of mechanics and electrodynamics to quantum physics. Mechanical quantities presuppose the concept 'particle'. Electrodynamics concepts presuppose 'field'. This paper is concerned with the application of the CP to the field concept. The classical concept of a field is a continuous distribution, such that the components have a value at every point. This, the authors insist, is an idealization. Electromagnetic quantities are quantized by the same procedure used for mechanical concepts. Set up Poisson brackets for components in Cartesian coordinates and then replace these brackets by commutators. These commutation relations lead to detailed conclusions concerning which field components can be simultaneously measured and to what degree of accuracy. The mathematical form of the results made an essential use of the Dirac delta function. The authors justified this by the physical significance they accorded it. The value of a field at a point is an extension of the classical idealization beyond the limits of its validity. Physical significance attaches only to space-time integrals of field components. The delta function is a tool for integration that effectively uses values defined over space-time intervals. Using the delta function, they computed average values of field components over different space-time regions and used this as a basis for predictions concerning measurability.

The averages of all field components over the same space-time region commute and, accordingly, should be independently measurable. The averages of two components of the same kind, such as E_x or H_y, over two spatially separate regions commute if the time intervals are identical. The averages of two components of different kinds over two arbitrary time intervals commute when the corresponding spatial regions coincide. However, average values of the same component, e.g. E_x, over different spatio-temporal regions (I and II) do not commute. Nor do the average values of one component, such as E_x in I, and a perpendicular component, such as H_y in II. Pauli, whose critical evaluation was regularly solicited, pointed out that vacuum fluctuations were not included. Rather than include them, the authors gave an epistemological justification for their omission.

To balance the debits with credits they considered, not actual measurements, but the most perfect measurements that could be conceived without contradiction. Again, the details are confusing, but the overall purpose is quite clear. Even idealized measurements of different field components can be broken down into two parts. The first is the actual measurement of a field quantity. The second is readjusting the 'machinery' so that it can perform another measurement. The analysis should include the actual measurement and any changes to other field values brought about by the measuring process, but exclude, or compensate for, the process of readjusting the machinery. The measurement of electromagnetic field components depends on the transfer of momentum to suitable electrical or magnetic test bodies placed in the field. Since measurements are of averages over space-time volumes, a suitable test body for measuring an electrical component must have a uniform charge distribution over a suitable volume. This is a classical charged particle, one whose atomic composition is ignored. Any direct measurement of the momentum this body acquires by using something like a radar gun, or by using the Doppler effect, would change the frequency. So for ideal measurements a more complicated device is needed.

Consider a collection of macroscopic uniformly charged rigid test particles, each with a fixed place in a rigid framework. To measure E_x in region I the test particle in I is disconnected from the framework. Then it is displaced by the value of the E_x field over the surface of the particle. Next, this displacement is compensated, e.g. by having the test particle attached to an oppositely charged body by magnetizable flexible threads. Then the test body is reattached to its original position. Since measurements are needed of different components in different regions and at different times the rigid framework must have a distributed series of detachable test particles each with its own compensating mechanism. The resulting apparatus is much more like a Rube Goldberg contraption than a feasible experimental arrangement. This is not significant. The postulation of the most perfect measurements compatible with the physical principles indispensable to the measurement in question supplies a clear basis for determining the overall consistency of measurements in QFT.

When Bohr and Rosenfeld developed these idealized measurements they were not trying to prove that The Copenhagen position was correct. Bohr always felt that learning where a system broke down is the best way to appraise its validity. What they found was that for the measurement of one or more components in different regions the results of idealized measurements did not coincide with the results obtained from the commutation relations. However, the analysis was not yet complete. The displacement of one test body changes the field at another test body. This change can be compensated by means of a third body hooked to the second by flexible springs. When this compensation is included, then the results are exactly the same as those obtained from the commutation relations. The debits and credits balance in precise detail. Hence the CP supplies a consistent basis for applying quantum mechanics to electromagnetic fields. This analysis does not establish, or even test, the consistency of quantum field **theory**. It simply shows that two different usages of classical field components, one using the CP to set up quantum analogs of classical components and the other in measuring fields, are consistent. This minimal consistency is a necessary but far from sufficient condition for any quantum field theory employing these concepts.

In their second paper (Bohr and Rosenfeld 1950), written when quantum electrodynamics (QED) dominated physics they extended their previous considerations from the measurement of fields to the measurement of charge currents. The goal was to show that second quantization is consistent in the treatment of 'matter waves' as well as photons. Again, they propose a highly idealized experiment, measuring current within a region by surrounding the region with a shell containing test bodies that absorb momentum, are moved, and then have the movement compensated. The first, or pre-QED, approximation presented no problems. In a second approximation they had to consider virtual pair production induced by displacement of test bodies. They gave a very non-technical argument to indicate that polarization of the vacuum would not influence their idealized measurements and that manipulations of a test body in one region would have a polarizing effect on other regions. When proper compensations are included the results are in accord with the commutation relations. The net result was a qualitative non-technical proof of what every one else assumed, that the way quantities are represented in QED is legitimate.

Bohr was finally convinced that his way of interpreting QM was consistent. One could use either the 'particle' or the 'field' cluster of concepts to interpret actual or ideal experiments and have a mathematical formulation that was consistent with the informal inferential structure used to report and extend experimental results. The final trial came from the challenge issued by Einstein, Podolsky, and Rosen. Since this has been exhaustively treated in the literature I will merely point out a divergence in the contrasting interpretative frameworks. The EPR paper argued that Copenhagen QM is incomplete on ontological grounds. There are elements of reality not included in the theory. Bohr defended QM as complete on epistemological grounds. It accommodates all the experimental information that can be used without introducing inconsistencies.

In his later analyses, Bohr gradually shifted from concepts used in individual experimental situations, to the supporting network of concepts, and finally to the language that made concepts possible. From about 1937 on Bohr advocated using 'phenomenon' as a general term covering the whole experimental situations, including the apparatus. Bohr was never concerned with the interpretation of quantum mechanics as a theory. He considered the mathematical formalism an inferential tool, not a theory. "Its physical content is exhausted by its power to formulate statistical laws governing observations obtained under conditions specified in plain language" (Bohr 1963, p. 12). With this background we may summarize the *Bohr Consistency Conditions*, the necessary conditions for the unambiguous communication of experimental information.

1. The meaning of classical concepts is rooted in ordinary language usage and its historical extension in the language of physics.
2. The doctrine of complementarity sets the limits to which classical concepts may be consistently extended.
3. Any use of classical concepts beyond these allowed limits may generate inconsistencies. Idealized thought experiments supply a vehicle for analyzing limits and exposing inconsistencies.
4. When concepts are used within their limits, then they support the normal inferences of experimental physics. Thus, predicting, or retrodicting, paths is valid in contexts where the classical particle concept is applicable.
5. The mathematical formulation based on the usual operator substitutions must be consistent with these conditions. This is a consistency relation between two inference supporting systems, a linguistic formulation and a mathematical formulation.

In introducing the dual inference model we used the simple example of how a dual inference system functions in the game of bridge. The informal ordinary-language inference system contains the physical content while an inferential system, like the Goren point-count system, functions as an inferential tool. In Bohr's position the extended ordinary language contains all the physical content, while the mathematical formalism is an inferential tool. The consistency conditions allow the dual-inference system to function without generating contradictions. The justification for imposing this is pragmatic. It works in atomic physics, nuclear physics, and quantum electrodynamics.

To see the significance of the Bohr Consistency Conditions we note that they disallow the standard formulation of Bell's theorem. Bell's original formulation specified the problem: "Consider a pair of spin one-half particles formed somehow in the singlet state and moving freely in opposite directions." This statement of the situation explicitly presupposes both the classical term 'particle' and a quantum specification of the state of the two-particle system. The fact that 'particle' is used in the classical sense of a localized body traveling in a trajectory is basic to every formulation of the problem. This problematic mixture of classical descriptive accounts, used to support inferences, and quantum state specifications carries over even to accounts given in purely quantum terms. Thus Redhead (1987, p. 73) says: "Consider a QM system consisting of two spin one-half particles, in the singlet state of the total spin, and widely separated, so that there is no significant overlap of the spatial wave functions of the two systems." In Bohrian semantics the term 'particle' serves as an apt designation and a basis for inference only in an experimental context set up to test for mechanical properties. Prior to such a measurement we are dealing with an entangled quantum mechanical system represented by one wave-function, not with two separated particles having separate wave functions. The Bohrian position supports the conclusion the QM correlations will always trump Bell limits. However, it does not explain, or even address, the distant correlations that Einstein labeled 'ghostly'. Bohr would argue that such questions are not properly formulated.

5.2 A Strict Measurement Interpretation of Quantum Mechanics

When quantum mechanics is developed on the basis of the Bohr consistency conditions it is not a theory. It uses the mathematical formalism of QM as a tool for extending classical concepts. Heisenberg[5] and Pauli[6] also interpreted quantum mechanics as a rational generalization of classical physics. Bohr repeatedly insisted

---

[5] "...the Copenhagen interpretation regards things and processes which are describable in terms of classical concepts, i.e., the actual, as the foundation of any physical interpretation" (Heisenberg 1958, p. 145).

[6] Pauli, Bohr's closest ally on interpretative issues, contrasted Reichenbach's attempt to formulate quantum mechanics as an axiomatic theory with his own position: "Quantum mechanics is a much less radical procedure. It can be considered the minimum generalization of the classical

that the complementarity interpretation is the only possible one.[7] In the view of many philosophers the Copenhagen patriarchs, like their Chalcedonian predecessors, were imposing orthodoxy by decreeing that no other position should be taught or held. What significance should be accorded Copenhagen orthodoxy?

Before answering that question we should consider the chief source of misunderstanding. David Bohm's presentation of a hidden variable interpretation of QM effectively changed the *status quaestionis*. Quantum mechanics was presented as a mathematical formalism that admitted of different interpretations. Heisenberg entered the fray arguing that the Copenhagen interpretation is the only viable interpretation. (See Heisenberg 1958, chaps. 3 and 8; Howard 2004) His defense effectively transformed the perception of the Copenhagen interpretation into an interpretation of quantum mechanics as a theory.[8] When Bohr's scattered comments were taken as the interpretation of QM *as a theory* then they seemed amateurish, outdated, and even perverse. Bohr never interpreted quantum mechanics as a theory. As the last footnote indicates what he regarded as necessary was the complementarity *description*. It is the only way to systematize experimental results without introducing inconsistencies. The mathematical formalism had to be used, and should be interpreted, in accord with these restrictions. Is this an adequate basis for an interpretation of QM? The three main objections can be labeled 'the Einstein objection', 'the Bohr objection', and the formalist objection. Einstein thought that Copenhagen QM is not what a fundamental theory should be. He realized that the only effective way to implement this criticism is to develop a better quantum theory. His 30 years of struggling to achieve this goal led only to frustration. Bohr thought that his way of interpreting QM should be rejected if it is inadequate to advances in physics. In 1930 he thought it might not be adequate to advances in RQM, QFT, and nuclear physics. The efforts previously summarized convinced him that these advances did not go beyond the limits his method allowed. The formalist objection is that a physical theory should be regarded as a mathematical formalism requiring a physical interpretation. This leaves no role for the dual-inference account that I summarized and which Bohr exemplifies.

We will focus on the Bohrian objection. Is the Copenhagen interpretation adequate to advances in physics since Bohr's death? This question cannot be answered by simply considering advances in physics. Creative physicists often rely on an 'Anything goes' methodology and deliberately go beyond accepted limits. The question can be rephrased. Does a systematic account of accepted advances go beyond

---

theory which is necessary to reach a self-consistent description of micro phenomena, in which the finiteness of the quantum of action is essential" (Pauli 1947, p. 1404).

[7] In an interview with Thomas Kuhn and others the day before his death Bohr claimed "There are all kinds of people, but I think it would be reasonable to say that no man who is called a philosopher really understands what one means by the complementary description.... They did not see that it was an objective description, and that it is the only possible objective description" (Bohr, AHQP, Interview 3, 5).

[8] See Gomatam (2007) for a clarification of the difference between Bohr's position and the standard Copenhagen interpretation.

the limits of the Bohr Consistency Conditions? Here we can take some guidance from the formalists. Explicit rules for theory interpretation have been developed for formal systems, such as symbolic logic. In an axiomatically formulated system there is: **a basis**, the axioms; **a method of extension**, the allowed rules of inference; and **a cutoff**. Only conclusions derived from the axioms by following the rules count as part of the system. Then a theory is a sharply delineated object of interpretation with clearly specified limits. A rigorous reformulation of QM could put it in this interpretative framework.

John von Neumann (1955 [1933]), who coined the term 'Hilbert space' extended Hilbert's axiomatic approach to quantum mechanics. J. Mackey (1963) gave a new axiomatic formulation of QM as a non-classical probability theory. Piron and the Geneva school developed axiomatic systems centered on the lattices of closed subspaces of a generalized Hilbert space.[9] Recent works generally rely on the semantic conception of theories rather than axiomatic models. We may schematize these formulations of QM in terms of a general structural form: $\mathcal{T} =< \mathcal{L}, \mathcal{A}, \mathcal{D}, \mathcal{K} >$, where $\mathcal{T}$ is a theory, $\mathcal{L}$ a formal language, $\mathcal{A}$ is a set of axioms expressed in $\mathcal{L}$, $\mathcal{D}$ is a set of inference rules, and $\mathcal{K}$ is a class of models of $\mathcal{A}$, or structures in which the axioms are true. In the semantic conception one dispenses with axioms and treats a Hilbert space as an abstract structure to be given an interpretation in terms of models.[10]

Bohr's methodology effectively reverses these methods of interpretation. Formal methods take a mathematical formalism as a foundation and then impose a physical interpretation on this foundation. Bohr takes a descriptive account of actual and possible measurements as foundational and then fits the mathematics to this foundation. This can be done in two ways. A *loose measurement interpretation* begins with the restrictions on the reporting of experimental data and then adapts the mathematical formalism to fit this basis. I am familiar with only five textbooks that take a basic consistency between the language used in experimental results and mathematical formulations as a basis for developing and interpreting quantum mechanics: Heisenberg 1930, Pauli 1947 [1930], Kramers 1957, Landau and Lifshitz 1965 [1956], and Gottfried 1966. There are undoubtedly more. This does not supply a basis for determining the limits of applicability of the method. A *strict measurement interpretation* relies on an analysis of quantum measurements to generate the mathematics of QM. In this case one can imitate the formal methodology and speak of the interpretation of QM in terms of a basis, the measurement analysis; a method of extension, the mathematical formulation, and a cutoff. Any conclusions incompatible with this methodology are not accepted. This methodology presents both theoretical and practical difficulties. As in the interpretation of classical physics, the theoretical difficulty is a reliance on sloppy mathematics. Since the physics is taken as foundational, physical considerations often replace

[9] Coecke et al. (2001) provides a good historical summary of the axiomatic approach.

[10] Healey (1989), Hughes (1989), and Van Fraassen (1991) have developed interpretations of QM using the semantic method of interpretation.

existence theorems and consistency considerations in justifying mathematical formulations. The practical difficulty is that this is an awkward, and often confusing, way of developing QM. Nevertheless, it seems to be the only method available for testing the limits of valid applicability of the Bohrian approach. I have presented the technical details elsewhere (MacKinnon 2008) and will present an informal summary here. As a preliminary point we should make a sharp distinction between the measurement *problem* and the measurement *interpretation*. The standard formulation of the measurement problem assumes the universal validity of QM. It should treat the apparatus as well as the system being analyzed. Consider an experimental situation where the state function, $|\psi\rangle$, representing the object plus the measuring apparatus is a superposition. In the linear dynamics of the Schrödinger equation a superposition of states evolves only into further superpositions. Measurement results require a mixture of states, which may be assigned different probabilities. How does a superposition become a mixture? In the von Neumann (or Wigner[11]) account one distinguishes two types of processes: the unitary evolution based on Schrödinger dynamics, and a non-unitary collapse proper to measurement situations. This has occasioned repeated criticism as an ad hoc postulate. When this postulate is rejected, then there are two interrelated problems. The first is the reduction problem, explaining how the superposition becomes a mixture. The second is the selection problem, explaining how the measurement selects one value from the mixture that has many values with differing probabilities.[12]

In a strict measurement interpretation the measurement problem does not arise. Instead of asking how the formalism yields measurement results one begins with measurements and asks how they can be represented mathematically. In a loose measurement interpretation one has the standard mathematical formalism and a form of the problem is treated in a reverse order. Thus, Landau and Lifshitz (pp. 21–24) claim that the measuring apparatus is represented by 'a quasi-classical wave function'. This means that one relies on a classical description of the apparatus and presupposes that there is a state function, or a large equivalence class of state functions, corresponding to this description. Gottfried (p. 186) insists that an experimental arrangement counts as a measurement device if and only if quasi-classical states are macroscopically distinguishable. This means that pure states and mixtures are indistinguishable in a measurement situation. He focuses on the conditions under which it is reasonable to replace a superposition of states by a mixture. This is not a consequence of the formalism of quantum mechanics; it is a necessary condition for a real measurement. The formalism of quantum mechanics does not yield real measurements.

---

[11] The account of measurement was developed in von Neumann (1955 [1932], chap. 6). In a conversation with Abner Shimony, Eugene Wigner claimed "I have learned much about quantum theory from Johnny, but the material in his Chapter Six Johnny learned all from me." (citation from Aczel 2001, p. 102)

[12] Bub (1997), chap. 7 gives a technical treatment that examines Bohr's position, the 'von Neuman-Dirac orthodoxy' and Bub's own development based on the Bub-Clifton theorem.

In his *Principles* Dirac generates the basic formalism of QM by analyzing idealized experiments. Messiah's (1964) well-known textbook helped make the Dirac formalism an established part of normal physics by presenting it with no reliance on Dirac's own development. As a result, Dirac's method of development is never considered. I will present his reasoning in its starkest form. Dirac justifies the representation of states by vectors through an analysis of measurements. A simplified recasting of his argument highlights the problematic features. Consider a beam of light consisting of a single photon plane-polarized at an oblique angle relative to the optic axis of a tourmaline crystal. Either the whole photon passes, in which case it is observed to be polarized perpendicular to the optic axis, or nothing goes through. The initial oblique polarization, accordingly, must be considered a superposition of states of parallel and perpendicular polarization. Again, consider another single-photon beam passed through an interferometer so that it gets split into two components that subsequently interfere. Prior to the interference, the photon must be considered to be in a translational state, which is a superposition of the translational states associated with the two components (Dirac 1958, pp. 4–14). Since particle states obey a superposition principle, they should be represented by mathematical quantities that also obey a superposition principle, vectors. The physics generates the mathematics.

This was a methodology that Dirac regularly relied on. In the second edition of *Principles* he introduced vectors "... in a suitable vector space with a sufficiently large number of dimensions" (Dirac 1935, p. 14). In the third edition he introduced his bra-ket notation and simply postulated a conjugate imaginary space with the needed properties. He assumed that the vector space he postulated must be more general than a Hilbert space, because it includes continuous vectors that cannot be normalized (Dirac 1958, p. 40, 48). He only spoke of the Hilbert-space formulation of quantum mechanics when he became convinced that it should be abandoned (Dirac 1964). Messiah developed a statistical interpretation of QM and did not apply the superposition principle to individual systems. In this context the Dirac argument from physical superposition of states of an individual system to a mathematical representation that also obeys a superposition principle has no foundation. Physicists generally learned the Dirac formulation through Messiah's elegant mathematical presentation and then failed to realize that Dirac's presentation represented better physics. The application of the superposition principle to individual states proved indispensable in particle physics. Schwinger described his early student years as "unknown to him, a student of Dirac's" (Schweber 1994, p. 278). Before beginning his freshman year at C.C.N.Y. he had studied Dirac's *Principles* and, at age 16, wrote his first paper, never published, "On the Interaction of Several Electrons", generalizing the Dirac-Fock-Podolsky many-time formulation of quantum electrodynamics. Schwinger explicitly puts QM on an epistemological basis: "Quantum mechanics is a symbolic expression of the laws of microscopic measurement" (Schwinger 1970b, p. 1). Accordingly, he begins with the distinctive features capturing these measurements. This, for Schwinger, is the fact that successive measurements can yield incompatible results. Since state preparations also capture this feature Schwinger actually uses state preparations, rather than complete measurements as his starting point. He begins by symbolizing a measurement, M, of a quantity, A, as an operation that sorts an ensemble into sub-ensembles characterized by their A values, $M(a_i)$ The paradigm case is a Stern-Gerlach filter sorting a beam of atoms into two or more beams. This is a type one measurement. An immediate repetition would yield the same results. Though Schwinger did not use quantum information theory, his point of departure in developing his measurement interpretation is a consideration of idealized measurements that yield Yes/No answers. There is no reduction of the wave packet or recording of numerical results. An idealization of successive measurements is used to characterize the distinguishing feature of these microscopic measurements. Symbolically

$$M(a')M(a'') = \delta(a', a'')M(a'). \tag{5.1}$$

This can be expanded into a complete measurement, $M(a') = \prod_{i=1}^{k} M(a_i')$ where a_i stands for a complete set of compatible physical quantities. Using A, B, C and D for complete sets of compatible quantities, a more general compound measurement is one in which systems are accepted only in the state $B = b_i$ and emerge in the state, $A = a_i$, e.g., an S-G filter that only accepts atoms with $\sigma_z = +1$ and only emits atoms with $\sigma_x = +1$. This is symbolized $M(a_i, b_i)$. If this is followed by another compound measurement $M(c_i, d_i)$, the net result is equivalent to an overall measurement that only accepts systems in state d_i and emits systems in state a_i. Symbolically,

$$M(a_i, b_i)M(c_i, d_i) =< b_i|c_i > M(a_i, d_i). \tag{5.2}$$

For this to be interpreted as a measurement $< b_i|c_i >$ must be a number characterizing systems with $C = c_i$ that are accepted as having $B = b_i$. The totality of such numbers, $< a'|b' >$, is called the transformation function, relating a description of a system in terms of the complete set of compatible physical quantities, B, to a description in terms of the complete compatible set, A. In the edition of Dirac's *Principles* that Schwinger studied, the transformation function was basic. A little manipulation reveals that N, the total number of states in a complete measurement, is independent of the particular choice of complete physical quantities. For N states the measurement symbols form an algebra of dimensionality N^2. These measurement operators form a set that is linear, associative, and non-commutative under multiplication.

To get a physical interpretation of this algebra consider the sequence of selective measurements $M(b')M(a')M(b')$. This differs from a simple or repeated measurement $M(b')$ in virtue of the disturbance produced by the intermediate $M(a')$ measurement. This suggests $M(b')M(a')M(b') = p(a', b')M(b')$, where $p(a', b') = < a'|b' >< b'|a' >$. Since this is invariant under the transformation, $< a'|b' >\rightarrow \lambda(a') < a'|b' > \lambda(b^{-1})$, where $\lambda(a'), \lambda(b')$ are arbitrary numbers, Schwinger argues that only the product, $p(a', b')$ should be accorded physical significance. Using $\sum_{a'} p(a', b') = 1$ Schwinger interprets this as a probability and imposes the restriction,

$$< b'|a' >=< a'|b' >^*. \tag{5.3}$$

The use of complex numbers in the measurement algebra implies the existence of a dual algebra in which all numbers are replaced by complex conjugate numbers. This algebra of measurement operators can be expanded into a geometry of states. Introduce the fictional null (or vacuum) state, 0, and then expand $M(a', b')$ as a product, $M(a', 0)M(0, b')$. Let $M(0, b') = \Phi(b')$, the annihilation of a system in state b', and $M(a', 0) = \Psi(a')$, the creation of a system in state a'. These play the role of the state vectors, $\Phi(b') =< b'|$ and $\Psi(a') = |a' >$. With the convenient fiction that every Hermitian operator symbolizes a property and every unit vector a state one can calculate standard expectation values. Like Dirac Schwinger relies on the complex space developed from his measurement algebra and never refers to Hilbert space. Accardi (1995) has shown that Schwinger's construction is equivalent to standard Hilbert space.

Schwinger extended this methodology to QED, where he was quite successful, and to QFT, where he was less successful. Standard QFT develops dynamics by introducing a classical Hamiltonian and substituting operators for dynamical variables. Schwinger relied on his methodology, rather than the Correspondence principle. He characterized his method as "... a phenomenological theory—a coherent account that it anabatic (from anabasis: going up)" (Schwinger 1983, p. 23, Flato et al. 1979). This anabatic methodology introduced two new steps. The first was a new dynamic principle (Schwinger 1959, p. xiv). The new dynamics is based on a unitary action principle whose justification hinges on the foundational role assigned measurement. A measurement-apparatus effectively defines a spatio-temporal coordinate system with respect to which physical properties are specified. A transformation function, $< a't_1|a''t_2 >$, relates two arbitrary complete descriptions. Physical properties and their spectra of values should not depend on which of equivalent descriptions are chosen. Hence, there must be a continuous unitary transformation leading from any given descriptive basis to equivalent bases. The continuous specification of a system in time gives the dynamics of the system (See Gottfried 1966, pp. 233–256). From this Schwinger infers that the properties of specific systems must be completely contained in a dynamical principle that characterizes the general transformation function.

Any infinitesimal alteration of the transformation function can be expressed as

$$\delta < a_1't_1|a_2''t_2 >= i < a_1't_1|\delta\mathbf{W_{12}}|a_2''t_2 >. \tag{5.4}$$

This suggests the fundamental dynamical postulate: There exists a special class of infinitesimal alterations for which the associated operators $\delta\mathbf{W_{12}}$ are obtained by appropriate variation of a single operator, the action operator $\mathbf{W_{12}}$, or $\delta\mathbf{W_{12}} = \delta[\mathbf{W_{12}}]$. Thus, quantum dynamics can be developed simply as an extension of the algebra of measurements without attaching any further ontological significance to state functions. The second advance was the introduction of operator fields. These dynamic variables, or operator fields, supply the theoretical concepts that replace the phenomenological concept 'particle'. This is the basic conceptual advance that Schwinger makes beyond Bohr's methodology. For Bohr all descriptions must be expressed exclusively in classical terms. Schwinger assumes that it is possible to use dynamical field variables to give a sub-microscopic descriptive account within the framework of his methodology. The spatial and temporal coordinates that function as parameters for operator fields are idealized extensions of the spatio-temporal framework of the measuring apparatus. "It is the introduction of operator variations that cuts the umbilical cord of the correspondence principle and brings quantum mechanics to full maturity" (Schwinger 1983, p. 343, Flato et al. 1979). In 1964 Gell-Mann and Zweig independently introduced the quark hypothesis, which Schwinger rejected. Schwinger's rejection had strong roots in his ideas of the proper relation between a phenomenological and depth level. 'Particle' functions on the phenomenological level. Speaking of a particle assumption he claimed: "But the essential point is embodied in the view that the observed physical world is the outcome of the dynamic play among underlying primary fields, and the relationship between these fundamental fields and the phenomenological particles can be comparatively remote, in contrast to the immediate connection that is commonly assumed" (Schwinger 1964, p. 189). The quark hypothesis entered on the wrong level, as part of an underlying theory rather than the phenomenology, and entered through a phenomenological classification, rather than through a depth theory. The standard model will be considered in the next chapter. Here we will simply indicate where it departs from Schwinger's anabatic methodology. For Schwinger the space-time framework of the measuring apparatus anchors all assignments of spatial and temporal values to fields. This supported universal gauge transformations, but not the local gauge transformations basic to the standard model. Finally, the standard model did not meet the requirement that Schwinger considered basic for a new theory. It should supply a theoretical basis for the coupling constants.

It may seem arbitrary to take the limits of Schwinger's anabatic advance as the limits of the measurement interpretation of QM. Yet, Schwinger's combination of awesome computational skill, profound knowledge of physics, and systematic development of a methodology supply a better guide then any alternative I might attempt. Furthermore, the consistent histories interpretation, which will also be treated in the next chapter, can easily be regarded as a replacement for the measurement interpretation. This grounds my evaluation. Orthodox quantum mechanics has had a success that is unprecedented in scope and precision. Yet, the usual formulations of orthodoxy, and the systematic misinterpretations, supply no clear basis for determining the limits of valid applicability. The measurement interpretation systematizes the orthodox interpretation. It supplies a basis, the distinctive features of quantum measurements and the algebra these generate; a method of extension, Schwinger's anabatic methodology; and a cutoff. This cutoff excludes the standard model of particle physics. In this respect Schwinger's position is similar to algebraic quantum field theory. This too relies on an epistemological foundation and a systematic method of advancement that also excludes the standard model of particle physics.[13]

We will rephrase the two basic objections to orthodoxy. A fundamental theory should be about reality at a fundamental level. Bohrian QM is grounded in classical physics and treats the mathematical formalism as a tool rather than a fundamental theory. Schwinger consciously went beyond this by using operator fields to give the equivalent of a sub-microscopic descriptive account. This did not go far enough. Orthodox QM does not answer the Einstein objection. Nor does it answer the Bohr objection. It is not empirically adequate to advances in quantum field theory. This evaluation does not suggest a change in the practice of fundamental physics. Creative theoreticians do not feel bound by, and rarely avert to, the restriction of a methodology. However, this evaluation does show where and why a revised interpretation of QM is needed.

I have appended schematic outlines of two different perspective on the interrelation of classical and quantum physics and, by an oversimplification, attached historical names to each. In the Einstein perspective, shared by many philosophers, theories are the basic units to be interpreted. In the Bohr perspective, our suspension in language plays a presuppositional role in the interrelation and interpretation of theories. A development of this unfamiliar interpretative perspective requires a clarification of the status of classical physics.

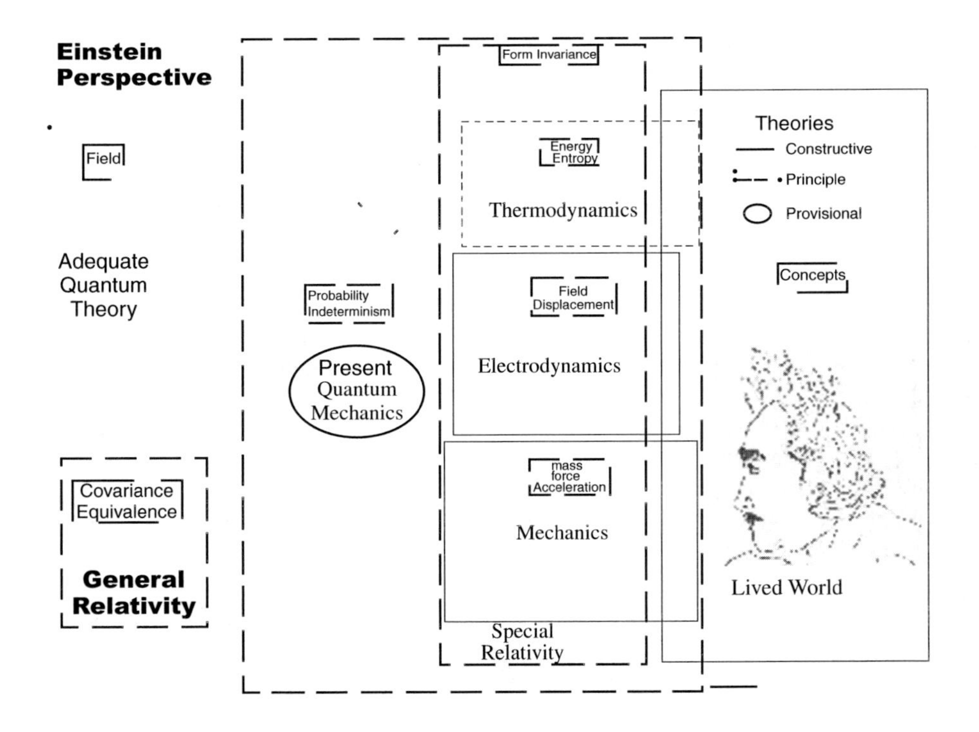

[13] Arguments supporting this evaluation are given in my 2007 and 2008 papers.

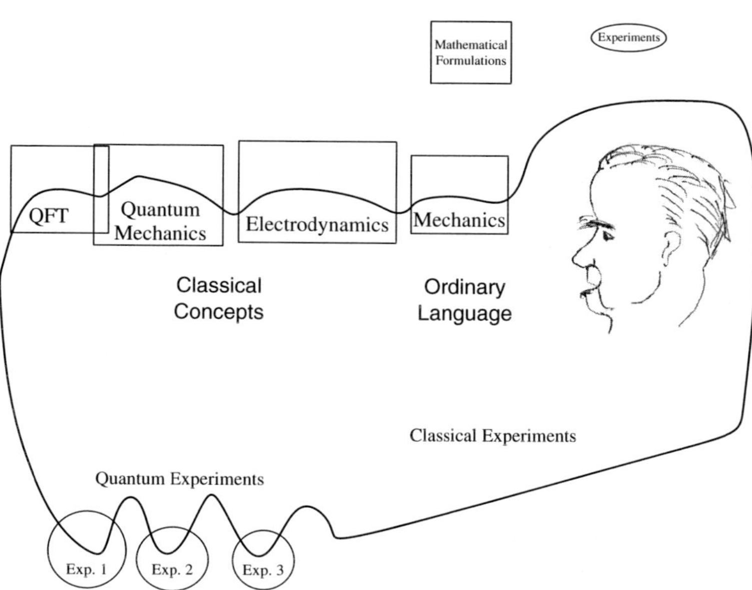

References

Accardi, L. 1995. Can Mathematics Help Solving the Interpretational Problems of Quantum Mechanics? *Il Nuovo Cimento, 110B*, 685–721.

Aczel, Amir D. 2001. *Entanglement: The Greatest Mystery in Physics*. New York, NY: Four Walls Eight Windows.

Bethe, Hans. 1999. Quantum Theory. *Reviews of Modern Physics, 71*, S1–S8.

Bohr, Niels. 1963. *Essays 1958–1962 on Atomic Physics and Human Knowledge*. New York, NY: Wiley.

Bohr, Niels, and Léon Rosenfeld. 1933. On the Question of the Measurability of Electromagnetic Field Quantities. In J. Wheeler, and W. Zurek (eds.), *Quantum Theory and Measurement* (pp. 478–522). Princeton, NJ: Princeton University Press.

Bohr, Niels, and Léon Rosenfeld. 1950. Field and Charge Measurements in Quantum Electrodynamics. *Physical Review, 78*, 794–798.

Born, Max. 1962. *Atomic Physics*. New York, NY: Hafner.

Bub, Jeffrey. 1997. *Interpreting the Quantum World*. Cambridge: Cambridge University Press.

Coecke, Bob, David Moore, and Alexander Wilce. 2001. *Operational Quantum Logic: An Overview*. ProCite field[12]: (2001).

Darrigol, Olivier. 1991. Coherence et complétude de la mécanique quantique: l'exemple de Bohr. *Review d'Histoire de Sciences, 44*, 137–179.

Dirac, Paul. 1928. *The Quantum Theory of the Electron, A117*, 610–625.

Dirac, P. A. M. (ed.) 1935. *The Principles of Quantum Mechanics* (2nd ed). Cambridge: Cambridge Universiwty Press.

Dirac, Paul. 1958. *The Principles of Quantum Mechanics*, (4th edn). Oxford: Clarendon Press.

Dirac, Paul. 1964. Foundations of Quantum Theory. *Lecture at Yeshiva University*. New York.

Flato, M., C. Fronsdal, and K. A. Milton. 1979. *Selected Papers (1937–1976) of Julian Schwinger*. Dordrecht: Holland: D. Reidel Publishing Company.

Gomatam, Ravi. 2007. Bohr's Interpretation and the Copenhagen Interpretation – Are The Two Incompatible? In Cristina Bicchieri, and Jason Alexander (eds.), *PSA06: Part I* (pp. 736–748). Chicago, IL: The University of Chicago Press.

Gottfried, Kurt. 1966. *Quantum Mechanics. Volume I: Fundamentals*. New York, NY: W. A. Benjamin.

Healey, Richard A. 1989. *The Philosophy of Quantum Mechanics: An Interactive Interpretation*. Cambridge: Cambridge University Press.

Heisenberg, Werner. 1930. *The Physical Principles of the Quantum Theory*. New York, NY: Dover.

Heisenberg, Werner. 1958. *Physics and Philosophy: The Revolution in Modern Science*. New York, NY: Harper's.

Howard, Don. 2004. Who Invented the "Copenhagen Interpretation": A Study in Mythology. In Sandra D. Mitchell (ed.), *Philosophy of Science: Proceedings of the 2002 Biennial Meeting* (pp. 669–682). East Lansing, MI: Philosophy of Science Association. ProCite field[8]: ed.

Hughes, R. I. G. 1989. *The Structure and Interpretation of Quantum Mechanics*. Cambridge: Harvard University Press.

Hund, Friedrich. 1974. *The History of Quantum Theory*. New York, NY: Harper & Row.

Jammer, Max. 1966. *The Conceptual Development of Quantum Mechanics*. New York, NY: McGraw-Hill.

Kramers, H. A. 1957. *Quantum Mechanics*. Amsterdam: North Holland.

Kuhn, Thomas, John Heilbron, Paul Forman (eds.) 1962. *AHQP: Archives for the History of Quantum Physics*. Berkeley, CA: Copenhagen.

Landau, L. D., and E. M. Lifshitz. 1965. *Quantum Mechanics: Non-Relativistic Theory* (2nd rev. edn). Reading, MA: Addison-Wesley.

Landau, Lev Davidovich, and Rudolf Peierls. 1931. Extension of the Uncertainty Principle to Relativistic Quantum Theory. In J. Wheeler, and W. Zurek (eds.), *Quantum Theory and Measurement* (pp. 465–476). Princeton, NJ: Princeton University Press.

Mackey, J. 1963. *The Mathematical Foundations of Quantum Mechanics*. New York, NY: Benjamin.

MacKinnon, Edward. 1982. *Scientific Explanation and Atomic Physics*. Chicago, IL: University of Chicago Press.

MacKinnon, Edward. 1985. Bohr on the Foundations of Quantum Theory. In A. P. French, and P. J. Kennedy (eds.), *Niels Bohr: A Centenary Volume* (pp. 101–120). Cambridge, MA: Harvard University Press.

MacKinnon, Edward. 1994. Bohr and the Realism Debates. In J. Faye, and H. Folse (eds.), *Niels Bohr and Contemporary Physics* (pp. 279–302). Dordrecht: Kluwer.

MacKinnon, Edward. 2008. The Standard Model as a Philosophical Challenge. *Philosophy of Science*, 75(4), 447–457.

Mehra, Jagdish, and Helmut Rechenberg. 1982. *The Historical Development of Quantum Mechanics*. New York, NY: Springer.

Messiah, Albert. 1964. *Quantum Mechanics: Vol. I*. Amsterdam: North Holland.

Pauli, Wolfgang. 1947. *Die Allgemeinen Prinzipien der Wellenmchanik*. Ann Arbor, MI: J. W. Edwards.

Petersen, Aage. 1968. *Quantum Physics and the Philosophical Tradition*. Cambridge: MIT Press.

Redhead, Michael. 1987. *Incompleteness, Nonlocality, and Realism*. Oxford: Clarendon Press.

Richardson, W. Mark, and Wesley J. Wildman. 1996. *Religion and Science: History, Method, Dialogue*. New York, NY: Routledge.

Rosenfeld, L. et al. 1972. *Niels Bohr: Collected Works*. Amsterdam: North Holland.

Schweber, Silvan S. 1994. *QED and the Men Who Made it*. Princeton, NJ: Princeton University Press.

Schwinger, Julian. 1970a. *Particles, Sources, and Fields*. Reading, MA: Addison-Wesley.

Schwinger, Julian. 1970b. *Quantum Kinematics and Dynamics*. New York, NY: W. A. Benjamin, Inc.
Schwinger, Julian. 1970c. *Selected Papers on Quantum Electrodynamics*. New York, NY: Dover.
Schwinger, Julian. 1959. The Algebra of Microscopic Measurement. *Proceedings of the National Academy of Sciences of the United States of America, 45*, 1542.
Schwinger, Julian. 1964. Field Theory of Matter. *Physical Review, 135*, B816–B830.
Schwinger, Julian. 1983. Renormalization theory of quantum electrodynamics. In Laurie Brown, and Lillian Hoddeson (eds.), *The Birth of Particle Physics* (pp. 329–353). Cambridge: Cambridge University Press.
Van Fraassen, Bas. 1991. *Quantum Mechanics: An Empiricist View*. Oxford: Clarendon Press.
von Neumann, John. (1955 [1933]). *Mathematical Foundations of Quantum Mechanics*, trans. Robert T. Beyer. Princeton, NJ: Princeton University Press.

Chapter 6
Beyond a Minimal Basis

Entia non sunt multiplicanda praeter necessitatem
John Duns Scotus

Quantum physics has gone beyond the material treated in the last chapter. Here we are primarily concerned with one type of advance going beyond the limits of the orthodox interpretation. This is something physicists routinely do. Many philosophers simply reject any limits imposed by orthodoxy. What is novel here is the methodology. In the last chapter we developed a measurement interpretation. This is essentially the orthodox, or Copenhagen, interpretation developed as a minimal epistemological interpretation of QM. The epistemological basis is a systematization of the language used to report actual and ideal measurements. The restrictions imposed are only those needed to avoid generating contradictions in reporting repeated measurements. The contradictions that must be avoided are those rooted in the material inferences proper to experimental physics and the dialog between experimentalists and theoreticians. The mathematical formalism was treated as a functional tool, not a theory to be interpreted. This does not give an ontology of quantum mechanics. In the spirit of Scotus, or of Davidson, ontological considerations are introduced only when they prove indispensable. We also indicated that the standard model of fundamental particles and quantum cosmology has gone beyond these limits. The operative question is not whether but how one goes beyond the limits of the orthodox interpretation. The more or less standard philosophical method is to put quantum theory on a rigorous foundation and then inquire into the ontology this theory supports.[1] Michel Bitbol has developed an interpretation of quantum mechanics quite similar to the one I am presenting.[2] However I am focusing on the actual practice of physics and its historical development, while he focuses on transcendental arguments from the mathematical formalism.

---

[1] See Healey (1989), Auyang (1995), and Kuhlmann et al. (2002) illustrate different ways of implementing such a program.

[2] This is developed in Bitbol (1996). A summary account may be found in my ISIS review, MacKinnon (1998).

E. MacKinnon, *Interpreting Physics*, Boston Studies in the Philosophy of Science 289,
DOI 10.1007/978-94-007-2369-6_6,

In the present chapter I am attempting a very different approach. The pertinent question is: What can we learn about the characteristic properties and processes of the quantum realm from the normal practice of quantum physics? A focus on the role of language highlights a peculiar interpretative problem. Language plays an indispensable role in describing and reporting experimental results involving QM. This reporting depends on informal inferences. These, in turn, depend on the categorial structure of the language used. Yet, as previous chapters indicated, this categorial structure is a systematically misleading guide to the ontology of the quantum realm. The concept of a spatio-temporal object with properties is a fundamental presupposition of ordinary language (OL) and the extension of ordinary language used in physics (EOL). Yet this concept cannot be systematically extended to the quantum realm without generating contradictions.

In one form or another various philosophers have addressed the question: How does one use language to speak beyond the limits of language? The two upper sections of Plato's divided line transcend the limits of language. The philosopher who leaves the cave to contemplate forms in their naked purity is unable to communicate his new knowledge when he returns to the cave. Aristotle, father of the language of physics, clearly recognized the problem of extending language to ultimate foundations. "Therefore the ultimate substratum is of itself neither a particular thing nor of a particular quantity nor otherwise positively characterized; nor yet is it a negation of these, for negations also will belong to it only by accident." (Metaphysics, 1029a, p. 23). As we saw in chapter 2, Moses Maimonides and Thomas Aquinas realized that discourse about God relies on categorial presuppositions that do not apply to God. Aquinas concluded: We do not know what God is, but only what he is not and how other things relate to him. Kant resorted to a split-level ontology. The categorial system treated in his system of twelve categories applies to phenomena, but not to Noumea. At the dawn of QM Whitehead attempted to develop a new metaphysics that would incorporate the advances or relativity and quantum mechanics. He was clear on the methodology required: "Philosophy will not regain its proper status until the gradual elaboration of categorial schemes, definitely stated at each stage of progress, is recognized as its proper objective" (Whitehead 1929, p. 12). The heroic efforts of Abner Shimony and others to develop a Whiteheadian metaphysics of QM have led to frustration.

Whitehead advocated replacing the ordinary language categorial system of objects with properties by a system of actual entities and processes, on the grounds that the established system is a misleading guide to an ultimate ontology. However, our ordinary language categorial system is finely tuned to our presence and action in the lived world. The challenge is to recognize that material inferences rely on an implicit relative ontology and finding some way of using this language and going beyond its limits. Heidegger's *Sein und Zeit* also came out at the dawn of the QM era. Though it was concerned with the problematic of being, rather than advances in physics, Heidegger clearly recognized the problem of using language to speak beyond the limits of language and suggested a method of treating it. Heidegger is concerned with the forgetfulness of being manifested in the history of Western philosophy. This forgetfulness is rooted in the lived world and its linguistic expression.

We are beings living in a world of beings. Yet our immediate experience of them and of ourselves is not an experience of beings as beings. We experience the world as a toolbox and of ourselves as users. We are aware of ourselves in the context of the surrounding others. Any questioning of the being of things entails some vague pre-understanding of being. Without this, the question of being would not arise. But, how should we go about answering this question?

Heidegger insists that we begin by questioning ourselves as beings. How can we phrase such questions. If we speak of ourselves as subjects we smuggle in the subject/object duality. To speak of ourselves as persons presupposes complex structures of properties, rights, and obligations. How can we inquire into ourselves as beings while suspending the presuppositions implicit in the categories used? Heidegger was well aware of the role of categories, having written his dissertation on Duns Scotus's treatment of categories. To disengage the normal presuppositions of the language in use Heidegger refers to the being being questioned as *Dasein*, there-being. Dasein is questioned as a process, not as an entity. The route to an answer begins with a phenomenological analysis of Dasein as a being-in-the-world. The others are beings, but are not experienced as beings. The ontological inquiry into beings as beings must be preceded by an *ontic* inquiry into beings as experienced. This highlights the ontological difference between being and beings and necessitates a further fracturing of ordinary language. Dasein's self-understanding is existential, grounded in a way of existing in the world. The understanding his inquiry seeks is *existentiell*, an analysis that uncovers Dasein's individual way of being. This analysis must begin by destroying the history of ontology, or the attempts to base ontological inquiry on established categories. A new non-categorical non-philosophical terminology supplies a basis for speaking of things while suspending the presuppositions implicit in the language of things. Heidegger distinguishes the *Vorhandenes*, the present at hand, from the *Zuhandenes*, the ready at hand. We encounter a hammer as a tool ready at hand for pounding nails. When it breaks it ceases to be a tool and may become something present at hand, something to be contemplated as an individual object. We will skip the details and quote Heidegger's appraisal of the role of his methodology of questioning being.

> One can never carry on researches into the source and the possibility of the 'idea' of Being in general simply by means of the 'abstractions' of formal logic—that is without any secure horizon for question and answer. One must seek a way of casting light on the fundamental question of ontology, and that is the *way* one must *go*. Whether this is the *only* or even the right one at all, can be decided only *after one has gone along it*. .... Something like 'Being' has been disclosed in the understanding-of-Being which belongs to existent Dasein as a way in which it understands. Being has been disclosed in a preliminary way, though non-conceptually. (Heidegger 1962, p. 437)

With a couple of exceptions analytic philosophers do not treat the problems involved in speaking beyond the limits of language. I can only think of two exceptions. Wittgenstein concluded his *Tractatus* with the claim: "What we cannot speak about we must pass over in silence." Wilfrid Sellars clearly recognized this problem in his analysis of the manifest and scientific images. The scientific image must be interpreted, he insisted, through a direct correspondence with reality, not through the mediation of the phenomenological concepts of the manifests image, though this language supports the inquiry. This, however, is not much of a guide. Sellars relied on the Tractatus as a functional placeholder for the scientific image, and never treated QM as a guide to the scientific image. The standard methods of theory interpretation in contemporary philosophy of physics cannot even formulate the problem. Interpretation is regarded as a model-mediated relation between a mathematical structure and aspects of reality. Language as such plays no role in interpretation.

We are focusing on functioning physics, rather than reconstructed theories, as a basis for interpretation. An epistemological analysis is anchored in measurement situations and the dialog between experimenters and theoreticians. Both the system studied and the measuring apparatus, including the inquirer, are quantum systems. I accept quantum mechanics as the basic science of reality. Yet the methodology of inquiry conceals the quantum nature of both systems. The language of discourse and experimental analysis is classical, grounded in the core of ordinary language. How can we analyze the advances so that they reveal the underlying quantum ontology rather than the classical presuppositions we bring to the inquiry? How do we use this language while bracketing its functional presuppositions?

There are two bridges stretching from classical physics to the quantum realm, mathematical formulations and symmetry principles. The mathematical formulations were erected on classical scaffoldings and projected into the quantum realm. Symmetry principles were developed in classical and semiclassical physics as external principles. In QFT they were transformed into internal principles. As projectors, these two bridges transcend classical categories. Yet to use them as guides for uncovering the ontology of the quantum realm we must discuss them in meaningful language. The meaningfulness of this language is rooted in the lived world and its extension to classical physics. These considerations motivate the quixotic selection of material in the present chapter. We begin with an analysis of the role of language in particle-physics experiments. This brings out the indispensable role of in describing experiments, reporting results, and supporting material inferences. Next we consider the neglected quantum theory, quantum electrodynamics. It has been systematically neglected by philosophers because it does not have a consistent mathematical formulation as a foundation for interpretation. Yet, it provides the clearest illustration of the dual inference model. The development of QED hinged on working out acceptable interpretations for mathematical expressions. Series expansions led to an indefinitely large collection of terms. A few had a clear significance. Many were ambiguous: actual infinities, potential infinities multiplied by very small coefficients, terms that could not be accorded any coherent physical interpretation. The convergence of the series was never established. Consistency was developed by physical, rather than mathematical, arguments. The series expansions probably diverge at very high energies. However, at these energies other physical processes should become prominent, an issue that will be treated later when we consider effective theories. In spite of such difficulties, the development of QED led to the most precise correspondence between theory and experiment ever achieved. If we are using mathematical forms to probe the ontology of the quantum realm we have to focus on the forms that fit reality, rather than those that fit a priori norms of mathematical propriety. Finally, we will make a limited examination of QFT and the standard model, focusing on the role of symmetry principles.

What follows, accordingly, does not have a neat methodological ordering. We will be focusing on aspects of quantum physics that have an ontological significance beyond that allowed by the minimal approach considered in the last chapter. This will help set criteria that an acceptable interpretation of QM must meet. Ontology will be treated on two levels, whose differentiation presents terminological problems that almost guarantee misunderstanding. The first level involves the functional ontology of EOL. I will follow recent trends and refer to this level as 'relative ontology'.[3] Using 'theory' in a loose sense we can say that a relative ontology is a clarification of the explanatory inferences licensed by the categorial basis of a theory (Seibt 2002). In an analysis of ordinary language, or its extension in EOL, 'object' plays a foundational role in supporting inferences. We are developing a dual inference system. We will call entities 'public objects' when they play a presuppositional role in physical inferences. This puts 'object' in an inference supporting network of concepts. This is extended to theories in the sense that one can speak of the relative ontology of chemistry or molecular biology. Since this is analytic ontology linguistic clarification is in order. For our purposes we will distinguish the theoretical concept, '$particle_t$', from the semiclassical concept, '$particle_c$'. The classical concept 'particle', is a concept of a sharply localized object. This concept is at the center of a cluster of concepts that support inferential relations. A particle travels in a sharp space-time trajectory, can collide with other particles, can penetrate or recoil from other objects, and obeys the laws of energy and momentum conservation. Two quantum effects are added to the classical concept to constitute '$particle_c$'. A $particle_c$ may decay into other particles. The distinctively quantum notion of a superposition of states is replaced by the quasiclassical notion of oscillation of state. The concept, '$particle_t$' has two components. The quantum component stems from field quantization. A creation operator, $a^+(k)|0\rangle = |k\rangle$ produces a particle with momentum k, while an annihilation operator, $a(k)$ destroys it. The group-theory component, stemming from Wigner, asks what kind of entity corresponds to an irreducible representation of the Poincaré group. It is an object characterized by mass, spin, parity, and various charges.

We will refer to the depth level as 'the ontology of the quantum realm'. This ontology is what the mathematical formulations that have proved successful say the quantum realm is. The peculiar problem is that any attempt to put this in words effectively makes it a part of EOL, the language of classical physics. We are assuming that both the mathematical expressions and any linguistic formulations we attach to it are pointers using the resources of EOL while bracketing its presuppositions. To paraphrase Heidegger: Something like 'Ontology of the quantum realm' has been disclosed through the understanding of relative ontology as a classical ontology of something that is really a quantum system. Through mathematical formulations the

[3] These two levels roughly correspond to Bitbol's (chap. 5) distinction between ontology in a descending mode and ontology in an ascending mode.

ontology of the quantum realm has been disclosed in a preliminary way, though non-conceptually. This need not entail contradictions if we regard the language of physics as an idiontology, a systematization of properties, and accord 'object' a presuppositional role.

6.1 The Role of Quantum Experiments

Our earlier analysis of experiments that played an important role in atomic physics, such as the Franck-Hertz and Stern-Gerlach experiments, focused on the necessity of using classical terms to describe the experimental situation and report the results. That carries over to the particle experiments we will consider. However, there is one significant way in which these later particle experiments go beyond this minimal epistemological basis. This concerns the necessity of including virtual processes in the experimental analysis. To illustrate the way these different requirements are met we begin with two experiments that played a decisive role in establishing the standard model of particle physics. The first experiment we will consider involves the discovery of intermediate vector bosons. We can begin with a summary statement by one member of the experimental crew. "In 1983 the intermediate vector bosons W^+, W^-, and Z^o were discovered at CERN, the European Center for Particle Physics in Geneva, Switzerland" (Kernan 1986, p. 21). In the style of experimentalists she simply presents this as a discovery about nature. Yet a study of this experiment probably illustrates the symbiotic interrelation of theory and experiment more clearly than any prior experiment. The idea that the electromagnetic and weak forces are unified at very small distances was suggested by Schwinger and developed by Glashow, among others, in 1961. The operative assumption was that Fermi's account of beta-decay, and subsequent variations, should be interpreted as phenomenological theories. On a deeper level beta decay, and other weak interactions, should be explained through the exchange of Intermediate Vector Bosons (IVBs), two charged and one neutral. These, together with photons, should form a family mediating electroweak interactions.

This suffered from various difficulties. The masses of the IVBs had to be supplied 'by hand' rather than deduced. The theory predicted results, such as the breakdown of neutral kaons into two muons, which were not observed. The neutral currents associated with the Z^o particle were not observed. Finally, the theory did not seem to be renormalizable. This precluded calculations that could be compared with experimental results. A series of developments changed this, some theoretical, some experimental, and some practical decisions. The crucial theoretical developments were: the separate reformulations of the theory by Weinberg and Salam, who introduced the masses of the IVBs by the Higgs mechanism; the proofs by G. 'tHooft and B. Lee that electroweak theory is renormalizable; and Glashow's postulation of charmed quarks as a way of explaining why some predicted reactions were not observed. The practical decision was a different way of treating the neutron background so that processes that had not counted as evidence for weak neutral currents were interpreted as supporting evidence. A crucial experimental advance was the detection of charmed particles, or more precisely of hidden charm. The J/ψ meson, discovered independently and announced on the same day by Richter and Ting, was interpreted as the union of a charmed quark and a charmed anti-quark.

The detection of IVBs required a more complicated and sophisticated experimental setup than anything preceding it. The experimental advances involved turning a synchrotron into an anti-proton storage ring, stochastic cooling of particles in the ring, and two detection apparatuses that required the labor of 190 physicists. Protons and anti-protons were made to collide in a region surrounded by two calorimeters, each consisting of alternating layers of scintillators and absorbers. The inner electromagnetic calorimeter absorbs electrons and photons. The outer hadronic calorimeter absorbs strongly acting particles. A uniform magnetic field allows the measurement of particles' momentum. It was anticipated that the IVBs would be so short-lived that any formed would break up into either a quark-anti-quark pair or a lepton anti-lepton pair before entering the detector. The quark pairs would form into π and heavier mesons. The experimental trick is to describe all the primary and secondary collision products and their behavior and then set a nested series of triggers that would select only the events that could be interpreted as decay products of a W or Z boson in the leptonic channel. Decay in the strong mode leads to π mesons that are difficult to separate from π mesons due to other collisions. Then there must be fast enough electronic processing so that the implementation of these criteria would activate the event recording mechanisms. Setting up the experiment and recording its results relies heavily on a network of informal inferences.

A crucial trigger for detecting the decay of charged IVBs was the search for high momentum electrons with no opposing jets. These could be interpreted as parts of electron-neutrino pairs. In this way the original run searching for charged IVBs selected five out of one billion events. The search for the Z^o used different triggers and also selected only five events. The selected events were seen through a multi-colored three-dimensional computer simulation of the events based on processing the information from the calorimeters and the magnetic spectrometers. Kernan summarizes the results: "The most dramatic confirmation that we were indeed seeing the carrier of the weak force was the direct observation of parity nonconservation (an absence of mirror symmetry) in the $W \rightarrow e + \mu$ decay." (Ibid., p. 27)

Here, as in the Franck-Hertz experiment, the interpretation of the results depended on a nested series of presuppositions and implications. There were the assumptions, well established by this time, concerning protons, anti-protons, electrons, neutrinos, and mesons as well as their behavior in electrical and magnetic fields and in collisions, and their decay products. A further assumption that played a crucial role in selecting and interpreting the results was that baryons are composed of quarks bound together by gluons. Mathematical theories are used in a supplementary way. However, the overall framework for inferences comes, not from formal theories, but from the inferential role of particle trajectories. As discussed in the last chapter, this depends on establishing an experimental environment in which the particle cluster of concepts can be consistently used. If any of the crucial trajectories involved passing a particle through a suitable crystal, then the inferential system clustering around the particle concept would not be reliable. Here the retroactive realism, to use Pickering's term, of functioning physics plays a crucial role. One does not treat the apparatus as a collection of observable objects and the protons, quarks, and leptons as theoretical entities. The development, deployment, and interpretation of the IVB experiments relied on a coherent detailed descriptive account of the behavior of the particles, of the apparatus, and of the intervention of various experimenters. All were simply accepted as public objects existing and interacting in precisely specifiable ways. The results were 'seen' through computer simulated three-dimensional representations.

From a philosopher's perspective all the crucial experimental 'observations' in this experiment are inferences. The crucial inferences concern unobservable processes. The high-energy collision of protons and anti-protons produces W and Z particles that immediately decay in the ways indicated. This fits the inferential network, because it relies on describing trajectories. However, inferences based on unobservable trajectories clearly go beyond the minimal epistemological basis of measurements, whether reported in ordinary language or QIT form.

Like the search for IVBs, the search for the Ω^- illustrated a new relation between experimenters and theoreticians. During the late 1950s and early 1960s experimenters kept discovering new particles and resonance states that supplied challenges for theoreticians. In the new style, theoreticians predicted singular events that could confirm a theory and supplied detailed directions on the type of collisions that might produce the desired results. After Gell-Mann made the prediction of the Ω^- particle at the 1962 CERN conference he had a discussion with Nicholas Samios, who directed high-energy experiments at Brookhaven. Gell-Mann wrote on a paper napkin the preferred production reaction.[4]

$$\begin{aligned}
K^- p &\longrightarrow \Omega^- K^+ (K^0)\\
\Omega^- &\rightarrow \Xi^0 \pi^-\\
\Xi^0 &\rightarrow \Lambda^0 \pi^0\\
\pi^0 &\rightarrow \gamma(\rightarrow e^+ e^-) + \gamma(\rightarrow e^+ e^-)\\
\Lambda^0 &\rightarrow p \pi^-
\end{aligned}$$

If a Ξ^- had been produced there would be a straight-line trajectory between its production point (the origin of the K^+ track) and the vertex of the $p \rightarrow \pi^-$ tracks. If an Ω^- is produced there is a slight displacement due to the intermediate steps indicated above. (See Samios 1997, p. 533). Samios and his Brookhaven team began the experimental search, developed thousands of bubble-chamber photographs, and even trained Long Island housewives to examine the photos for the slight deflection inferred in the production of an Ω^-. After 97,024 negative results they finally produced a photograph that Samios interpreted as the detection of the Ω^-. A copy of

[4] The experimental search is described in Samios (1997). The changed relation between experimenters and theoreticians is discussed in Pickering (1984, p. 16), and Galison (1987).

this photo adorns the paper cover of Johnson's 1999 biography of Gell-Mann. Here again, the experimental analysis hinges on using particle trajectories as an inferential basis and on the inclusion of unobservable trajectories.

6.2 QED and Virtual Processes

An excursion into virtual processes leads through an ontological swampland that most philosophers systematically avoid. If reasons were offered for this avoidance, three would be prominent. First, the interpretation of theories depends on properly formulated theories. QED, where the interpretation of virtual processes became a crucial issue, never achieved the status of a properly developed theory. Second, interpretation hinges on observable results, not unobservable virtual processes. Finally, Feynman diagrams emerged as the indispensable tool for treating virtual processes. In the development of QED Feynman's path-integral method paralleled Schwinger's method of canonical transformations. In further developments of QFT the Feynman method was followed because it left physicists free to choose whatever gauge they needed (See Kaku 1993, p. 298). Yet, any literal interpretation of Feynman diagrams seems untenable. We begin with the key problem treated in QED, the Lamb shift.

Figure 6.1 illustrates the Lamb shift. The Schrödinger equation for the hydrogen atom leads to the Bohr-Schrödinger atom where energy levels depend only on n. The Dirac solution removes the degeneracy of states with different j values. The Lamb shift indicated a slight displacement of energy levels and removed the degeneracy of the $2S_{1/2}$ and $2P_{1/2}$ energy levels. In terms of the energy diagram this is a correction to a correction. This accounts for the fantastic accuracy attributed to Lamb shift calculations. They effectively begin at the sixth decimal place.

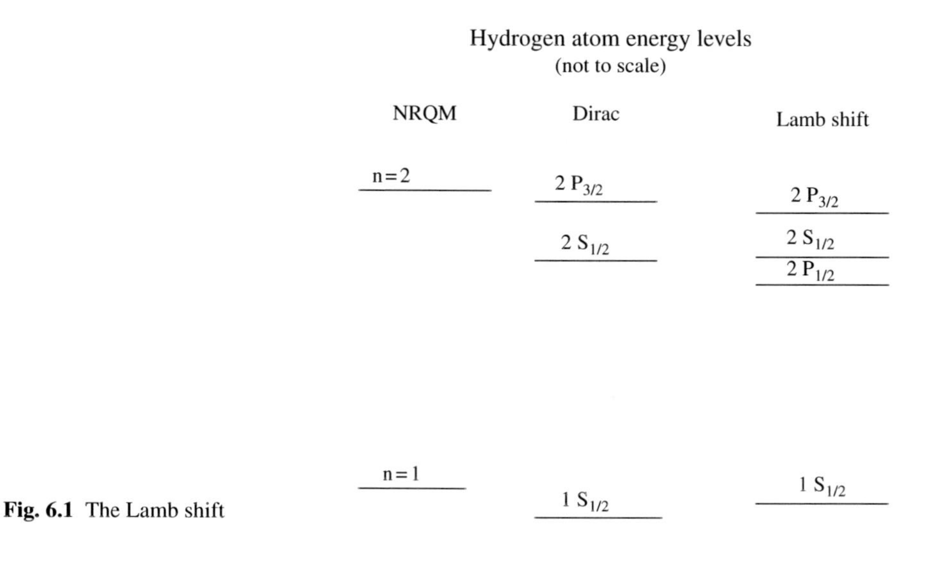

Fig. 6.1 The Lamb shift

The calculations involved have become part of the lore of modern physics. After the Shelter Island conference on 2–4 June, 1947, where Willis Lamb announced his experimental results, Hans Bethe made a hasty, essentially non-relativistic calculation on the train back to Ithaca. It agreed with Lamb's results remarkably well. Bethe communicated his results and precipitated a widespread recognition that a relativistic calculation was needed, as well as more precise experiments. A large number of people made very detailed calculations. After a preliminary clarification, I will indicate the three methods that achieved a dominant position.[5]

Two types of approximations are involved in these calculations. First, the successive approximations in the treatment of the virtual radiation field of the electron give an expansion in powers of α. Second, successive approximations in the treatment of the Coulomb field give an expansion in powers of $Z\alpha$. To keep these distinct, the Z is retained, even though it has a value of 1 for hydrogen. Schwinger totally dominated the early theoretical discussions. His program was to begin with a covariant relativistic formulation. Then he introduced the interaction representation and a series of canonical transformations to order, α^2 and higher orders. Once the individual terms were separated out in a covariant fashion, then they were calculated by whatever approximations seemed physically justified. Oppenheimer's famous appraisal indicates the difficulty people had in following Schwinger's long intricate calculations: "Others calculate to show how it is done; Julie calculates to show that only he can do it." The second path was Feynman's and the initiation of Feynman diagrams. The calculations were simpler, at least for Feynman, but the underlying assumption that a positron is an electron going backwards in time was one that most physicists found literally incredible. Schwinger and Feynman used to compare their results term by term, but neither understood the other's method. Finally, Freeman Dyson established the equivalence of the Schwinger and Feynman methods. Then he showed that a perturbation expansion of the S-matrix could lead to the Feynman terms and utilize Feynman diagrams without any explicit reliance on the Feynman ontology. This emerged as the canonical form of QED.

The point that chiefly concerns us is a methodological one. In the previous chapter we considered the need for a dual inference system and mixed reasoning. This was illustrated by some elementary examples showing how physical reasoning often guides the introduction and interpretation of the mathematics used. In the last chapter we illustrated Bohr's use of a dual-inference methodology. Prior to quantum mechanics he used physicalistic reasoning to arrangements of electrons in atoms that would explain the periodic table. Other physicists were amazed to learn that he achieved his results without reliance on a mathematical theory. In scattering accounts he use physicalistic reasoning to guess how electrons of different speeds would react with different types of atoms, Then he set up a mathematical formula to infer numerical values. His liquid drop and collective models of the nucleus were inferred from speculations about how neutrons and protons would behave

[5] A complete history of this development is given in Schweber (1994).

inside a nucleus. Here again, a mathematics was imposed. The Bohr consistency conditions were developed as a guide to using and interpreting the mathematical formalism of QM. In each case subsequent developments followed the familiar pattern. After physicists accepted the results of the physical speculation they were, through retroactive realism, regarded as facts. Then one could simply use the formulas.

The Lamb shift calculations probably supply the clearest example in physics of such mixed reasoning. Physicists had assumed that the $2S_{1/2}$ and $2P_{1/2}$ levels in the hydrogen atom should be slightly different. This motivated the Lamb-Retheford experiment. The problem was finding the mathematical formula that fit the observed difference. The three types of expansions just considered all led to an ordered series of correction terms. The calculation of each individual term was guided by the physical interpretation accorded the term. This was not the interpretation of a theory, but the physical interpretation of individual terms in an expansion that lacked mathematical justification.

To put some flesh on these bare bones it is necessary to review to review some background ideas. The first, weak divergence, is illustrated by integrals of the form, $\int_a^b dx/x = log(b/a)$. This diverges if $a = 0$. The lower limit of 0 generally results from treating electrons, or other particles, as point particles. If a is assigned a small value, corresponding to a finite size particle, and the term is part of a perturbation expansion with a coefficient of, e.g., α^4, then the overall term has a small value. A Fourier transform to a momentum space representation requires a corresponding high-energy cutoff for the momentum. In his original calculations Bethe used a value of mc for the cutoff, a reasonable but arbitrary choice. The second new idea is *r*enormalization, a term R. Serber introduced in 1936. H. Kramers developed this as an extension of Bohr's correspondence principle and presented a summary account at the Shelter Island conference. Kramers developed this on a classical basis, but this proved inadequate. The mass of the electron may be represented by the form, $m = m_0 + \delta m$. Here m_0 is the theoretical bare mass. Since the electron interacts with the radiation field there is an electromagnetic contribution to the mass, δm, a weakly divergent term. The term m stands for the measured mass. If the theoretical calculations are to fit measured results, then the appropriate formulas should use $m_0 + \delta m$, instead of assuming the equivalence between the bare and measured mass. In practice, calculations used the bare mass term and developed a weakly divergent electromagnetic mass contribution. The divergent terms were present in the expansions for both bound and free electrons. Subtracting one from the other yields a finite electromagnetic mass contribution which is added to the bare mass to give the measured mass. Thus, the corrections of QED could all be interpreted as mass renormalization and charge renormalization.

For a more systematic treatment of virtual processes, we can distinguish two types. The first kind involves virtual transitions, and is a feature of NRQM. The second involves the virtual emission and absorption of particles or photons. These are dubbed 'virtual' because they cannot be observed. However, virtual electrons are assigned the same properties as real electrons. We will consider examples of each kind. Bethe's original calculation (Bethe 1947) led to a formula for the energy shift,

$$W_n' = 8/3\pi \left(e^2/\hbar c\right)^3 \; RyZ^4/n^3 \, ln \frac{Z}{< E_n - E_m >_{Ave}}, \tag{6.1}$$

where R_y is the Rydberg constant. The interesting term here is $< E_n - E_m >_{Ave}$, the averaged energy for virtual transitions between energy levels. The assumption is that an electron in a particular state continually makes virtual transitions to higher states, including continuum states, and then a transition to the final state. This is a feature of quantum mechanics not QED. It was implicitly present from the beginning. As I showed is some detail elsewhere (MacKinnon 1977) Heisenberg's introduction of non-commuting variables was based on his use of the virtual oscillator model. His treatment of a transition from an initial to a final state included a summation of transitions to virtual intermediate states and from the intermediate state to the final state. An insistence on the proper temporal ordering of these virtual transitions introduced non-commuting variables. Bethe could make the initial calculation on the train because he had effectively memorized tables of virtual oscillator strengths. Subsequently, he secured the aid of a proto-computer to calculate the averaged virtual transition energies for the $2S_{1/2}$ and $2P_{1/2}$ states and later refined these initial calculations.

To get at the physical significance of virtual processes we will consider some basic Feynman diagrams for the Lamb shift.

In a review article E. Salpeter (1953) summarized the principal correction terms. There are terms due to: the shift of the $2S_{1/2}$ level; the shift of the $2P_{1/2}$ level; vacuum polarization, which must be calculated for each succeeding order; terms involving two virtual photons; terms for the effect of the electron's anomalous magnetic moment; corrections to include the finite mass of the proton; and corrections to include the structure of the proton, which was not yet known. There are also corrections introduced to compensate for approximations used in calculations. Figure 6.2 contains Feynman diagrams for two significant corrections. Figure 6.1 is the emission and absorption of a photon. Because of energy conservation, a free electron cannot simply emit a photon. Virtual emission and absorption involves uncertainty in energy and time that fits the Heisenberg uncertainty principle, and leaves the electron in the same final state. Higher order corrections may be introduced by having the electromagnetic field of the proton act once (1a) or twice (1b). Figure 6.2 is vacuum polarization, the creation and annihilation of an electron positron pair. This modifies the effective potential of the proton and is ultimately absorbed in charge renormalization.

Earlier we indicated that the physical significance accorded each correction term guides the mathematical formulation. The treatment of vacuum polarization supplies an apt illustration. The energy shift for the $2S_{1/2}$ state is:

$$\Delta E = (\alpha/3\pi) \left\{ ln \frac{m}{2k_0(2s)} + [\cdots] \right\} \left(n_0\, 0 \left| \frac{\nabla^2}{m^2} \, V \right| n_0\, 0 \right). \tag{6.2}$$

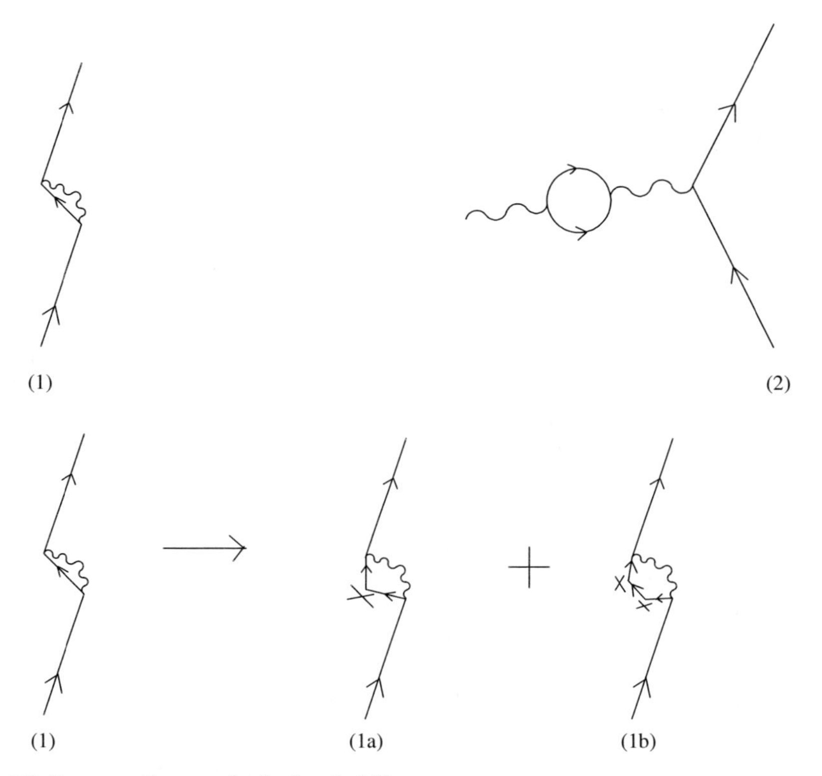

Fig. 6.2 Feynman diagrams for the Lamb shift

Here $[\cdots]$ stands for correction terms. All these correction terms are multiplied by the matrix element, $\left(n_0\, 0 \left| \frac{\nabla^2}{m^2}\, V \right| n_0\, 0\right)$, where $n_0\, 0$ means the wave function for the $n = 1, l = 0$ state. For a point proton the potential term is

$$\nabla^2\, V = -4\pi\rho(\mathbf{r}) = -4\pi e\delta^3(\mathbf{r}) \tag{6.3}$$

Since the Dirac wave equation supplies the basis for the splitting, one should use Dirac wave functions in Eq. (6.3). This, however, was not done. The Dirac wave function causes a divergence. The Schrödinger wave function does not cause a divergence. The reason for this can be seen by noting that the asymptotic form of the radial part of the Dirac $1S$ wave function near the origin has the approximate form,

$$|\psi_{dir}(0)|^2 \approx |\psi_{sch}(0)|^2 \left(\frac{a_0}{r}\right)^{\alpha^2} \approx \left(1 + \alpha^2\, ln\, \frac{a_0}{r}\right)\, |\psi_{sch}(0)|^2 \tag{6.4}$$

where α the fine-structure constant $\approx 1/137$ and a_0 is the radius of the first Bohr orbit. To avoid the divergence, calculations used the Schrödinger wave functions and estimated the corrections needed for the relativistic contribution. The divergence here stems from treating the proton as a point particle. If the $\delta^3(\mathbf{r})$ in Eq. (6.3) is replaced by a spatial distribution, then, as Eq. (6.4) indicates there is a relative correction of order α^2. The larger the spatial distribution the smaller the correction. When this integral was evaluated for the vacuum polarization contribution (Kroll and Lamb 1949) what was used for the charge distribution $\rho(\mathbf{r})$ was a form factor roughly characterized by the electron's Compton wavelength, $2{,}400 \times 10^{-13}$ cm. The convergence factors used in the self-energy calculations were smeared out over a similar volume. Thus, the integration is carried out for a volume approximately 40×10^6 larger than that given by the form factor for the proton. For this volume, the relativistic correction would be roughly 0.025 Mc./s. How is the use of such a relatively large volume justified? Partially by the physical interpretation of the processes involved. Consider the vacuum polarization term, which modifies the Coulomb potential of the proton and leads to charge renormalization. If this is attributed to the creation of an electron-positron pair followed by their mutual annihilation, then these processes require a spatial volume around the proton. Any precise specification would be misleading, and would have minimal significance for the correction term. However, the physical significance accorded virtual polarization suggests the limits of integration.

For the final illustration of the physical significance of the virtual processes represented in Feynman diagrams we will consider, and update, the second major contribution of QED, the calculation of the anomalous magnetic moment of the electron. In NRQM the magnetic moment of the electron has a value given by its formula, $\hbar^2/2m_e c$. If we assign this the value 1 then experiments gave a corrected value of 1.00118 ± 0.00003. Schwinger's laborious calculations led to a correction term of $\alpha/2\pi = 0.00116$. This spectacular success was followed by a spectacular failure in calculating the magnetic moment of the proton. Experiment and theory differed by a factor of 3. Much later it was followed by a high, but perhaps limited, success in calculating the anomalous magnetic moment of the muon.

Figure 6.3 indicates the first order Feynman diagrams used in these calculations. Diagrams (a)–(d) give the first order corrections that Schwinger calculated by his own methods. Subsequently calculations were carried out to second and third order in α. This gave an agreement to nine decimal places, making it the most accurate calculation on record. The total failure of the similar proton calculation was interpreted as indicating that the proton has a structure. The muon is in the same family as the electron and is considered an elementary particle. It should have the same Feynman diagrams, though with a different mass in the calculations. Initial measurements showed strong agreement between theory and prediction. Because of the extreme importance attached to a precise fit between theory and experiment-more detailed experimental results were sought. A Phys. Rev. letter, signed by 70 researchers (Bennett et al. 2002), gave an experimental value accurate to 8 places. Extensive theoretical calculations led to a result that indicates a discrepancy of the order of 10^{-9} between the measured and predicted values. Slight as this is, it seems to indicate a need to go beyond QED (considered as a part of the standard model). The expectation is that a calculation of all the Feynman diagrams will yield the

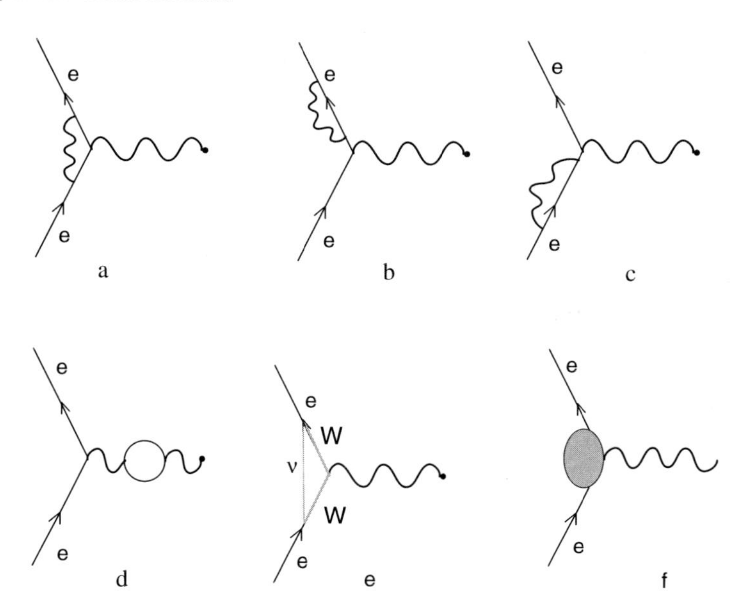

Fig. 6.3 Anomalous magnetic moment calculations

correct results. I have indicated this schematically in diagram (e) and (f). When QED is treated as a part of electroweak unification, then (e) represents an allowed virtual process. Because of the large mass of the W meson this process should make a negligible contribution. We will return to this point in considering effective theories. In (f) the shaded area indicates presently unknown higher energy interactions. The operative assumption is that these are not renormalizable but that they make only a minimal contribution due to the effective weakness of the coupling constants. However, if some such processes make a measurable contribution, then they could point to physics beyond the standard model, such as the virtual creation of supersymmetric particles, or an indication of some internal structure in leptons (See Calmet et al. 2001). This illustrates the way Feynman graphs of virtual processes guide a physical interpretation.

The physical interpretation accorded these terms emerged from protracted interactions between theoreticians and experimenters. Because of the intense competition between followers of the different methodologies, any proposed interpretation of an individual term in the expansion was subjected to severe criticism. The consensus that emerged and is now part of standard physics exemplifies the way functioning physics treats the interpretation of selected parts of quantum mechanics. There is still the problem of the interpretation to be accorded Feynman diagrams. As an internal question, this is clear an unambiguous. There are well-known rules for developing these diagrams and assigning mathematical expressions to every element in a diagram. The individual mathematical expressions are correlated with distinct quantum processes. Within the formalism virtual processes are qualitatively on an equal footing with real processes. Quantitatively, there are vastly more virtual processes than real processes. The external question is not so clear. Do these diagrams represent processes that really happen in the vacuum? For what it is worth, I will present my answer to this external question. The Feynman diagrams are clear explicit representations of particles traveling in trajectories. This is the way *particles*$_c$ behave. Virtual *particles*$_c$ traveling in trajectories have the same status as real *particles*$_c$ traveling in trajectories. This diagrammatic and linguistic representation is a part of quasi-classical physics, the physical component of the dual inference system in use. Can these diagrams be accorded any realistic significance? My view is that as descriptive accounts they have a status quite similar to the observable processes interpreted as particle trajectories. They support context-dependent inferences. In this regard they are like the elliptical orbits of the Bohr-Sommerfeld atomic theory. Even after these were denied any precise descriptive significance, they still played a valuable role in supporting calculations and in the models used in atomic physics and chemistry.

Do the details of these mathematical formulations correspond to details of reality? Here we should let Dyson have the final word:

> I always felt it was a miracle that electrons actually behaved the way the theory said. To me it was always an amazing experimental fact that the perturbation series was somehow real, and everything the perturbation series said turned our right. I never felt that we really understood the theory in the philosophical sense-where by understand I mean having a well defined and consistent mathematical scheme. [Nonetheless] I always felt it was obviously true, true even with a big T. Truth to me means agreeing with the experiments . . . For a theory to be true it has to describe accurately what really happens in the experiments. (cited from Schweber 1994, p. 568)

6.3 The Standard Model

The standard model is the outcome of the Lagrangian field theory approach and of unprecedented close cooperation between theoreticians and experimenters. It is a difficult topic, but one well covered in many texts.[6] When judged by the standards set by either Schwinger's or algebraic quantum field theory, the standard model has some serious shortcomings. Schwinger insisted that the ultimate theory should explain the coupling constants. The standard model does not do this. It is now widely regarded, not as the ultimate theory, but as a relatively low-energy approximation to a deeper theory at a much higher energy level. Also, where Schwinger relied on a step-by-step advance and restricted postulation to the minimum necessary for advance, both components of the standard model rely on postulation. Algebraic and

---

[6] My account is chiefly based on the texts by Kaku (1993) and Weinberg (1995), the historical accounts given in Hoddeson et al. (1997) and Cao (1998, Part III), and on various articles. For philosophical appraisals of QFT see Auyang (1995), Teller (1995), Brown and Harré (1988), the symposia papers in Wayne (1996), and MacKinnon (2008).

axiomatic QFT prize the virtues of mathematical rigor and consistency.[7] A strong defense of the axiomatic approach may be found in Fraser (2009). The algebraic methodology is also relatively clear on the epistemology/ontology circle. Instead of beginning with specific models one begins with the general principles of quantum mechanics and relativity. In this context 'field' does not have any basic ontological significance. Measured fields correspond to smeared out averages over a spatio-temporal volume. The mathematical representation of fields should be regarded as operator valued distributions, rather than physical values at a point. The basic things treated are observables, rather than fields. In this context 'observable' is only distantly related to a process of observing or measuring. For each compact space-time region one constructs a local algebra (a von Neumann algebra) of observables. Relativity and QM enter with the assumption that operators in causally separate regions commute. Then the physical significance of the theory is contained in the net of maps from space-time regions to algebras. This formulation supplies a basis for defining numerical values proper to detectors and coincidence counters. The major shortcoming of this approach is that it treats free particles, and only interactions that can be generated from free particle states, but not or the standard model. There is also a mathematical difficulty that admits of various solutions. The point of departure for the algebraic approach is the Stone-von Neumann theorem. Quantum commutation relations generate the structure of a Hilbert space in a way that is unique up to a unitary transformation. This does not hold for the infinite dimensional representations that QFT uses. Nevertheless, defenders of AQFT hope that these difficulties can be overcome without compromising the integrity of the theory. This is not an impediment if one switches the epistemological base from 'observables' to 'measurables'. In any experimental situation measurements record ranges. Only a finite number of measurements are required to cover any experimental situation. In principle, a finite Hilbert space is adequate to represent this. Then the switch to an infinite-dimensional Hilbert space is an idealization that lacks physical significance. The chief complaint that members of the AQFT community lodge against mainstream QFT tradition is that the theory is mathematically unsound.

In terms of scope and precision, the standard model is the most successful theory in the history of science. It is capable, in principle, of handling the particles, and all the basic interactions, involved in strong, weak, and electromagnetic interactions. For our purposes, we do not need the technical details of this model, which I have not mastered, but only two basic aspects. The first is ontological. The family of particles proper to the standard model is a permanent part of physics. As a theory of matter, it is essentially immune to revision by any foreseeable advances in physics.[8] The second point is epistemological. The success of the standard model stems from its fusion of basic principles of: quantum mechanics, relativity, and

---

[7] My evaluation of this approach is chiefly based on Haag (1992), Redhead and Wagner (1998), Buchholz and Haag (1999), and Buchholz (2000). The axioms used are given in Redhead and Wagner, p. 1, and are analyzed in Haag (1992, Section II).

[8] This appraisal stems from evaluations given by Wilczek (1999, 2000, 2002a, 2002b, 2004a, 2004b).

local symmetry. The standard model does not accord with the norms proper to a formal theory. This is not a shortcoming. It is an inevitable consequence of the way the theory is constructed. Physical arguments play an essential role in setting up the mathematical formulations. This may be unacceptable in an interpretative framework in which a theory is developed as an *uninterpreted* mathematical formalism and then given a physical interpretation. But, it is the normal practice of physicists and fits the dual inference model of explanation. For this reason we will focus on aspects of the standard model that diverge from philosophical anticipations. I will fill in the minimal background needed for a non-technical account of two key aspects, local gauge symmetry and the renormalization program. After presenting the basic ideas in a qualitative way, I will add a bit of flesh to the bare bones.

We have already considered external symmetries proper to the Poincaré group. These include symmetries under rotation. The trick here is to extend rotational symmetry from external to internal degrees of freedom. Heisenberg introduced isotopic spin space as a simple device for treating protons and neutrons as different states of nucleons. Talk of the z-component of a nucleon's isotopic spin was treated as a metaphorical extension of particle spin. Thus, the different spin components of atomic electrons with the same n and l quantum numbers may be degenerate until a magnetic field is turned on to cause an energy splitting. The fiction was that the energy differences between protons and neutrons would vanish if the electrical field were shut off. This is fictional, because the proton's electric field is internal. There was no a priori reason to expect that rotational symmetry could be applied to isotopic spin space. Local gauge symmetry is just such an extension, for isotopic spin space in electroweak unification, and for color space in quantum chromodynamics. The fact that it works so well indicates that the extension of a mathematical structure into the quantum realm reveals something of the ontology of that realm. We have already considered renormalization in QED. The general program of renormalization supports the concept of *effective theories*. The basic idea is that theories in atomic and particle physics apply to ranges characterized by values of energy or distance. The correlation is important. Intuitively one might think that high energies would overwhelm smaller energies, as they do with competing radio stations or track teams. In the theories we are considering higher energy interactions are correlated with smaller interaction lengths. The basic ideas may be illustrated by Fig. 6.4. One point about the table should be noted. The standard custom, which we will generally follow is to use natural units ($\hbar = 1,\ c = 1$), so that energy and momentum are both measured in mass units.

The boxes correspond in a rough fashion to specialized fields that rely on different presuppositions. A theory at a particular level accepts input parameters that should be explained at a deeper level. Thus chemistry and atomic physics overlap at the energy level proper to valence electrons. Chemists generally presuppose atoms and molecules with a mass, spatial configuration, and special properties for valence electrons, e.g., to explain covalent bonding. The energy states of inner electrons, for multi-electron atoms, and the nucleus are not considered. Consider the NRQM treatment of the hydrogen atom. This takes as input parameters the mass and charge

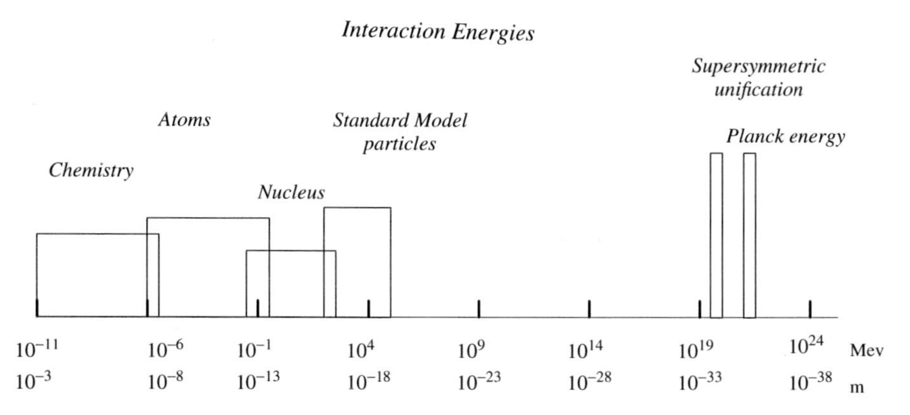

Fig. 6.4 Interaction Energies

of the proton. For more detailed calculations, e.g. of hyperfine splitting, the magnetic moment of the proton is used. However, it is never necessary to consider the quark-gluon structure of the proton. Chemistry and atomic physics remain effective theories on their proper energy levels, totally shielded from modification due to advances at a much deeper level, e.g., string theory.

We have been considering theories proper to adjoining or overlapping energy levels. The situation is much clearer for widely separated levels. A study of ocean currents treats water as a continuous fluid. An effective theory of fluid flow is far removed from its ultimate base. It follows that it supplies no information about its ultimate constituents. The equations of fluid flow can be adapted to fit collections of water molecules, the flow of money in banking systems, the migrations of peoples over time, or information in a computer. In terms of orders of magnitude the separation of the standard model from the length characterizing string theory and GUT theories is greater than that between ocean currents and hydrogen molecules. The general principle is that a theory can be regarded as an effective theory up to a proper high-energy cutoff. Above that level new physics is needed. Beyond 10^6 Mev (or a few Tev) new physics may explain the masses of the W and Z bosons, but would not introduce new standard model particles. This gives strong support to the contention that the standard model is essentially immune to revision by further advances in physics. Such advances should explain the parameters the standard model assumes, especially coupling constants. They will not dissolve the table of particles and properties of the standard model.

To get at the crucial feature of gauge field theory in its simplest form, we will begin with a classical formulation of electrodynamics and the non-relativistic Schrödinger equation. The electrical field strength, **E**, and the magnetic flux density, **B**, may be expressed in terms of the vector potential, **A**, and the scalar potential, V, as

$$\mathbf{E} = -\nabla V - \partial \mathbf{A}/\partial t, \ \ \mathbf{B} = \nabla\, x\, \mathbf{A} \tag{6.5}$$

If we introduce a gauge, f such that $\mathbf{A}' = \mathbf{A} - \nabla f$, $V' = V + \partial f/\partial t$, then the values of **E** and **B** remain the same. This is a global gauge transformation, applying to all of space.

The phase of a wave function has a peculiar status. The phase is not measurable and does not enter into the calculation of probability densities, since $P = \int \psi^* \psi d\mathbf{x}$. However, phase differences have empirical consequences. Consider the *local* phase transformation,

$$\psi(\mathbf{x}, t)' = e^{-\imath\theta(\mathbf{x})}\, \psi(\mathbf{x}, t). \tag{6.6}$$

This is local in the sense that the value of $\theta(\mathbf{x})$ can be different for different positions. To relate this to non-relativistic quantum mechanics as well as electrodynamics, use the Schrödinger wave equation,

$$-\, 1/2m\, \nabla^2\, \psi(\mathbf{x}, t) = \imath \partial \psi(\mathbf{x}, t)/\partial t. \tag{6.7}$$

If in Eq. (6.7) we substitute $\psi(\mathbf{x}, t)'$ then we get

$$-\, 1/2m \left[\nabla^2(-\imath \nabla\theta\right]\psi e^{-\imath\theta(\mathbf{x},t)}\big] = \imath \left[\partial/\partial t - \imath \partial\, \theta(\mathbf{x}, t)/\partial\, t\right] e^{-\imath\theta(\mathbf{x})}. \tag{6.8}$$

This definitely does not have the same form as Eq. (6.7). However, we may supplement Eq. (6.6) with the local electromagnetic gauge transformation,

$$\mathbf{A}' = \mathbf{A} + 1/q\, \nabla\theta(\mathbf{x}), \;\; V' = V - 1/q\, \partial\, \theta(\mathbf{x})/\partial\, t. \tag{6.9}$$

For a particle of charge, q, the conjugate momentum is $\mathbf{p} \rightarrow \mathbf{p} - q\,\mathbf{A}(\mathbf{x})$. The standard quantization procedure is to replace $\mathbf{p}$ by $-\imath\, \nabla$. So,

$$-\, \imath\, \nabla \rightarrow -\imath[\nabla - \imath q \mathbf{A}(\mathbf{x})] \equiv -\imath \mathbf{D}. \tag{6.10}$$

D is the covariant derivative. When this is substituted for the ordinary derivative and the two sets of phase changes are introduced, then the transformed Schrödinger equation has the same form as the untransformed equation. This covariant derivate gives the coupling between a classical electromagnetic field and a charged particle. The standard model couples the Weinberg-Salam model of electroweak interactions to the SU(3) color model of strong interactions. Here, the covariant derivative can be written in symbolic form as:

$$\partial/\partial x_\mu \rightarrow D/Dx_\mu = \partial/\partial x_\mu - \imath g/\hbar\, L_k\, A^k_\mu, \tag{6.11}$$

where L_k are the generators of the gauge group, the A^k_μ are the gauge fields, and g is the matching coupling constant. This covers all basic interactions: hardron-hardron, hardron-lepton, and lepton-lepton, and fits all empirical results until the recent (2002) discovery that neutrinos have mass (not allowed by the standard model).

All couplings are effectively represented by covariant derivatives. The values of the couplings are adjusted to fit the empirical data. There is no plausible way to reach these couplings by a series of anabatic steps from a measurement basis. Here again, the fact that these mathematical forms work reveals something about the ontology of the quantum realm. One further aspect should be treated, the relation between symmetries and invariances. We begin with some general relations between measurements, symmetry operations, and conservation laws proper to classical physics and NRQM. If a system is invariant under linear displacement then linear momentum is conserved. This is the momentum proper to an inertial system specified by the measuring apparatus. Absolute momentum is unmeasurable. Similarly invariance under: time translation implies energy conservation; rotation implies angular momentum conservation. These are external symmetries included in the Poincaré group. Local gauge theory extends this general approach in two ways. First, it adapts the machinery developed for external symmetries to internal symmetries. Second, it goes beyond a measurement basis by postulating that symmetry principles apply to functional spaces characterizing internal degrees of freedom. I will indicate how the principles proper to rotational symmetry and quantization are extended to internal symmetries. Rotations in a plane involve the group of transformations that keeps $r^2 = x^2 + y^2$ invariant. They can be represented by the one-parameter continuous group, $r' = re^{\iota\theta}$, or through the matrix representation $r' = R(\theta)r$ where

$$r = \begin{pmatrix} x \\ y \end{pmatrix}, \; r' = \begin{pmatrix} x' \\ y' \end{pmatrix}, \quad R(\theta) = \begin{pmatrix} \cos\theta & \sin\theta \\ -\sin\theta & \cos\theta \end{pmatrix}$$

This is the group $SO(2)$. It is the rotational part of the group $O(2)$, which includes inversions. $SO(2)$ is an irreducible representation. For a reducible representation describe the same system in 3-dimensional space as a rotation about the z-axis. The rotation matrix is

$$R(\theta) = \begin{pmatrix} \cos\theta & \sin\theta & 0 \\ -\sin\theta & \cos\theta & 0 \\ 0 & 0 & 1 \end{pmatrix} \tag{6.12}$$

This can be reduced to 2 components, a rotation matrix in the $x - y$ plane and a 'do nothing' action along z. Any matrix that has the form

$$T = \begin{pmatrix} T^{q_1} & 0 & 0 & \cdots & 0 \\ 0 & T^{q_2} & 0 & \cdots & 0 \\ \cdots & \cdots & \cdots & \cdots & \cdots \\ 0 & 0 & 0 & 0 & T^{q_n} \end{pmatrix}$$

where T^{q_i} is a rectangular matrix in a space of q_i dimensions and the '0's stand for block matrices of 0's, is reducible to the rectangular blocks.

To relate this to symmetry we note that we can generate the $SO(2)$ group from its infinitesimal elements. For infinitesimal rotations $\cos\theta = 1$, $\sin\theta = \theta$, and

$$R(\theta) = \lim_{n\to\infty}\left(1 + \frac{\imath\,\theta}{n}\,\tau_3\right)^n = e^{\imath\,\theta\,\tau_3}, \qquad \text{where } \tau_3 = \begin{pmatrix} 0 & 1 \\ -1 & 0 \end{pmatrix}$$

Instead of limited rotations about the z-axis we may consider general 3-dimensional rotations. This leads to the group $O(3)$ and the sub-group of continuous rotations, $SO(3)$. The infinitesimal generator for this group is

$$R(\theta) = \sum_{j=1}^{3} \lim_{n\to\infty}\left(1 + \frac{\imath\,\theta^j}{n}\,\tau^j\right)^n = e^{\sum_{j=1}^{3}\theta^j\,\tau^j},$$

A Lie group is characterized by topological as well as group properties. In a Lie group there exists a neighborhood of the identity element, e, in which the inverse of any element and the product of any two elements are continuous and continuously differentiable functions of the parameters. The key theorem that Lie developed is that the infinitesimal generators of a Lie group obey the commutation relations

$$[I_j,\, I_k] = \sum_{l=1}^{r} c^l_{jk}\, I_l, \tag{6.13}$$

These commutators constitute a *Lie algebra*, where c^l_{jk} are the structure constants. Because the parameters are all continuous and continuously differentiable, one can work out the properties of infinitesimal group elements and then integrate to get finite group elements. For a compact Lie group, or one with a finite volume, any representation is equivalent to a representation by unitary operators. So, only they need be considered. If we follow the Gell-Mann specification of the τ operators

$$\tau^1 = -\imath \begin{pmatrix} 0 & 0 & 0 \\ 0 & 0 & 1 \\ 0 & -1 & 0 \end{pmatrix} \quad \tau^2 = -\imath \begin{pmatrix} 0 & 0 & -1 \\ 0 & 0 & 0 \\ 1 & 0 & 0 \end{pmatrix} \quad \tau^3 = -\imath \begin{pmatrix} 0 & 1 & 0 \\ -1 & 0 & 0 \\ 0 & 0 & 0 \end{pmatrix}$$

These antisymmetric matrices obey the anti-commutation relations

$$[\tau^i,\, \tau^j] = \imath\, \varepsilon^{ijk}\, \tau^k, \tag{6.14}$$

where $\varepsilon^{ijk} = 1$ for ε^{123} and is completely anti-symmetric. Equation (6.14) is a simple example of a 3 parameter Lie algebra with the structure constants ε^{ijk}.

To facilitate the transition from external to internal symmetries in a complex Hilbert space we should consider how external symmetries are represented in a complex space. Consider the complex object, $w = x + \imath y$. Rotations in the complex plane are given by the group of unitary matrices, $U(1)$, where $U' = U(\theta)\, w = e^{\imath\,\theta} w$. This gives a correspondence between rotation operators defined in real space and unitary operators defined in a complex space.

$$SO(2) \sim U(1)$$

This may be extended to get the relation between $SO(3)$ and the special unitary group $SU(2)$. $SU(2)$ is represented by the set of unitary 2×2 matrices with unitary determinant Any element of this group can be written in the form, $U = e^{\imath\theta^i\sigma^i/2}$, where the Pauli matrices, σ^i satisfy the relation

$$\left[\sigma^i/2, \sigma^j/2\right] = \imath\, \varepsilon^{ijk}\, \sigma^k/2 \tag{6.15}$$

Since Eqs. (6.14) and (6.15) define the same Lie algebra we have the correspondence

$$SO(3) \sim SU(2).$$

To see this correspondence in a particular case we note that the rotations, $\mathbf{x}' = \mathbf{O(3)} \cdot \mathbf{x}$ given by Eq. (6.12) could also be given by $h' = U\, h\, U^{-1}$, where

$$h(\mathbf{x}) = \sigma \cdot \mathbf{x} = \begin{pmatrix} z & x - \imath y \\ x + \imath y & z \end{pmatrix}$$

Different $SU(n)$ groups supply the basic tools used for symmetry operations in the standard model. The ontology is determined by the combination of three principles: quantum mechanics, relativity, and internal symmetry. From an a priori perspective the idea that internal symmetries should play such a role is doubly implausible. The first implausibility is an old one. When Sommerfeld introduced the j and m quantum numbers they were bookkeeping devices for spectral lines. Neither he, nor apparently anyone else with the possible exception of Pauli, believed in the reality of space quantization. Then Stern and Gerlach demonstrated that space quantization is real. Consider one aspect of this that gets adapted to $SU(n)$ internal symmetries. Consider joining together 2 spin-1 systems. Each system, considered separately has 3 components along an axis, set by the magnetic field. In quantum numbers, m takes $2j + 1$ values. What of the composite? Diagrams bring out the m components of the 3 different spin state the composite can have.

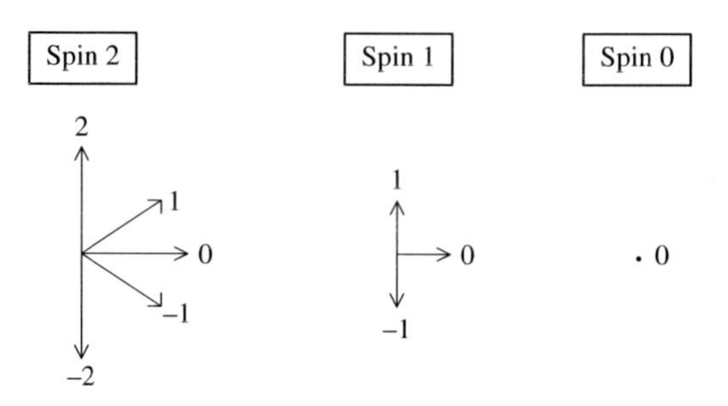

This can be formulated as a tensor decomposition rule

$$3 \otimes 3 \rightarrow 5 \oplus 3 \oplus 1 \tag{6.16}$$

The second a priori implausibility is the extension of rules like Eq. (6.16) to inner symmetries. For external symmetries, the preferred direction is picked out by some physical means, such as the direction of a magnetic field or aspects of a measuring apparatus. How does one set a preferred direction in isospin space or color space? The assumption that compositions in these internal symmetry spaces admit of tensor decomposition and that the components can be accorded physical significance has a profound ontological significance. This can best be seen by beginning with the example that led to the use of Lie algebras in particle physics. The quantum numbers of Q (charge), S (strangeness), and B (baryon number), and I_3 (the projection of isospin along a preferred axis a re related by the formula $Q = I_3 + (S + B)/2$. Gell-Mann introduced the quark hypothesis with the assumptions that baryons are composed of three quarks, mesons are composed of a quark and an anti-quark, that up quarks have a charge of 2/3, down and strange quarks have a charge of $-1/3$, that up and down quarks have isospin values of 1/2, while strange quarks have an isospin value of 0. He introduced the further assumption that the component of isospin along an arbitrary z-axis is $+1/2$ for the up quark and $-1/2$ for the down quark. Then the allowed composition of mesons and baryons led to the tensor decomposition

$$\begin{array}{lll} \text{meson} & q\bar{q}: & \mathbf{3} \otimes \mathbf{3} \rightarrow \mathbf{8} \oplus \mathbf{1} \\ \text{baryon} & qqq: & \mathbf{3} \otimes \mathbf{3} \otimes \mathbf{3} \rightarrow \mathbf{10} \oplus \mathbf{8} \oplus \mathbf{8} \oplus \mathbf{1} \end{array}$$

The known mesons and baryons fit into this systematization. The decuplet had only 9 members, rather than the 10, given by the tensor decomposition. How could symmetry in an abstract isospace supply a basis for predicting a new particle? This hinges on a physical assumption. On a deep level, below the symmetry breaking that accounts for the mass differences, there is only one system. Then, all the allowed states of this system should be realizable, at least in principle. Gell-Mann used the decuplet decomposition for baryons with spin, $j = 3/2$, plus the Gell-Mann Okubo mass relation to predict the existence of a new particle, the Ω^- with an isospin value of -2, a strangeness of -3, and a mass of 1675 Mev. This was in 1962, prior to the 1964 introduction of quarks and based on the phenomenological eightfold way. We have already considered the experimental confirmation of this prediction.

The prediction of particles was based on according internal symmetry principles an ontological significance. Many physicists, including Gell-Mann and Feynman, were originally reluctant to accept quarks as real. In addition to the novelty of fractionally charged particles there were two basic difficulties. Quarks were not observed and, as fermions, they violated the Pauli exclusion principle. The introduction of color solved the statistics problem and led to quantum chromodynamics (QCD) and the standard model. The proof that non-Abelian gauge theories have the unique property of asymptotic freedom explained why individual quarks could not be observed. The evidence that convinced the scientific community of the reality of quarks stemmed from the Stanford-MIT experiments scattering high-energy electrons off protons. The experimental setup separated elastic, inelastic, and deep inelastic components, while the theoreticians tried to fit the results into a model of the proton as a point mass surrounded by virtual mesons. The deep inelastic component did not fit. In a famous analysis Feynman showed that this component could be explained as elastic scattering from parts of the proton.[9] The parts of his parton model were quickly identified with quarks and gluons. The 1974 discovery of the J/ψ particle led to the charmed quark. When the bottom quark was discovered, symmetry suggested that there must be a corresponding top quark and guided the search that eventually succeeded. Schwinger's anabatic QFT and AQFT concur on one point that stems from the epistemological primacy accorded measurements. The ongoing process of explaining composites in terms of ever more elementary units should not terminate in particles, but in some more fundamental entities postulated by quantum field theory. The success of the quark hypothesis and the standard model should be taken as conclusive proof that the progress of quantum physics had gone beyond the limits proper to the strict measurement interpretation and its anabatic and algebraic extensions.

The prediction and interpretation of these particles was based on symmetry principles, not on formal theories. This raises the question of the status of the standard model and its two components, electroweak unification and quantum chromodynamics, as theories. The theorists involved in the development of LQFT were not trying to meet the criteria proper to the formal conception of theories. Yet, they did have criteria of acceptability. Paramount among these criteria was the requirement that a theory must supply a basis for calculation and prediction.

6.3.1 Renormalization and Effective Theories

Renormalization involves complex calculations, which we will skip. It also draws a line separating the formalists from the pragmatists. This is the focus of our concern. For a formalist, a theory should be developed, or at least be recast, as an uninterpreted mathematical formalism that is given a physical interpretation. Using 'pragmatist' loosely for want of a better term, a pragmatist regards a functioning theory as an inferential tool not an uninterpreted formalism. A crucial difference between the two perspectives concerns justification. If the mathematical formulation is treated as an uninterpreted formalism, then the inferential process must have a mathematical

[9] Feynman (1974) has a non-technical summary of these calculations. Feynman replaced electrons scattered off stationary protons by high-energy collisions between protons and electrons moving in opposite directions at relativistic velocities. In this framework the transverse movements of the proton are negligible and the momentum distribution of the backscattered electrons gives the distribution of the charged parts of the proton.

justification. A pragmatist is willing to rely on physical justification for key steps. This fits the dual inference model of scientific explanation, but not the formalist conception of theories. Neither component of the standard model, electroweak unification or quantum chromodynamics, meets basic standards for mathematical rigor. They rely on series expansions that have not been shown to meet the standards of Cauchy convergence, and probably do not converge. In a broader context the term 'pragmatist' might be replaced by 'theory learner'. Creators of original theories generally begin theory construction with a priori norms of what a theory should be and later learn that the norms must be modified or abandoned to construct an adequate theory in a new domain. As was shown in earlier chapters Newton thought of theory of mechanics as a three-stage process: mathematics, concerned primarily with the formulation of force laws and their consequences; physics, concerned with the application of these laws to terrestrial and celestial data; and philosophy, concerned with an explanation through causes. He never succeeded in developing the third state. Subsequent developments showed that a science of mechanics could flourish without it. After developing his eponymous laws Maxwell retired to his family estate and spent his time writing his *Treatise* attempting to ground his laws in a depth understanding of what electricity, magnetism, current, and light really are. By his own standards he failed. Here again, subsequent developments showed that electrodynamics did not require such a foundation. Dirac initiated quantum electrodynamics and subsequently rejected the successful development of QED because it did not meet his standards for a proper theory. Weinberg summarized the development of the standard model as involving "a slow change in the attitude that defines what we take as plausible and implausible in scientific theories" (Weinberg 1995, p. 36). A pragmatist appraises a theory more by it success in accommodating actual and potential data than by its conformity to a priori norms. In QFT this means that the theory must supply a basis for unambiguous calculations and predictions. Thanks to renormalization, both components of the standard model supply such a basis. Renormalization hinges on the physical interpretation accorded mathematical terms.

Needless to say, I support the pragmatist perspective To put renormalization in a pragmatist perspective we begin with the presuppositions that generate the problem.[10] Classical electrodynamics introduced the idealization of point charges and soon encountered infinities in calculations, such as the self-energy of an electron, that depended on this idealization. Both QED and QFT began with a correspondence principle approach, using and quantizing classical electrodynamics. So, the idealization and problems of point particles was built in. In QED Schwinger was very clear on the physical significance of renormalization. QED is valid for reasonably small

[10] The historical development of renormalization is summarized in Cao, pp. 185–207. See 't Hooft (1997) for the problems involved in developing a renormalized electroweak theory.

distances or reasonably large momenta. For values below some distance cutoff, a, or above some momenta cutoff, Λ, where $a \propto 1/\Lambda$ a new unknown physics must enter. The series expansions involve terms with increasingly higher values of the coupling constant (α in QED) and the energy. As explained earlier, renormalization in QED effectively absorbs the problematic terms in the charge and mass coupling constants and justifies the neglect of high energy terms by the sharply diminishing values of the coupling constants raised to higher and higher powers. Feynman achieved the same effect by a regularization scheme. Regularization involves replacing the actual theory by a modified theory with a built-in cutoff. Feynman's trick was to replace the mass term involved in integrals by $(m^2 - \lambda^2)^{1/2}$, where λ is a fictitious mass assigned to the photon. In both approaches, renormalization involved showing that if one let the cutoff go to 0 (or ∞) the results remained finite.

Regularization and renormalization presented formidable problems in the physics beyond QED. Again, I will skip the details and relate the established results to the ontological issues we are considering. The electroweak theories developed independently by Glashow, Weinberg, and Salam were not taken seriously until 't Hooft established the renormalizability of non-Abelian Yang-Mills theory in 1972. QCD, developed in 1973, presented a different problem. An expansion in terms of increasing powers of the coupling constant led to a divergent series, because of the large value of the strong coupling constant. The assumption of asymptotic freedom supplied an elegant solution to this difficulty. As quarks move closer together the force between them decreases leading to a cutoff at very small distances. The standard model as a theory is given by a Lagrangian with many terms. In terms of gauge properties it is $SU(3)$ (color); $SU(2)$ (weak), and $U(1)$ (electromagnetic). We can extend the earlier summary to consider the particles proper to the standard model.

The Weinberg-Salam theory of electroweak interactions presents a unification of the electromagnetic, $U(1)$, and weak, $SU(2)$ fields. The weak field in not parity conserving and requires a left-handed neutrino. For the unification one needs a left-handed component,

$$L \equiv \begin{pmatrix} \nu_e \\ e \end{pmatrix}_L,$$

and a right-hand component

$$R \equiv (e)_R$$

Separate gauge transformations must be introduced for each component under $SU(2)$ and $O(1)$. To get massive vector bosons one must also include spontaneous symmetry breaking and the Higgs mechanism. The standard model adds to these the $SU(3)$ color component for strong interactions. The basic particles of the standard model are given in the following table (adapted from Wevers 2001).

Particle	spin	B	L	I	I_3	charge	m_0 (MeV)	Antipart.
u	1/2	1/3	0	1/2	1/2	+2/3	5	$\overline{\mathrm{u}}$
d	1/2	1/3	0	1/2	−1/2	−1/3	9	$\overline{\mathrm{d}}$
s	1/2	1/3	0	0	0	−1/3	175	$\overline{\mathrm{s}}$
c	1/2	1/3	0	0	0	+2/3	1350	$\overline{\mathrm{c}}$
b	1/2	1/3	0	0	0	−1/3	4500	$\overline{\mathrm{b}}$
t	1/2	1/3	0	0	0	+2/3	173000	$\overline{\mathrm{t}}$
e^-	1/2	0	1	0	0	−1	0.511	e^+
μ^-	1/2	0	1	0	0	−1	105.658	μ^+
τ^-	1/2	0	1	0	0	−1	1777.1	τ^+
ν_e	1/2	0	1	0	0	0	$< 1 \times 10^{-8}$	$\overline{\nu}_e$
ν_μ	1/2	0	1	0	0	0	< 0.0002	$\overline{\nu}_\mu$
ν_τ	1/2	0	1	0	0	0	< 0.02	$\overline{\nu}_\tau$
γ	1	0	0	0	0	0	0	γ
gluon	1	0	0	0	0	0	0	$\overline{\text{gluon}}$
W^+	1	0	0	0	0	+1	80400	W^-
Z^0	1	0	0	0	0	0	91187	Z^0

Here B is the baryon number and L the lepton number and there are three different lepton numbers, for e, μ and τ, which are separately conserved. I is the isospin, with I_3 the projection of the isospin on the third axis. The antiparticles have quantum numbers with the opposite sign except for the isospin I. In addition to these basic particles there are hypothetical particles postulated by the standard model. The two most significant are the Higgs particles, postulated to confer mass on the W and Z bosons, and the axion, postulated to explain CP violating terms in QCD. Neither has yet been observed. The axion is a candidate for the dark matter astronomers have inferred. The novel idea that the value of the coupling constant is a function of distance received theoretical support with the idea of a renormalization group, developed by Wilson and others in condensed matter physics. The idea of running (or variable) coupling constants initially seems counter-intuitive. For almost 100 years α, the fine structure constant, has been regarded as one of the fundamental *constants* of physics. Calculations of increasing accuracy led to an established result, $\alpha = 1/137.03599911$. To see how this 'constant' could be treated as a variable we recall its definition, $\alpha = e^2/2hc\varepsilon_0$. Ideally, a direct measurement of α might be made by bringing two electrons increasingly close together and measuring the strength of their interaction, the e^2 term in the definition. In practice, one relies on high energy scattering experiments. In QFT a charged particle is represented as a bare charge surrounded by a cloud of virtual pairs, the loops in the Feynman diagrams. The distinction between bare and clothed particles plays a basic role in renormalization. If this conception is valid, then electrons at very close distances would penetrate each other's virtual clouds and experience an electrical force stronger than that experienced by more distant particles. The virtual cloud has a screening effect. For the strong coupling constant the virtual cloud has an anti-screening effect, because interactions get weaker as quarks get closer. This hand-waving physical argument can be given a more precise mathematical expression.

Thus, the strong-coupling constant, α_s is expressed as a function of momentum, μ by an equation, $\mu\partial\alpha_s/\partial\mu = 2\beta(\alpha_s)$. The solution of this equation involves a series expansion and a constant of integration, whose value must be determined by experiment. This is generally done by determining α_s at some fixed reference scale, such as the mass of the Z boson. Then $\alpha_s \rightarrow 0$ as $\mu \rightarrow \infty$ (Schmelling 1997). This idea of running coupling constants supports the idea of *effective theories*. This leads to simplified calculation methods, relying on finite cutoffs in theories while ignoring limiting behavior as cutoffs approach 0 or infinity. It also leads to an appraisal of the physics that lies beyond the standard model. The leading candidate for physics beyond the standard model is supersymmetry. (See Kane 2000, for a non-technical account). Just as a proton and neutron are regarded as states of a primitive undifferentiated nucleon, so fermions and bosons may be states of some primitive undifferentiated superparticle.

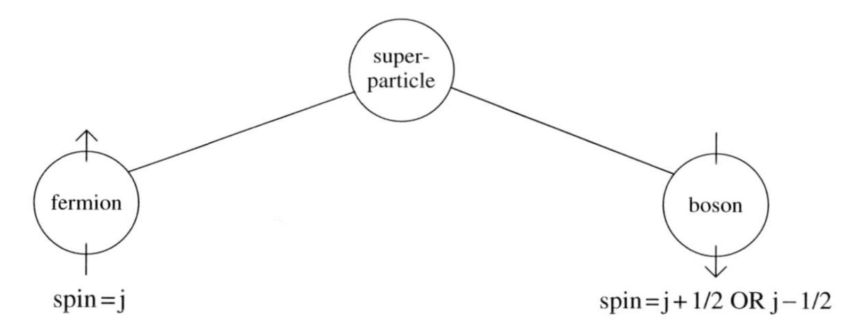

All components of the standard model are renormalized. This suggests that it is accurate for calculations up to much higher energy levels. Though it lacks the high precision of QED, the standard model explains a wider class of phenomena and does so with a high degree of accuracy. It also has a basic role in cosmological accounts of the state of the universe immediately after the big bang. There are, nevertheless, compelling reasons for regarding the standard model as an effective theory, a low energy approximation to a deeper theory. The electroweak component postulates, but does not explain, the spontaneous symmetry breaking that leads to the separation of the electrical and weak components. It requires a massless neutrino. The observed neutrino oscillation requires assigning neutrinos the masses indicated on the preceding table of particles. The hierarchy problem concerns the enormous gap between the energy level of the standard model and that of supersymmetric unification or string theory. Some theories postulate a new physics just a couple of orders of magnitude beyond the standard model. The default position is that there is no new physics between the standard model and the level of supersymmetric unification. Detection of Higgs particles should help to resolve these disputes (See Pokorski 2004). The significance of this for running coupling constants is given by Wilczek's diagrams.

Figure 6.5 shows the near convergence of the strong (α_1), electromagnetic (α_2), and weak (α_3) coupling constants near an energy level of 10^{15} GeV. Fig. 6.6 shows a much sharper convergence at a somewhat higher energy level with the assumption

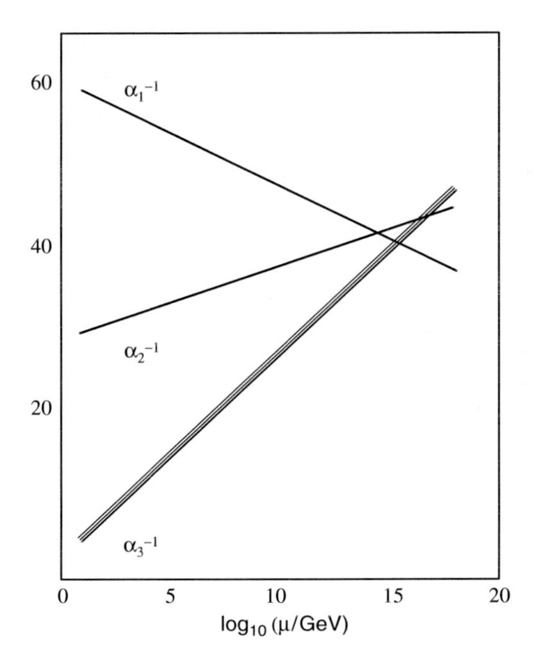

Fig. 6.5 Unification without supersymmetry

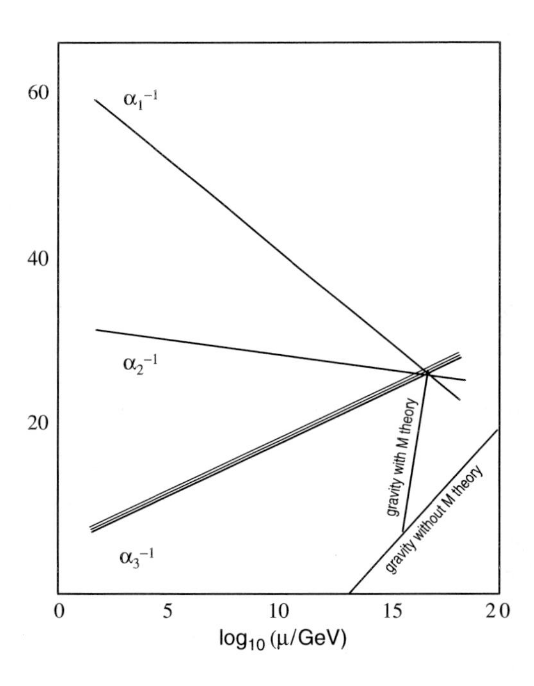

Fig. 6.6 Unification with supersymmetry

of supersymmetric particles. Witten claims that the M version of string theory supports the conclusion that the gravitational coupling constant also converges to this point (See Greene 1999, p. 363). I have included this in Fig. 6.6 as a bit of extra information. However, I will make no further use of string theory. Physics at the string-theory level would not change the standard model or its list of basic entities. It should explain some of the parameters that are presupposed in the standard model.

6.4 Idiontology of the Quantum Realm

Earlier we indicated the need for a split level ontology. The phenomenological level is ground in ordinary language and its extension to the language of physics. The inference-supporting base of ordinary language is its categorical structure. This is rooted, not in a theory of objective reality, but in our way of existing as agents and knowers in a physical world structured by a social order. The extension of this grounds a relative ontology of public objects. Extended ordinary language is indispensable. However, we must maintain a critical awareness of its limits. If we use 'description' in a broad sense of specifying the features characterizing a system, then most classical systems admit, at least in principle, of a precise description. Quantum systems may only be described in a complementarity sense. Here description is relative to an experimental situation involving decisions of experimenters regarding which features of a quantum system they wish to investigate. The functional ontology presupposed in such descriptive accounts is a streamlined extension of a lived-world ontology. It is not an ontology of the quantum realm. The measurement interpretation, a streamlined version of the Copenhagen interpretation functions on this phenomenological level. It does not present an ontology of the quantum realm. Since it is inadequate to contemporary advances in quantum physics, it seems reasonable to speculate about such an ontology.

Before attempting this we should situate the effort in the context of similar attempts to develop an ontology of QM or QFT. I will not summarize other positions, but merely indicate why I follow a different path. There have been many attempt to develop an ontology of QM. These generally take the form of presenting, or redeveloping non-relativistic quantum mechanics as a formal system. This entails excluding the reduction postulate and introducing the measurement problem. Ontology enters as an answer to the question: What must the world be like if this theory is true of it? If the answers are not compelling then one can fall back on an antirealist or an epistemological interpretation. Auletta 2000, *Sect. X* has summarized this oscillation. If an epistemological approach is followed, then observables are fundamental. The state of a system can only be precisely specified for an isolated system, which is unobservable. So, the state of a system is defined as positive normalized linear functionals of observables and is not accorded an ontological status. This approach leads from algebras of observables to correlations, but not to distinct individuals. When correlations are basic, individuation is a form of noise, breaking the interconnections. In a more ontological approach, the state of a system is taken as basic and is characterized by the superposition principle. This supports individuality and the particle concept. Here, however, correlations appear as a form of noise, a weakening of individuality. Yet, measurements depend on correlations. I am assuming that an attempt to go beyond the measurement interpretation must reproduce the successful features of the measurement, or orthodox, interpretation. This means that the suggested replacement must also include some version of the reduction postulate. Such a replacement will be considered in the next chapter. The present concern is with the issue of whether QM supports an ontology, or whether one should be satisfied with an idiontology, a systematic account of the properties characterizing quantum systems.

The current standard for theory interpretation entails arguing from a theory to the reality it represents. The present linguistic approach suggests a methodological reversal. Bohr insisted that the orthodox interpretation represents the limits of what can be meaningfully said about quantum systems. Now we are confronted with the need to go beyond orthodoxy. The preliminary question concerns how we can say something meaningful beyond these limits. Superficially one can easily go beyond by adding new terms: ‘quark’, ‘gluon’, ‘intermediate vector boson’, ‘axion’, ‘Higgs particle’, and terms not yet coined. These are embedded in the language of physics and do not involve a modification, or abandonment, of the categorial system. Like ‘tree’, ‘ghost’, ‘rock’ and ‘angel’, they have the status of referring to things that may or may not exist. The key terms characterizing the distinctively non-classical *properties* of quantum systems, ‘superposition’, ‘interference’ ‘distributed probability’ and ‘entanglement’ have a different status. Consider ‘superposition’. It has an acceptable sense. Radio signals involve the superposition of sound waves on electromagnetic waves. ‘Superposition’ does not apply to objects and their properties. It is incoherent to represent a man weighing 175 pounds as a superposition of two men, one weighing 150 pounds and the other 200 pounds, with the assumption that the process of measuring the man’s weight might catch either of the two superimposed men. Dirac introduced the superposition principle with the physical argument that a photon polarized at a 45° angle may be consider a superposition of a photon polarized at a 0° angle and one polarized at a 90° angle with the assumption that an appropriate measurement could catch either the 0° or the 90° photon. The resolution of the solar neutrino problem hinged on accepting a neutrino as a superposition of 3 types of neutrinos: ν_e, ν_μ, and ν_τ. When the term is used in a quantum context it is an oblique way of referring to mathematical expressions of the form, $\|\psi\rangle = \sum |\psi_i\rangle$. One is using a classical term while rejecting the presuppositions that ground its normal usage. Tossing three coins in a fountain leads to a simple display of wave interference. Water waves in the ripple tanks formerly used for high-school physics demonstrations manifest wave behavior whether they pass through one or two slits. The idealized one-slit, two-slit thought experiment illustrates quantum interference. Whether an electron manifests wave behavior and interference depends on the experimenter’s choice of opening the second slit. A tossed coin that is not yet seen is assigned equal probability of being either head or tails. This is an epistemic probability referring to a state of knowledge. Objectively the coin is either H or T. This epistemic probability does not support counter factual conditionals such as: If the coin observed to be T were not observed it might be H. An X-ray propagating from a radioactive source has equal probability of being in either half of a sphere surrounding the source. This is not an epistemic probability. It does support counter factual conditionals. If the X-ray detected in the upper half of the sphere had not been detected it might have been detected in the lower half. 'Entanglement' when used in a classical sense presupposes locality. Two strings are entangled when one is wound around the other. When 'entanglement' is used in quantum contexts it refers to a non-local, or holistic, property. Two photons that are produced together remain entangled until one of them is measured no matter how large the separation between them. If we take the terms listed as characterizing quantum systems then they must be recognized as analogous extensions of classical terms. They function without the presuppositions that grounds the classical usage and, accordingly, do not support the material inferences that hinge on these presuppositions and the network of concepts in which they are embedded. The meanings of these terms in a quantum context depends on the mathematical forms and quantum uses they presuppose. One is using language beyond the limits of normal usage by suppressing presuppositions and using mathematics as a vehicle for moving usage beyond normal limits. This supports an idiontology, a specification of the distinctive properties of quantum systems. It does not yield an ontology, either in the sense of a specification of basic entities or a categorial system for the quantum realm. This analysis was based on quantum mechanics considered as a set of principles present in different quantum theories: non-relativistic quantum mechanics, quantum electrodynamics, quantum field theory, the standard model, and string theory. Can something more be said of the ontology of QED or QFT? We begin with a basic point. QED, QFT and the standard model are best regarded as effective theories. They supply a basis for representing interactions within a certain energy range. None can be considered the ultimate theory. Therefore none of them can supply a fundamental ontology. In each case we are dealing with relative ontologies. A relative ontology is basically a clarification of the things a theory treats. In this minimal sense the functional ontology of the standard model, including QED, is given by table 6.3.1. Philosophers want to go beyond this minimal functional basis and ask what the physics reveals about the reality treated. I think that such questions are best answered by analyzing what physics says to the degree that this can be rendered in meaningful language.

We will begin on a simple descriptive level and then assess its limitations. Hydrogen atoms, weakly excited by Lamb and Retherford's microwave radiation, reveal an indefinitely large number of virtual transitions and virtual processes. This privileged access to the microworld opens up a new vista of ubiquitous, incessant, ultra fast non-classical processes. It should also induce a change in philosophical perspective on two basic issues. The first is what counts as a theory. QED is an extremely successful physical theory. It does not accord with the criteria philosophers impose as a condition for theory interpretation. I am taking functioning physics, rather than philosophical reconstructions of selected theories, as a basis for interpretation. QED so interpreted suggests a shift in ontological perspective. Traditionally ontological systems has been concerned with individual entities, beings, substances, monads, events. Attempts to extend such systems to the quantum realm have to develop accounts of individuation that accommodate Fermi-Dirac and Bose-Einstein particles. Yet the 4 emphasis remains on individuals. Some philosophers, such as Richard Healey, stress holism as characterizing the quantum realm. Even this advance centers on observable systems. QED draws attention to the ubiquity of virtual processes. Within QED they function on a par with observable processes. However, for every observable process there are an indefinitely large number of virtual processes. Feynman path integrals treat observables as outcomes of virtual processes. According virtual processes a foundational role upsets all traditional ontologies. To handle this in a proper fashion we need a formulation of quantum mechanics that gives virtual processes a foundational role. This will be treated in the next chapter. Bohrian semantics set norms for what could be meaningfully said about atoms and particles. The development of QFT involved two probes extending beyond these limits. Do they reveal anything that can be meaningfully communicated about the ontology of the quantum realm? The first probe, mathematics can, for our present limited purposes, be divided into two parts. The first part is correspondence principle mathematics. One begins with classical expressions for dynamic variables, Lagrangians, Hamiltonians, and Poisson brackets and then follows operational rules for transforming these into quantum operators, Lagrangians, Hamiltonians, and anticommutation relations. This admits of alternatives, such as Schwinger's variational methods, and amplifications. The non-correspondence principle mathematics includes Feynman path integrals, and the postulation of local gauge invariance. Many philosophical analyses dismiss Feynman path integrals as a basis for determining a quantum ontology on the grounds that they do not admit of a realistic interpretation. I think that they should be used for precisely that reason. As Auyang (1995, p. 73) puts it: "Feynman diagrams are not visualizations, but informative symbols of microscopic processes". The path integrals and the symbolic diagrams represent the most that physics can say about submicroscopic processes.

The symmetry principles used in QFT can be divided into spatio-temporal principles and internal principles. We can speak meaningfully about the former in terms of properties of a system being invariant under translation, rotation, or inversion. When mathematics is used to express internal symmetries then the language we use to speak about them is analogous, not literal. What does it mean to speak of rotating a system in color space or isospin space? The general conclusion I draw from these observations is more concerned with the role of philosophy than with the ontology of quantum field theory. I believe that a philosophical analysis can put the accomplishments of physics in a different interpretative framework. This serves to relate it to the questions philosophers ask. But it does not add anything substantive to what physics says. Perhaps such questions can be best addressed if quantum physics is given a better formulation.

The measurement interpretation of QM, like the orthodox interpretation it systematizes, supplies a consistent fit to the observational basis and the inference-supporting conceptual structures used to express it. It has been remarkably successful in atomic, nuclear, and particle physics. Yet, it has severe limitations. It is not adequate to advances in QFT and to quantum cosmology. It says nothing positive about the ontology of the quantum realm. Can an interpretation of QM be developed that accommodates the pragmatic success of the measurement interpretation and also explicitly recognizes the primacy of the quantum realm? This would require an interpretation that accommodates actual measurements, rather than dissolving them in the measurement problem. It should also incorporate the distinctive quantum features just discussed. Superposition, interference, distributed probability, and virtual processes should be explicitly present in the formalism. An attempt to present such an interpretation is the burden of the next chapter.

References

Auletta, Gennaro. (ed.) 2000. *Foundations and Interpretation of Quantum Mechanics in the Light of a Critical-Historical Analysis of the Problems and of a Synthesis of the Results*. Singapore: World Scientific.

Auyang, Sunny. 1995. *How is Quantum Field Theory Possible?* New York, NY: Oxford University Press.

Bennett, G. W. et al. 2002. Measurement of the Positive Muon Anomalous Magnetic Moment to 0.7 ppm. *Physical Review Letters, 89*, 101.

Bethe, Hans. 1947. The Electromagnetic Shift of Energy Levels. *Physical Review, 72*, 339–341.

Bitbol, Michel. 1996. *Mécanique Quantique*. Paris: Flammarion.

Brown, Harvey, and Rom Harré. 1988. *Philosophical Foundations of Quantum Field Theory*. Oxford: Clarendon Press.

Buchholz, Detlev, and Rudolf Haag. 1999. The Quest for Understanding in Relativistic Quantum Physics.

Calmet, Xavier, Harald Fritzsch, and Dirk Holtmannspoeter. 2001. Anomalous Magnetic Moment of the Muon Andradiative Lepton Decay. *Physical Review D, 64*, 64.

Cao, Tian Yu. 1998. *Conceptual Developments of 20th Century Field Theories*. Cambridge: Cambridge University Press.

Fraser, Doreen. 2009. Quantum Field Theory: Underdetermination, Inconsistency, and Idealization. *Philosophy of Science, 76*, 536–567.

Galison, Peter. 1987. *How Experiments End*. Chicago, IL: University of Chicago Press.

Greene, Brian. 1999. *The Elegant Universe: Superstrings, Hidden Dimensions, and the Quest for the Ultimate Theory*. New York, NY: Vintage Books.

Haag, Rudolf. 1992. *Local Quantum Physics: Fields, Particles, Algebras*. Berlin: Springer.

Healey, Richard A. 1989. *The Philosophy of Quantum Mechanics: An Interactive Interpretation*. Cambridge: Cambridge University Press.

Heidegger, Martin. 1962. *Being and Time*. New York, NY: Harper & Row.

Hoddeson, Lillian, Laurie Brown, Michael Riordan, and Max Dresden. 1997. *The Rise of the Standard Model: Particle Physics in the 1960s and 1970s*. Cambridge: Cambridge University Press.

Johnson, George. 1999. *Strange Beauty: Murray Gell-Mann and the Revolution in Twentieth-Century Physics*. New York, NY: Alfred Knopf.

Kaku, Michio. 1993. *Quantum Field Theory: A Modern Introduction*. New York, NY: Oxford University Press.

Kane, Gordon. 2000. *Supersymmetry: Squarks, Photinos, and the Unveiling of the Ultimate Laws of Nature*. Cambridge, MA: Perseus Publishing.

Kernan, Ann. 1986. The Discovery of Intermediate Vector Bosons. *American Scientist, 74*, 21–24.

Kroll, Norman, and Willis Lamb. 1949. On the Self-energy of a Bound Electron. *Physical Review, 75*, 388–398.

Kuhlmann, Meinard, Holger Lyre, and Andrew Wayne. 2002. *Ontological Aspects of Quantum Field Theory*. River Edge, NJ: World Scientific.

MacKinnon, Edward. 1977. Heisenberg, Models, and the Rise of Matrix Mechanics. *Historical Studies in the Physical Sciences, 8*, 137–188.

MacKinnon, Edward. 1998. Review of Michel Bitbol's Mecanique Quantique: Une Introduction Philosophique. *Isis, 89*, 360–361.

MacKinnon, Edward. 2008. The Standard Model as a Philosophical Challenge. *Philosophy of Science, 75*, 447–457.

Pickering, Andrew. 1984. *Constructing Quarks: A Sociological History of Particle Physics*. Chicago, IL: University of Chicago Press.

Pokorski, Stefan. 2004. Phenomenological Guide to Physics Beyond the Standard Model. *arXiv/hep-ph*, 0502132.

Redhead, Michael L. G., and Fabian Wagner. 1998. Unified Treatment of EPR and Bell Arguments in Algebraic Quantum Field Theory. ArXiv:09083.2844v1, quant-ph/9802010.

Salpeter, Edwin. 1953. The Lamb Shift for Hydrogen adn Deuterium. *Physical Review, 89*, 92–99.

Samios, Nicholas. 1997. Early Baryon and Meson Spectroscopy Culminating in the Discovery of the Omega-Minus and Charmed Baryons. In Hoddeson et al. (ed.), *The Rise of the Standard Model* (pp. 525–541). Cambridge: Cambridge University Press.

Schmelling, Michael. 1997. Status of the Strong Coupling Constant. http://arXiv:hep-th, 9701002.

Schweber, Silvan S. 1994. *QED and the Men Who Made it*. Princeton, NJ: Princeton University Press.

Seibt, Johanna. 2002. Quanta, Tropes, or Processes: Ontologies for QFT Beyond the Myth of Substance. In Meinrad Kuhlmann, et al. (eds.), *Ontological Aspects of Quantum Field Theory*. Singapore: World Scientific.

't Hooft, G. 1997. *In Search of the Ultimate Building Blocks*. Cambridge: Cambridge University Press.

Teller, Paul. 1995. *An Interpretive Introduction to Quantum Field Theory*. Princeton, NJ: Princeton University Press.

Wayne, Andrew. 1998. Conceptual Foundations of Field Theories in Physics. In Don A. Howard (ed.), *PSA98*. Part II; Symposia Papers (pp. S466–S522). Kansas City, MO: Philosophy of Science Association.

Weinberg, Steven. 1995. *The Quantum Theory of Fields: Vol. I*. Cambridge, NY: Cambridge University Press.

Wevers, ir. J. C. A. 2001. Physics Formulary. www.xs4all.nl/~johanw/index.html

Whitehead, Alfred North. 1929. *Process and Reality: An Essay in Cosmology*. New York, NY: Macmillan.

Wilczek, Frank. 1999. Quantum Field Theory. *Reviews of Modern Physics, 71*, S85–S95.

Wilczek, Frank. 2000. Future Summary. *http://arXiv:hep-th*, 0101087.

Wilczek, Frank. 2002a. QCD and Natural Philosophy. *http://arXiv:physics*, 0212025.

Wilczek, Frank. 2002b. Inventory and Outlook of High Energy Physics. *http://arXiv:hep-th*, 0202128.

Wilczek, Frank. 2004a. Asymptotic Freedom: From Paradox to Paradigm. *http://arXiv:hep-ph*, 0502113.

Wilczek, Frank. 2004b. A Model of Anthropic Reasoning: Addressing the Dark to Ordinary Matter Coincidence. *http://arXiv:hep-ph*, 0408167.

Chapter 7
Interpreting Quantum Mechanics

... in taking a state the conqueror must arrange to commit all his cruelties at once, so as not to have to recur to them every day, and so as to be able, by not making fresh changes, to reassure people and win them over by benefiting them. Whoever acts otherwise, either through timidity or bad counsels, is always obliged to stand with knife in hand, and can never depend on his subjects, because they, owing to continual fresh injuries, are unable to depend on him. For injuries should be done all together, so that being less tasted, they will give less offense. Benefits should be granted little by little, so that they may be better enjoyed.
Machiavelli, The Prince, chap. VIII

Many studies have been written on the interpretation of quantum mechanics. Most share an implicit assumption. Interpretation is a matter of analyzing theories, particularly quantum mechanics and quantum field theory. I believe that this is a radically inadequate basis for analyzing the significance of the quantum realm, or the subject matter treated by quantum mechanics. As a preliminary to developing an alternate position we begin by reflecting on some of the issues treated in the preceding chapters.

The original Copenhagen interpretation, stemming chiefly from Bohr, was essentially a guide for the practice of quantum physics. Bohr's analyses focused on idealized thought experiments. The common theme was one of clarifying the type of experimental information that could be reported in an unambiguous manner. He treated the mathematical formalism as a tool for insuring consistency and enabling predictions. By the time interpreting QM became a growth industry Bohr's perspective was lost. His scattered reflections were generally treated as interpretations of a mathematically formulated theory. As such, they seemed amateurish, outdated, even bizarre. Historically, Bohr's reply to the EPR paper was taken as a definitive refutation by the physics community. In the altered perspective of philosophical interpretation, Bohr's paper was taken as missing the point, because he only treated the mathematical formalism in a footnote.

E. MacKinnon, *Interpreting Physics*, Boston Studies in the Philosophy of Science 289, DOI 10.1007/978-94-007-2369-6_7,

In this theory-alone perspective experiments are relegated to the role of supplying data that confirm, falsify, or somehow test theoretical predictions. The experimental inferences that Bohr focused on are no longer accorded a role in interpretation. This introduces an inconsistency that is much stronger now than in the physics Bohr treated. Theories in particle physics are *never* tested by observations. They are tested by *inferences* drawn from experimental analyses. If these inferences are invalid, then the theories are not tested. A philosophical analysis of quantum theories leads to the conclusion that most of these inferences rely on a foundation that is incompatible with QM. To relate the theoretical and experimental perspectives we will consider how each represents the motion of a free particle, i.e., a particle not subject to any force.

In non-relativistic quantum mechanics (NRQM) a particle with momentum **p** is represented by a plane wave. A plane wave is distributed over all space. So there is no localization. A purely quantum consideration leads to the same conclusion. Since the particle has a definite momentum, **p**, Heisenberg's uncertainty principle prohibits any localization. Accordingly, it is customary to represent a particle moving in the x direction by a wave packet. For a trajectory we need the maximum values for both position and momentum. This is given by a Gaussian wave packet, with initial wave packet diameter, $\sigma_x(0)$, and later wave packet diameter, $\sigma_x(t)$

$$\psi(x) = A\, e^{\frac{-x^2}{2\sigma_x{}^2}}\, e^{ik_0 x}, \tag{7.1}$$

$$\sigma_x(t) = \sigma_x(0)\left(1 + \left[\frac{\hbar t}{2m\sigma_x(0)^2}\right]^2\right)^{1/2} \tag{7.2}$$

Since we will be considering protons moving through accelerators, we will take the initial wave packet as describing a proton with an initial value of $\sigma_x(0) = 10^{-2}$ cm. The packet width would increase by a factor of 3 in one second, and by a factor of 17 in 100 s.

Relativistic quantum mechanics (RQM) does not support a position variable. One may repeat the above considerations of plane waves and wave packets (See Björken and Drell 1964, chap. 3). The results are very unsatisfactory. Since the velocity operator is $c\alpha$ each component has the average value c. The attempt to form a wave packet including both positive and negative energy solutions leads to the notorious *zitterbewegung* oscillation between positive and negative energy solutions. The Newton-Wigner FAPP solutions to this difficulty was to construct a wave-packet using only positive-energy solutions of the Dirac equation. Foldy and Wouthuysen (1950) generalized this by introducing a transformation that systematically eliminated the odd operators in an expansion of the Dirac equation. The same transformation led to the replacement of the position (**x**) and velocity **x** by complex expressions. However, the mean position is $X' = x$, while the mean velocity is $\dot{X}' = \beta p/E_p$, where β is a diagonal matrix. This yields the conventional velocity operator for positive energy states. In effect the location is effectively represented by a small region that grows with time. Halvorson and Clifton (2002) argue from the no-go theorems of Malament and Hegerfeldt that there is no place for an ontology of localized particles in quantum field theory (QFT). They accommodate experimenter's reliance on localization by photographic tracks by showing that the observables characterizing a system can be approximately localized under certain conditions.

Thus NRQM, RQM, and QFT can all be interpreted as supporting the conclusion that quantum mechanics is incompatible with, an ontology of localized particles traveling in sharp trajectories. The original Feynman formalism was based on particles traveling in trajectories. These, however, could not be interpreted as trajectories of a classical particle ($particle_c s$). In the Feynman representation a particle ravels all possible paths between the origin and the terminus. As indicated in our earlier discussions of particular experiments, particle accelerators and detectors rely on particle trajectories as an inferential basis. Both the design of particle accelerators and the related detectors rely on '$particle_c$' rather than '$particle_t$' and on classical mechanics and electrodynamics, rather than quantum mechanics or QFT. The progress from the original cyclotron to the Large Hadron Collider (LHC) required the production and guidance of ever sharper particle trajectories. Basically the guidance system has three components: dipole magnets that curve the particle's trajectory; paired quadrapole magnets that produce the sharp focusing and drift tubes, where particles are accelerated. The length and sharpness of the trajectories are notable. The LHC uses five linked proton accelerators: a linear accelerator: a Proton-Synchrotron Booster; a Proton-Synchrotron; a Super Proton Synchrotron; and the Large Hadron Collider (Lincoln 2009, chap. 3). A switching mechanism allows the introduction of successive bunches of about 10^{11} protons rotating in opposite directions in the LHC. When, or if, the LHC is operating at full capacity it will keep particles rotating in the LHC for 10–20 h to record collisions. In this time the particles travel more than 60 times the earth-sun distance. The sharp focusing required to prevent unintended collisions must be replaced by sharper focusing of opposing beams to produce collisions in the detectors.

The detectors now used represent the fusion of two experimental traditions (Galison, 1987, 1997). The image tradition, which relied on cloud chambers, nuclear emulsions, and bubble chambers, produced pictures of trajectories. It was often necessary to process thousands of images to find one significant, or sought for, event. The logic tradition counted clicks from Geiger-Müller counters, spark chambers, and various wire detectors. This led to statistics rather than pictures and allowed for triggering mechanisms. The fusion of the two traditions paralleled the switch from aiming particles at fixed targets to collisions between particles rotating in opposite directions. In this case the motion of the center of mass is 0. So, the particles produced in the collision can scatter in all directions. Tracking these requires multistage detectors in concentric shells surrounding the point of collision. The current state of the art detectors are those designed for the LHC. If it is ever operating at full capacity there should be about one hundred million proton-proton collisions per second in each of the four detectors (See Lincoln 2009, chap. 4). The experimental analysis of events recorded in these detectors has two stages. The first selects the few events that may have interesting physics. This stage analyzes the debris from many secondary, tertiary, etc., collisions and infers whether the originating collision merits analysis. This involves identifying the various secondary particles produced: $e^+, e^-, \gamma, \mu\, \pi$, their energy and momenta. The largest of the two main detectors are ATLAS, which is 45 m long and 22.5 m wide, and the smaller, but more complex, Compact Muon Solenoid (CMS). ındexcompact muon solenoid. The CMS is 19.8 m long, 14.6 m in diameter, and weighs 12,500 t. It is divided into 5 layers that utilize magnetic bending, ionization, showering (the multiple collisions produced in a solid), transition radiation, and Cerenkov radiation to infer energy, momentum, mass, charge, and point of origin of secondary decay products. When the individual units are counted, the CMS contains about 830,000 detectors. This selects approximately 1 out of 100,000 originating collisions by processing the multiple inputs through high-speed computers. The second stage analyzes the records produced in the first stage. The information selected is passed on to members of a research team consisting of approximately 3,800 people in 38 countries (CMS 2005).

The final result of all this processing are 'observables' that test theoretical predictions or occasionally produce unanticipated results. These observations clearly depend on a complex collection of inferences. The overall framework interrelating these inferences is what Bohr labeled 'a phenomenon', a descriptive account of the accelerator, the detector chamber, and a host of actual and virtual processes. Both classical and quantum theories may be used, but the overall framework is not a theory. It is a descriptive account that includes both the apparatus and accounts of actual and virtual processes. When appropriate, it also uses accounts of particles traveling in trajectories. This informal inferential system relates to a formal, or theoretical deductive system in a particular measurement context. The two representations meet by identifying p_c and p_t in the context of an actual or possible experiment. This may be indicated schematically.

In a measurement situation the semiclassical concept, p_c, and the theoretical concept, p_t, have a common referent. Each concept has a central role in an inference-supporting cluster of concepts. The p_c cluster relies on particle trajectories as a foundation for informal inferences and uses mathematical formulas, e.g., energy and momentum conservation, as tools. Each has surplus structure not represented in the other system. The p_t cluster now relies on QFT as a deductive system. The loose correlation between these two inferential systems is illustrated by the Ω^- detection treated in the last chapter. On the basis of the eightfold way, a phenomenological account later subsumed in QFT, Gell-Mann predicted that the Ω^- would decay into a Ξ^0 and a π^-, and that these secondary particles would follow familiar decay patterns. Samios inferred that the difference between the production of a Ξ^- particle and a very short0-lived Ω^- particle would be manifested by a slight displacement in the start of the Ξ^0 track stemming from Ω^- production.

7.1 Formulations and Interpretations

Before presenting yet another interpretation of QM we should clarify the difference between what we are attempting and more familiar interpretations. In the semantic model interpreting a theory is essentially a matter of asking what the world is like if the theory is true of it. An individual theory, at least a fundamental one, supplies a basis for an ontology. If the theory-ontology inference is not persuasive, then either anti-realism or an epistemological interpretation supplies a fall-back position. I find this methodology unreliable and particularly misleading when applied to QM. There are two reasons for the assessment of general unreliability. The first stems from the earlier historical analysis. The mathematical formulation of a general physical theory presupposes an idiontology. The mathematical formalism relates to a systematization of properties and processes. Objects, as bearers of these properties, have a presuppositional role.

The second reason stems from the concept of effective theories. An individual theory fits into a hierarchical network of theories and is interpreted through a descriptive account of the interactions proper to some energy range. The objects involved in the interactions enter through the descriptive account, not through an interpretation of the formalism. The EFT tower of theories scenario suggests that the ultimate theory will supply the ontological foundation supporting the tower. This suggestion, however, should be accompanied by two reservations. The first is that the ultimate ontological foundation still remains a matter of speculation.[1] The second reservation is that the development of a 'theory of everything' is not expected to change higher-level theories and their relative ontologies.

For these reasons I am not presenting the interpretation of QM *as a theory*, where 'theory' denotes a deductively organized system. Rather, 'quantum mechanics' will be used as a general term to cover principles common to NRQM, QED, and especially QFT, the most basic quantum theory. Interpreting QM as an idiontology means focusing on the properties that distinguish QM from classical physics and accepting them as characterizing reality. Adapting Gell-Mann and Hartle, who will be treated later, and relying on the analysis presented in the end of the last chapter, I take three distinctively quantum mechanical properties as characterizing

[1] A brief non-technical survey of such speculations is given in Susskind (2006), 348–356, who supports superstrings. 't Hooft (1997, chap. 26) suggests that miniature black holes are the ultimate entities, while Wilczek (2008, chap. 21) postulates that 'empty space' is really a multilayered, multicolored superconductor.

reality at a deeper level than classical physics: superposition, interference, and non-locality. Tests of Bell's theorem highlighted the significance of non-locality. A basic manifestation of non-locality is distributed probability. 'Distributed probability', or 'propensity' as Teller (1995, p. 8) defines the term, is a form of quantum non-locality that requires clarification. I will present a simple example that brings out the non-classical character of distributed probability. Consider a variation of an apparatus often used in high-schools classes as an illustration of probabilities (or of the binomial theorem).

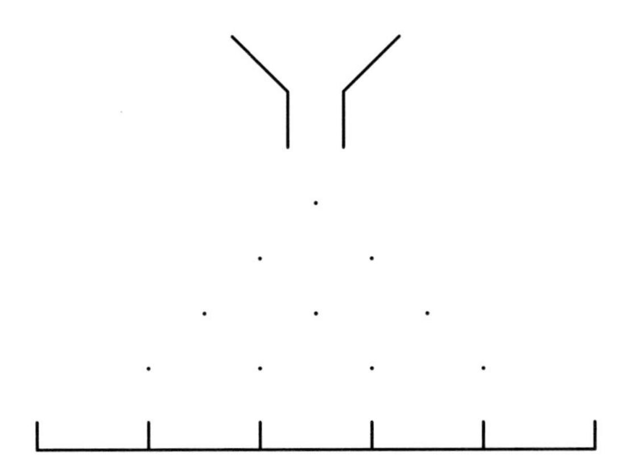

A steel ball whose diameter is the width of the tube is dropped into the tube. The steel pegs are spaced so that a ball bouncing off a peg has an equal chance of hitting the top of either the lower right or the lower left peg. The variation is that the apparatus is enclosed so that the final destination of the ball cannot be seen until the bottom is opened. A simple counting of trajectories leads to the probabilities that it is in each box: 1/16, 4/16, 6/16, 4/16, 1/16, reading left to right. These are epistemic probabilities. The ball is in only one box. If a large number of balls are dropped successively, then the distribution in the bottom boxes generally approximates the epistemic probabilities.

Now consider the familiar one-slit experiment. For the ideal case assume the electron gun shoots one electron that passes through the slit and strikes one of the five sections of the photographic plate. If we cannot see the plate we could assign probabilities to each section that, like the dropped ball case, would be highest in the center and symmetrically lower as we move from the center. These too would be epistemic probabilities. The spot is on only one of the five sections. Now we remove the photographic plate and consider the probabilities that if a measurement were performed the electron would be found to be in one of these five sections. In this case the epistemic probabilities express real propensities for possible localizations. The calculation of $\int \psi^{+}\psi$ over the volume of each box would have a positive value. The transition from distributed probabilities to a unique measured position is an example of wave-packet reduction. We will return to this point shortly.

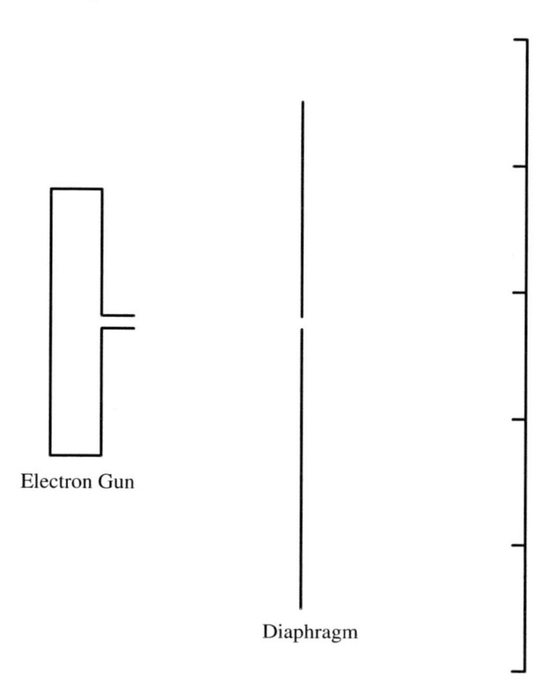

Two more reservations should be noted. The terms 'superposition', 'interference', 'non-locality' and 'distributed probability' have established meanings through their use in quantum contexts. Inasmuch as this language is EOL it is a quasi-classical specification of these properties. The basis specification comes from the mathematical formulation. Second, we have omitted some distinctively quantum holistic properties, such as 'entanglement' and holistic properties of quantum fields suggested by holonomic loop representations of gauge fields (Healey 2004). Such properties are developed as consequences of QM. They are not implicit in the formalism.

If QM is accepted as the fundamental science of reality, then it should be formulated as a fundamental theory, using 'theory' in a loose enough sense to cover non-rigorous formulations. The ideal new formulation is the one Einstein attempted, developing a new quantum theory that replaces the established theory. Assuming that this is not a viable project, something less ambitious is sought. Present QM has an unprecedented success in solving problems, calculating probabilities, and making predictions. However, it is not adequate to foundational problems, to advances in QFT, and to quantum cosmology. A new formulation should be used to treat these problems. Before considering possible candidates we should consider the requirements that such a formulation should meet, as suggested by the preceding analysis.

1. The formulation should be purely quantum, not semi-classical.
2. The new formulation should be downwardly compatible with the orthodox interpretation.

3. The formulation should reflect the complementarity between theory and experiment as sources of information.
4. The formulation should treat real and virtual processes equally.

The term 'downwardly compatible' is taken from computer software. A new version of a program must be able to do everything the preceding version did. The requirement of compatibility with the orthodox interpretation relates to the third requirement. Earlier we distinguished a measurement interpretation from the measurement problem. In a strict measurement interpretation, the specifications of distinctively quantum measurements plays a determining role in developing the mathematical formalism. In a loose measurement interpretation the acceptability of measurements is determined by experimental expertise. This is a minimal functional formulation of QM. The formalism is adjusted to accommodate such results, e.g., by accepting a projection postulate or the collapse of a wave function. Many rigorous reconstructions of NRQM as a theory attempt to eliminate the projection postulate. This entails eliminating any real role for experimental inferences. I accept the projection postulate as a feature of orthodox QM.

The attempt to meet these requirements leads to the elimination of many formulation/interpretations of QM. They are not rejected as wrong, but on the grounds that they do not meet the methodological requirements of the interpretative perspective developed here. To begin the necessary cruelties we first exclude interpretations based on foundational reconstructions of NRQM or QFT. None that I am familiar with accords experimental inferences an interpretative role. Most try to eliminate, or somehow get around, the projection postulate. The modification of QM proposed by Ghirardi, Rimini, and Weber (1986) is not downwardly compatible with established QM. It modifies the superposition principle by postulating spontaneous random localization processes. The interpretation given by Bitbol does meet these requirements. However, this is a philosophical interpretation, not a reformulation.

Feynman initiated a tradition of taking quantum mechanics seriously, of accepting quantum descriptions when they are incompatible with classical intuitions. This initiated a serious acceptance of the first requirement. Here we should distinguish the Feynman formalism from the Feynman *interpretation* of this formalism. He interpreted antiparticles as normal particles going backwards in time. Both the many-worlds interpretation and the consistent histories interpretation can be developed as extensions of the Feynman formalism. The many worlds interpretation stemming from Everett (1957) and Dewitt and Graham (1973) is usually interpreted as entailing an uncontrollably profligate ontological reduplication. Tegmark (1997) and Wallace (2006) have presented defenses of Everett's interpretation that minimize these ontological difficulties. I find even the residual ontological profligacy unacceptable. However, I believe that it would be possible to develop a many-worlds formulation that meets these requirements. There remains only one contender that meets these constraints as an reformulation/interpretation that takes quantum mechanics seriously, the consistent histories (CH) formulation/interpretation.

7.2 The Consistent Histories Interpretation of Quantum Mechanics

We begin with a preliminary issue. What is a quantum history a history of? Adapting (Zee 2003, pp. 7–10) consider the traditional single-slit double-slit thought experiment. This would be depicted by the previous diagram, but with two slits. Electrons are shot out of a projector, pass through a slit in a diaphragm and then strike a photographic plate. If only one slit is open the electron passes through that slit and there are no interference patterns on the plate. If two slits are open the electron passes through both slits and there are interference patterns. The claim that the electron passes through both slits relies on EOL and electrons as public objects to make a paradoxical claim that corresponds to the QM formulation of the electron system as being in a superposition of states. Suppose that the diaphragm has area *A* and that there are n square slits of area *A/n*, then the electron path is a superposition of *n* paths. The diaphragm, however, has effectively disappeared, yielding an experimental basis for the Feynman formulation of QM as path integrals. According a foundational role to such paths gives a distinctively quantum account a foundational role.

The Gell-Mann–Hartle (G-H) formulation presents consistent histories as an extension of Feynman path integrals. The original formulation of consistent histories given by Griffiths (1986) was presented as an independent formulation of QM, compatible with but not dependent on Feynman. The Griffiths formulation broke with the orthodox interpretation by treating closed systems, by not assigning measurement a foundational role, and by insisting that quantum mechanics supply an account of all basic processes including measurements.[2] The Griffiths formulation has three basic features. First, there is the specification of a closed system at particular times by a series of events. An event is the specification of the properties of a system through a projection operator for the Hilbert sub-space representing the property. Second, there is a stochastic dynamics. Though Griffiths relied on Schrödinger dynamics, he treated it as going from event to event, rather than as a unitary evolution of a system prior to measurement and collapse. The events could be steps in a uniform evolution, measurements, interaction with the environment, or whatever. At this stage there is no distinction between real and virtual processes. A history is a time-ordered sequence of events. It is represented by projectors on a tensor product of the Hilbert spaces of the events. Third, a consistency condition is imposed on histories, or families of histories. Only those that may be assigned classical probabilities are given a physical interpretation.

A comparison with classical physics clarifies the status accorded quantum histories. Consider classical statistical mechanics, where the state of a system is represented by a point in phase space and the evolution of the system, or its history, by

---

[2] This is based on Griffiths (1984, 1996, 2002a, b), Griffiths and Hartle (1997) and on Griffiths's helpful comments on an earlier draft of this material. I have given a more detailed summary in MacKinnon (2009a).

the trajectory of this point. The phase space may be coarse-grained by dividing it into a set of cells of arbitrary size that are mutually exclusive and jointly exhaustive. A cell will be assigned a value 1 if the point representing the system is in the cell, and has the value 0 otherwise.. We introduce a variable, B_i for these 0 and 1 values, where the subscript, i, indexes the cells. These variables satisfy

$$\sum_i B_i = 1 \qquad\qquad B_i B_j = \delta_{ij} B_j$$

This assignment of 0 and 1 values supports a Boolean algebra. To represent a history, construct a Cartesian product of copies of the phase space and let them represent the system at times $t_0,\ t_1,\ \ldots,\ t_n$. Then the product of the variables, $\{B_i\}$, for these time slices represents a history. The relation to classical probabilities can be given an intuitive expression. The tensor product of the successive phase spaces has a volume with an a priori probability of 1. Each history is like a hole dug by a phase-space worm through this volume. Its a priori probability is the ratio of the volume of the worm hole to the total volume. The probability of two histories is additive provided the worm holes don't overlap. In the limit the total volume is the sum of a set of worm holes that are mutually exclusive and jointly exhaustive.

Quantum mechanics uses Hilbert space, rather than phase space and represents properties by sub-spaces. The correlate to dividing phase space into cells is a decomposition of the identity, dividing Hilbert space into mutually exclusive and jointly exhaustive subspaces whose projectors satisfy:

$$\sum_i B_i = 1 \qquad B_i^\dagger = B_i \qquad B_i B_j = \delta_{ij} B_j \tag{7.3}$$

Corresponding to the intuitive idea of a wormhole volume the *weight operator* for a history is

$$K(Y) = E_1 T(t_1, t_2) E_2 T(t_2, t_3) \cdots T(t_{n-1}, t_n) E_n, \tag{7.4}$$

where E stands for an event or its orthogonal projection operator, $T(t_1, t_2)$ is the operator for the evolution of the system from t_l to t_2. Equation (7.4) can be simplified by using the Heisenberg projection operators

$$\hat{E}_j = T(t_r, t_j) E_j T(t_j, t_r) \tag{7.5}$$

leading to

$$\hat{K}(Y) = \hat{E}_1 \hat{E}_2 \cdots \hat{E}_n. \tag{7.6}$$

Then the weight of a history may be defined in terms of an inner product

$$W(Y) = \langle K(Y), K(Y') \rangle = \langle \hat{K}, \hat{K}' \rangle. \tag{7.7}$$

The significance of this equation, defined on the vector space of operators, may be seen by the phase-space comparison used earlier. Classical weights used to assign probabilities are additive functions on the sample space. If E and F are two disjoint collections of phase-space histories, then $W(E \bigcup F) = W(E) + W(F)$. Quantum weights should also satisfy this requirement, since they yield classical probabilities and must be non-negative. As Griffiths (2002a, 121–124) shows, Eq. (7.7) achieves this. Quantum histories behave like classical histories to the degree that mutual interference is negligible. This is the key idea behind the varying formulations of a consistency condition. If two histories are sufficiently orthogonal, $\langle K(Y), K(Y')\rangle \approx 0$, then their weights are additive and can be interpreted as relative probabilities. This idea of mutual compatibility may be extended to a *family* of histories. Such a family is represented by a consistent Boolean algebra of history projectors. This may be extended from a family of projectors, $\mathcal{F}$ to a refinement, $\mathcal{G}$, that contains every projector in $\mathcal{F}$.

This consistency requirement concerns pairs of histories that may be assigned probabilities. Essentially, it is the requirement that interference between two histories is negligible. Interference and superposition are not eliminated. They are essential features of the formulation. This consideration introduces the basic unit for interpretative consistency, a *framework*, a single Boolean algebra of commuting projectors based upon a particular decomposition of the identity.[3] A framework supplies the basis for quantum reasoning in CH. Almost all the objections to the CH interpretation are countered by showing they violate the single framework rule, or by a straightforward extension, the single family rule. This notion, accordingly, requires critical analysis.

There are two aspects to consider: the relation between a framework and quantum reasoning, and whether the framework rule is an ad hoc imposition. The first point is developed in different ways by Omnès and Griffiths. Omnès develops what he calls consistent (or sensible) logics. The logic is standard; the way it is applied is not. A consistent logic applies to a framework and by extension to families of histories. If two families differ in any detail, then they have different logics (Omnès 1992, p. 155). A specific logic that is consistent may become inconsistent by changing the framework, e.g., using a larger radius (Ibid, p. 174). In the standard philosophical application of logic to theories, one first develops a logic system, or syntax, and then applies it. The content to which it is applied does not alter the logic. Omnès uses 'logic' for an interpreted set of propositions. This terminology does not imply a non-standard logic. However, it may occasion misunderstanding.

Griffiths focuses on frameworks and develops the logic of frameworks by considering simple examples and using them as a springboard to general rules The distinctive features of this reasoning confined to a framework can be seen by contrast with more familiar reasoning. Consider a system that may be characterized by two

[3] This idea of a distinctive form of quantum reasoning was developed by Omnès (1994 chaps. 9 and 12), and in Griffiths (1999, 2002a, chap. 10).

or more complete sets of compatible properties. The Hilbert space representing the system may be decomposed into different sets of subspaces corresponding to the different sets of compatible properties. To simplify the issue take σ_x^+ and σ_z^+ as the properties. Can one attach a significance or assign a probability to 'σ_x^+ AND σ_z^+'? In CH propositions are represented by projectors of Hilbert subspaces. The representation of σ_x requires a two-dimensional subspace with states $\mid X^+\rangle$ and $\mid X^-\rangle$, projectors $X^{\pm} = \mid X^{\pm}\rangle\langle X^{\pm} \mid$, and the identity, $I = X^+ + X^-$. One cannot represent 'σ_x^+ AND σ_z^+' in any of the allowed subspaces. Accordingly it is dismissed as 'meaningless'.

Griffiths has answered the technical objections brought against his formulation. I am more concerned with the philosophical problems concerning truth and meaning. These problems arise from the way the formalism is deployed. So, we begin with technical difficulties and move to philosophical problems. The distinctive features and associated difficulties of this framework reasoning are illustrated by Griffiths's reworking of Wheeler's (1983) delayed choice experiment. Both Wheeler and Griffiths (1998) consider a highly idealized Mach-Zehender interferometer (Fig. 7.1).

The classical description in terms of the interference of light waves may be extended to an idealized situation where the intensity of the laser is reduced so low that only one photon goes through at a time. Here S and L are beam-splitters, M_1 and M_2 are perfect mirrors, and C, D, E, and F are detectors. If D registers, one infers path d; if C registers, then the path is c. If C and D are removed, then the detectors E and F can be used to determine whether the photon is in a superposition of states. Wheeler's delayed choice was based on the idealization that detectors C and D could be removed after the photon had passed through S. It is now possible to implement such delayed choice experiments, though not in the simplistic fashion depicted.

To see the resulting paradox assume that detectors C and D are removed and that the first beam splitter leads to the superposition, which can be symbolized in abbreviated notation as

$$\mid a\rangle \mapsto \mid s\rangle = (\mid c\rangle + \mid d\rangle)/\sqrt{2}, \tag{7.8}$$

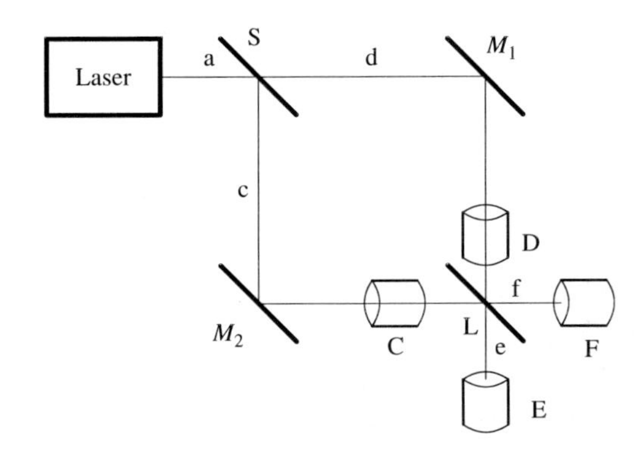

Fig. 7.1 A Mach-Zehender interferometer

where $| a\rangle$, $| c\rangle$, and $| d\rangle$ are wave packets at the entrance and in the indicated arms. Assume that the second beam splitter L leads to a unitary transformation

$$| c\rangle \mapsto | u\rangle = (| e\rangle + | f\rangle)/\sqrt{2}, \quad | d\rangle \mapsto | v\rangle = (- | e\rangle + | f\rangle)/\sqrt{2}, \tag{7.9}$$

with the net result that

$$| a\rangle \mapsto | s\rangle \mapsto | f\rangle. \tag{7.10}$$

Equations (7.8) and (7.10) bring out the paradox. If the detectors, C and D were in place, then the photon would have been detected by either C or D. If it is detected by C, then it must have been in the c arm. If the detectors are removed and the F detector registers, then it is reasonable to assume that the photon passed through the interferometer in the superposition of states given by Eq. (7.8). The detectors were removed while the photon was already in the interferometer. It may seem reasonable to ask what state the photon was in before the detectors were removed. Here, however, intuition is a misleading guide to the proper formulation of questions in a quantum context.

Griffiths treats this paradox by considering different families of possible histories. Using C and D for the ready state of detectors, considered as quantum systems, and C* and D* for triggered states then one consistent family for the combined photon-detector system is

$$|a\rangle|CD\rangle \longrightarrow \begin{pmatrix} |c\rangle|CD\rangle \longrightarrow |C^*D\rangle \\ |d\rangle|CD\rangle \longrightarrow |CD^*\rangle \end{pmatrix} \tag{7.11}$$

Here $|a\rangle|CD\rangle$ represents a tensor product of the Hilbert spaces of the photon and the detector. Equation (7.11) represents a situation in which the photon enters the interferometer and then proceeds either along the c arm, triggering C^* or along the d arm, triggering D^*. These paths and outcomes are mutually exclusive.

For the superposition alternative, treated in Eqs. (7.8), (7.9), and (7.10), there is a different consistent family of histories,

$$|a\rangle|EF\rangle \longrightarrow |s\rangle|EF\rangle \longrightarrow \begin{pmatrix} |e\rangle|EF\rangle \longrightarrow |E^*F\rangle \\ |f\rangle|EF\rangle \longrightarrow |EF^*\rangle \end{pmatrix} \tag{7.12}$$

Equation (7.12) represents superposition inside the interferometer and exclusive alternatives after the photon leaves the interferometer. In accord with Eq. (7.10) the upper history in Eq. (7.12) has a probability of 0 and F^* is triggered.

Suppose that we replace the situation represented in Eq. (7.12) by one in which the photon is in either the c or d arms. There is no superposition within the interferometer, but there is when the photon leaves the interferometer. This can be represented by another consistent family of histories,

$$|a\rangle|EF\rangle \longrightarrow \begin{pmatrix} |c\rangle|EF\rangle \longrightarrow |u\rangle|EF\rangle \longrightarrow |U\rangle \\ |d\rangle|EF\rangle \longrightarrow |v\rangle|EF\rangle \longrightarrow |V\rangle \end{pmatrix}, \tag{7.13}$$

where

$$|U\rangle = (|E^*F\rangle + |EF^*\rangle)/\sqrt{2},$$
$$|V\rangle = (-|E^*F\rangle + |EF^*\rangle)/\sqrt{2}.$$

Both $|U\rangle$ and $|F\rangle$ are Macroscopic Quantum States (MQS), or Schrödinger cat states. The formalism allows for such states. However, they are not observed and do not represent measurement outcomes. This delayed choice example represents the way traditional quantum paradoxes are dissolved in CH. Reasoning is confined to a framework. Truth is framework-relative. The framework is selected by the questions the physicist imposes on nature. If a measurement has an outcome, then one must choose a framework that includes the outcome. Within a particular framework, there is no contradiction. One is dealing with consistent histories. The traditional paradoxes all involve combining elements drawn from incompatible frameworks.

This and other examples show how the formalism is explicitly adapted to a measurement situation. Adapting the formalism to fit an observed outcome, such as a particle location or interference in a Mach-Zehender interferometer, plays the same role as the projection postulate in orthodox QM. This differs from the measurement interpretation in precisely the way Omnès (1994, chap. 2, 1999, p. 80) and Griffiths (2002a, Preface) indicated. In the measurement interpretation experimental results as reported supply the basis and the formalism is used as a tool. In the CH formulation one begins with families of consistent histories and selects the one that fits the observed results. Experimental results are expressed in the language of the theory. In no case does the formalism give a reason for selecting only one of various possibilities. The experiment is taken as a source of information. This clearly illustrates how the CH interpretation is downwardly compatible with the measurement interpretation in the domain of this interpretation's established validity.

These experimental analyses lead to two general principles:

1. *A quantum mechanical description of a measurement with particular outcomes must employ a framework in which these outcomes are represented.*
2. *The framework used to describe the measuring process must include the measured properties at a time before the measurement took place.* This embodies the experimental practice of interpreting a pointer reading in the apparatus after the measurement as recording a property value characterizing a system before the measurement.

Here my evaluation differs from Griffiths and Omnès. The CH analysis of actual and idealized experiments relies on quasiclassical state functions like $|C^*D\rangle$, indicating that the C detector has been triggered and the D detector was not. These are needed to formulate CH as a closed system. There is no outside observer. However, these are place holders for equivalence classes of state functions, that will never be specified in purely quantum terms. In treating only closed systems one treats measurement *as a process*, but not measurement *as a measurement*. In an actual measurement one does not rely on $|C^*D\rangle$, but on a description of a measurement situation in the standard language of physics. Experimental values are given by an outside observer performing measurements on a system. This treats measurement *as a measurement*, not as a quantum process. This put us back in the realm where the Copenhagen interpretation has a well established success. In the laboratory one carries on with physics as usual. Because of the way it is constructed the CH formulation parallels the Copenhagen interpretation with a projection postulate, or the measurement interpretation. The CH formulation/interpretation is not a stand alone interpretation in this practical sense. However, it can function as a replacement in answering theoretical questions about QM as a foundational system.

7.2.1 Criticisms of Consistent Histories

The objections brought against the CH interpretation cluster around the border separating physics from philosophy. The technical physical objections have been answered largely by showing that confining quantum reasoning to a framework eliminates contradictions.[4] Here we will focus on the more philosophical issues of truth and meaning. The basic objection is that the CH interpretation makes meaning and truth framework relative.

Adrian Kent has brought the issue of meaning to the forefront.[5] Consider two histories with the same initial and final states and intermediate states σ_x and σ_z, respectively. In each history one can infer the intermediate state with probability 1. A simple conjunction of two true propositions yields 'σ_x AND σ_z'. Griiffiths and Hartle contend, and Kent concedes, that there is no formal contradiction since the intermediate states are in separate histories. Kent finds this defense arbitrary and counter-intuitive. Our concepts of logical contradiction and inference are established prior to and independent of their application of quantum histories. If each intermediate state can be inferred, then their conjunction is meaningful.

The issue of truth arises when one considers the ontological significance of assigning values to properties. In classical physics assigning a value to a property means that the property *possesses* the value. Copenhagen quantum physics fudges this issue by speaking only of the *assignment* of properties in particular experimental situations. The CH interpretation exacerbates the difficulty. A realistic interpretation of projectors take them as representing the properties a system *possesses* at a time. This does not fit the Griffiths treatment of the delayed choice experiment when one asks what position the photon really had at time t_2. Thus, d'Espagnat (1995, chap. 11) argues that the CH interpretation involves inconsistent

[4] See Griffiths (2002b, chaps. 20–25) for a detailed treatment of objections and quantum paradoxes.

[5] Kent (1996) was answered by Griffiths and Hartle (1997), which was answered by Kent (1998).

property assignments. In a similar vein Bub (1997, p. 236) expressed the objection that if there are two quasiclassical histories of Schrödinger's cat, then one does not really know whether the cat is alive or dead. Bassi and Ghirardi (1999) make the issue of truth explicit. The attribution of properties to a system is true if and only if the system actually possesses the properties. They find Griffiths's reasoning "shifty and weak", implying the coexistence of physically senseless decoherent families. This criticism extends to probabilities. From an ontological perspective probabilities of properties must refer to objective and intrinsic properties of physical systems. There is, they claim, no other reasonable alternative. If they referred to the possibilities of measurement results, then this would be a measurement interpretation, not a replacement for it. Goldstein (1998) argues that the CH interpretation cannot be true, since it contradicts established no-go theorems.

To treat the framework relevance of truth we should distinguish 'truth' and 'true'. Correspondence theories of truth begin with obvious examples like

> "The cat is on the mat" is true iff the cat is on the mat.

This looks unproblematic in the context of someone who sees the cat and understands the claim. It becomes highly problematic when one argues from the acceptance of a theory as true to what the world must be like to make it true. This has been treated in detail in preceding chapters. I will summarize the pertinent aspects. Adapting Davidson's semantics, we took the normal use of true as a semantic primitive. Acceptance of a claim as true implicitly presupposes the acceptance of a vast but amorphous collection of truths. Consider our old example: "This shirt is yellow" is true iff the shirt is yellow. This presupposes the normal assimilation of the language of objects, properties, and color terms. One might impose an ordinary language version of the Bassi-Ghiradi criterion and assert that "This shirt is yellow" is true iff the shirt possesses the property of being yellow. But, the traditional argument goes, yellow is not a property of objects. So the claim is objectively false. As many philosophers have pointed out, this argument is ultimately self-destructive. Acceptance of claims, such as attributing colors to objects, as true is a necessary condition for learning how to use the language of color terms. In the terminology of the present debate we could say that the truth of the claim is framework relevant. Here, however, the framework is the ordinary language that extensions presuppose.

Earlier we showed how this the extension of ordinary language in the development of physics is controlled by constraints. The classical/quantum divide presented a crisis for the normal process of language extension. Thus, the observation of electrons traveling in trajectories did not support the general claim: "Electrons travel in trajectories". This relates to both meaning and truth. The meaning of terms like 'trajectory' is set by their use in ordinary language and its extension to the language of physics. A reliance on these established meanings is a necessary condition for the unambiguous communication of experimental information and for standard material inferences. This entails that the truth of the preceding claim is framework relevant. It applies to a single slit experiment. It does not apply to the scattering of electrons off nickel crystals. Bohr eventually crystallized his position on this by an idiosyncratic use of 'phenomenon', to include the total experimental situation in which a claim is made. Wheeler's treatment of the delayed choice experiment explicitly relied on Bohr's use of 'phenomenon'. Within a phenomenon one uses the classical extension of ordinary language. Hence one can rely on standard logic.

The CH use of 'framework' is downwardly compatible with Bohr's use of 'phenomenon'. This downward compatibility carries over to the issues of meaning, truth, and implication. Within the CH formulation the usage of these terms is framework relevant. If a claim can not be represented through the projectors proper to a particular framework, then the claim is not meaningful in that framework. Similarly, in a particular framework in which one asserts that the photon traveled along path c, one could make the corresponding claim: "The photon traveled through path c" is true. Such claims support material inferences. If only the C detector registers one infers that the photon traveled along path c. When such claims are transported to an ontological context buttressed by a correspondence theory of truth, then they may seem perverse. They do not address questions about where the photon **really** was before it was detected. When these claims are related to the semantics governing the normal use of language in quantum contexts, then they seem uniquely reasonable. An adequate defense of the CH formulation against such philosophical criticisms requires a recognition of the role of Bohrian semantics and of 'true' as a semantic primitive.

I doubt if the CH formulation/interpretation will ever replace standard QM. In dealing with actual, rather than thought, experiments one needs an outside observer performing measurements. A closed system does not treat measurement as measurement. However, the CH formulation clearly shows that there is a formulation of QM that treats QM processes as foundational, and is downwardly compatible with orthodox QM. This supplies the best currently available basis for investigating the consequences of accepting QM as the basic science of reality. To see how far it can be extended we switch from Griffiths and Omnès to Gell-Mann and Hartle.

7.3 The Gell-Mann–Hartle Project

Gell-Mann and Hartle developed an independent formulation of the consistent histories interpretation and expanded it to include quantum cosmology and a program for treating reductionism and the emergence of complexity.[6] This is a very ambitious project. It entails a radical modification of our understanding of physical reality. I will begin with a summary of the project, focusing more on the new reductionism than on the technical details. Then I will consider objections brought against this project.

It is now common to distinguish three different types of reductionism: ontological, epistemological, and methodological. Ontological reductionism is traditionally

[6] The project was formulated in Gell-Mann and Hartle (1990, 1993, 1995, 1996). Hartle (1993b) is a monograph presenting a clear summary. Gell-Mann (1994), Part II presents a non-technical summary. See also Cowan et al. (1999).

concerned with explaining macroscopic entities in terms of microscopic or submicroscopic entities. Epistemological reductionism, at least in it simpler forms, aims at reducing all knowledge to one kind of knowledge, e.g. sense impressions for the empirical tradition. Methodological reductionism is concerned with reducing, or at least explaining, phenomenological theories in terms of depth theories. The clash between ontological and epistemological reductionism has been clear since ancient times.[7] The new reductionism straddles these categories The G-H project involves explaining the emergence of the semi-classical realm from the quantum realm and then relating the semi-classical realm to classical reality. The G-H project treats quantum mechanics as the fundamental science, but does not specify the properties of ultimate objects.[8]

Before getting into the details of this project it is important to consider the significance of this or any similar program with the same general goal. Traditional ontological reductionism, from Democritos through Kim, is concerned with explaining complex entities and their distinctive properties in terms of the properties of the building blocks presumed to be basic. In spite of charges of materialism or atheism one is dealing with the familiar world. When quantum mechanics is taken as the basic science of reality, then one is taking as foundational something that is non-classical, highly counterintuitive. The familiar cliché is: Quantum mechanics is not only stranger than we imagine; it is stranger than anything we can imagine. It took 13.7 billion years to go from a pure quantum realm to our familiar world. No program can reproduce this. A more mane gable goal is to accept reality as depicted by quantum mechanics and then develop a program that shows the possibility of explaining features that characterize the reality depicted in classical physics on a quantum basis. This requires an abstract mathematical program.

The G-H project is not a matter of explaining properties of complex wholes in terms of the properties of their ultimate constituents. It is concerned with explaining the classical realm in terms of quantum mechanics as the court of last appeal in all matters physical. This presents an immediate and formidable difficulty. The quantum realm is characterized by superposition, interference, and distributed probability. Deterministic laws characterize the classical realm. The strategy for tackling this problem is one of using the resources of the quantum realm to construct a *quasi-classical realm*, something that supports structures and relations that have an approximate isomorphism to those found in the reality depicted by classical physics. This is done through a process of simplifying and systematizing selected aspects of the quantum realm. The tools used are decoherence, the substitution of hydrodynamic variables for the average values of energy, momentum and other dynamic variables, and further coarse graining. This is not the mereological reduction of

---

[7] Galen, the famous second century physician, attributed to Democritos the statement, "...wretched mind, do you who get your evidence from us [the senses], yet seek to overthrow us? Our overthrow will be your downfall. (Cited from Kirk and Raven (1962), fragment 593, p. 424). This conflict played a basic role in the transition from positivistic interpretations of science, emphasizing observation sentences, to post- positivist interpretations, emphasizing theories.

[8] A more detailed comparison of the old and new reductionisms is given in MacKinnon (2009b).

wholes to constitutive parts found in traditional reductionism. A philosophical analysis of realms and reductionism will be postponed to the next chapter.

The universe is the ultimate closed system. Now it is characterized by formidable complexity, of which we have only a very fragmentary knowledge. The big bang hypothesis confers plausibility on the assumption that in the instant of its origin the universe was a simple unified quantum system. If we sidestep the problem of a state function and boundary conditions characterizing the earliest stages,[9] we may skip to slightly later stages where space-time was effectively decoupled. Then the problem of quantum gravity may be postponed. The universe branched into subsystems. Even when the background perspective recedes over the horizon, a methodological residue remains, the treatment of closed, rather than open systems. To present the basic idea in the simplest form, consider a closed system characterized by a single scalar field, $\phi(x)$. The dynamic evolution of the system through a sequence of space-like surfaces is generated by a Hamiltonian labeled by the time at each surface. This Hamiltonian is a function of $\phi(\mathbf{x}, t)$ and the conjugate momentum, $\pi(\mathbf{x}, t)$. On a spacelike surface these obey the commutation relations, $[\phi(\mathbf{x}, t), \pi(\mathbf{x}', t)] = \imath\delta(\mathbf{x}, \mathbf{x}')$ (with $\hbar, c = 1$). Various field quantities (aka observables) can be generated by ϕ and π. To simplify we consider only non-fuzzy 'yes-no' observables. These can be represented by projection operators, $P(t)$. In the Heisenberg representation, $P(t) = e^{\imath Ht}\, P(t_0)\, e^{-\imath Ht}$.

A sum over histories formulation of QM allows different histories. Using the index, k, to distinguish histories and the subscript, α, to distinguish observables, an exhaustive set of 'yes-no' observables at one time is given by the set of projection operators, $\{P^k{}_{\alpha_k}(t_k)\}$. Since these are exhaustive and mutually exclusive,

$$\begin{aligned} P^k{}_{\alpha_k}(t_k)\, P^k{}_{\alpha'_k}(t_k) &= \delta_{\alpha_k \alpha'_k}\, P^k{}_{\alpha_k}(t) \\ \sum_{\alpha_k} P^k{}_{\alpha_k}(t_k) &= 1 \end{aligned} \tag{7.14}$$

A particular history can be represented by a chain of projection operators,

$$C_\alpha = P^n_{\alpha_n}(t_n) \cdots P^1_{\alpha_1}(t_1) \tag{7.15}$$

This is essentially the same as the Griffiths's formula, presented earlier. The novel factor introduced here is a coarse graining of histories. Coarse graining begins by selecting only certain times and by collecting chains into classes. The decoherence functional is defined as

$$D(\alpha', \alpha) = Tr[C'_\alpha\, \rho\, C^\dagger_\alpha], \tag{7.16}$$

where ρ is the density matrix representing the initial conditions. In this context 'decoherence' has a special meaning. It refers to a complex functional defined

[9] This is treated in Hartle (2002a, 2002b).

over pairs of chains of historical projectors. The basic idea is the one we have already seen. Two coarse grained histories decohere if there is negligible interference between them. Only decoherent histories can be assigned probabilities. Different decoherence conditions can be set (Gell-Mann and Hartle 1995). We will consider two.

$$\begin{aligned} \textit{Weak}: \quad Re\,Tr[C'_\alpha\, \rho\, C^\dagger_\alpha] &= \delta(\alpha'\alpha)P(\alpha) && (7.17)\\ \textit{Medium}: \quad Tr[C'_\alpha\, \rho\, C^\dagger_\alpha] &= \delta(\alpha'\alpha)P(\alpha) && (7.18)\end{aligned}$$

Weak decoherence is the necessary condition for assigning probabilities to histories. When it obtains the probability of a history, abbreviated as α is $P(\alpha) = D(\alpha\alpha)$. Medium decoherence relates to the possibility of generalized records. Here is the gist of the argument. Consider a pure initial state, $|\psi\rangle$ with $\rho = |\psi\rangle\langle\psi|$. Alternative histories obeying exact medium decoherence can be resolved into branches that are orthogonal, $|\psi\rangle = \sum_\alpha C_\alpha|\psi\rangle$. Only when this condition is met are the corresponding projectors unique. If the projectors did not form a complete set, as in weak decoherence, then the past is not fixed. Other decompositions are possible. This relates to the more familiar notion of records when the wave function is split into two parts, one representing a system and the other representing the environment, $R_\alpha(t)$. These could not count as environmental records of the state of a system if the past could be changed by selecting a different decomposition. Thus, medium decoherence, or a stricter condition such as strong decoherence, is a necessary condition for the emergence of a quasiclassical order.

It is far from a sufficient condition. The order represented in classical physics presupposes deterministic laws obtaining over vast stretches of time and space. The G-H project must show that it has the resources required to produce a quasiclassical order in which there are very high approximations to such large scale deterministic laws. At the present time the operative issue is the possibility of deducing such quasi-deterministic laws. The deduction of detailed laws from first principles is much too complex. Zurek, Feynman and Vernon, Caldeira and Leggett, and others initiated the process by considering simplified linear models. The G-H project puts these efforts into a cosmological framework and develops methods for going beyond linear models. The standard implementation of a linear model represents the environment, or a thermal bath, by a collection of simple harmonic oscillators. In an appropriate model the action can be split into two parts: a distinguished observable, q^i, and the other variables, Q_i, the ignored variables that are summed over.

The G-H project extends this to non-linear models, at least in a programmatic way. I will indicate the methods and the conclusions. As a first step we introduce new variables for the average and difference of the arguments used in the decoherence function:

$$\begin{aligned} X(t) &= 1/2(x'(t) + x(t))\\ \xi(t) &= x'(t) - x(t)\\ D(\alpha', \alpha) &= f(X, \xi) && (7.19)\end{aligned}$$

The rhs of Eq. (7.19) is small except when $\xi(t) \approx 0$. This means that the histories with the largest probabilities are those whose average values are correlated with classical equations of motion. Classical behavior requires sufficient coarse graining and interaction for decoherence, but sufficient inertia to resist the deviations from predictability that the coarse graining and interactions provide. This is effectively handled by an analog of the classical equation of motion. In the simple linear models, and in the first step beyond these, it is possible to separate a distinguished variable, and the other variables that are summed over. In such cases, the analog of the equation of motion has a term corresponding to the classical equation of motion, and a further series of terms corresponding to interference, noise and dissipation. The factors that produce decoherence also produce noise and dissipation. This is handled, in the case of particular models, by tradeoffs between these conflicting requirements. The goal is to produce an optimum characteristic scale for the emergence of classical action. In more realistic cases, where this isolation of a distinguished variable is not possible, they develop a coarse graining with respect to hydrodynamic variables, such as average values of energy, momentum, and other conserved, or approximately conserved, quantities. A considerable amount of coarse graining is needed to approximate classical deterministic laws. Further complications, such as the branching of a system into subsystems, present complications not yet explored in a detailed way. Nevertheless the authors argue that they could be handled by further extensions of the methods just outlined. Since this is an ongoing project, it is reasonable to assume that such extensions will be developed.

Before considering objections it is helpful to consider what kind of a project we are dealing with. I take it as a kind of abstract reverse engineering.

> In a universe governed at a fundamental level by quantum-mechanical laws, characterized by indeterminacy and distributed probabilities, what is the origin of the phenomenological, deterministic laws that approximately govern the quasiclassical domain of everyday experience? What features of classical laws can be traced to their underlying quantum-mechanical origin? (Gell-Mann and Hartle (1993), p. 3345)

The G-H project is essentially a form of methodological reductionism. The goal is to show how a consistent histories formulation of quantum mechanics plus the G-H project supplies a possibility for explaining the already established form of classical laws. It relates to ontological reductionism in a somewhat anticipatory fashion. The ultimate ingredients of the universe, e.g., superstrings, branes, loop gravity, space-time foam, or something yet unknown, is presumed to supply the basic building blocks. Regardless of what these ultimate constituents are, they are presumed to be quantum systems. Quantum mechanics, in some formulation, is assumed to be the basic science of material reality. There is another significant difference. Traditional reductionism is essentially synchronic. The new reductionism is diachronic, or evolutionary. The programmatic point of departure is the state of the universe at the moment of its inception. Though this is not known, it seems reasonable to assume that it is a quantum state, perhaps a pure state. Subsequent evolution involves multiple branching and 'frozen accidents'. The form that galaxies, organic molecules, DNA, mammalian prototypes, etc., took is partially due to accidental features in evolutionary history. Yet, these features are passed on to later members of the branch.

If a consistent histories formulation of quantum mechanics is accepted as a fundamental account of reality, then it should have the resources required to reproduce classical physics, at least in its essential features. In a methodological perspective this involves an interrelation of two realms, the quantum realm and the quasiclassical realm. Classical reality and the lived world are not treated as realms, but as phenomenologicl manifestations of the quasiclassical realm. The problems this generates will be treated in the next chapter. In the quantum realm the basic units considered are triples, $\{\{C_\alpha\}, H, \rho\}$, representing: a set of alternative coarse-grained histories of a closed system; a Hamiltonian, connecting field operators at different times through Heisenberg equations of motion; and a density matrix, representing the initial conditions. This is an abstract schematism that does not have physical content until H is specified in terms of fundamental fields. For programmatic purposes, however, such a specification is not required. The program involves decoherence and further coarsegraining through considerations of hydrodynamic variables, the treatment of inertia, friction, and dissipation. This should lead to a quasiclassical realm. This is taken to be a set of histories (or a class of nearly equivalent sets) maximally refined, consistent with obeying a realistic principle of decoherence, and exhibiting patterns of approximately deterministically correlations governed by phenomenological classical laws connecting similar operations at different times. The basic task, accordingly, is to show the possibility of deducing a quasiclassical realm from the basis and project just summarized. To be complete, this must include the conditions for the possibility of IGUSes, information gathering and utilizing systems. Without this the quasiclassical realm cannot treat the functions observers play in the classical realm. The term 'IGUS' is defined in a broad enough way to encompass human or alien observers, computers, animals, bacteria, and immune systems.

In this perspective a fundamental problem is the possibility of multiple inequivalent quasiclassical realms. This requires a more detailed specification. From the to-be-implemented specification of basic fields, it should be possible to define quasiclassical operators through sets of orthogonal projectors, $\{P_\alpha(t)\}$. A quasiclassical operator is a local operator averaged over small regions of space at a sequence of times. An example is an operator representing the center of mass of a set of decoherent histories. The projectors of this operator would have smeared out average values, and would be represented by a Hilbert subspace. The problem of inequivalent representations stems from the fact that many such sets of projectors are possible.

As a first step in tackling this problem Gell-Mann and Hartle (1996) clarify what is meant by equivalent sets. A reassignment of time intervals would lead to an equivalent set. A transformation from old to new field variables, $\tilde{\phi}(x) = \tilde{\phi}(x; \phi(y), \pi(y))$; $\tilde{\pi}(x) = \tilde{\pi}(x, \phi(y), \pi(y))$, would lead to a physically equivalent description, and a physically equivalent triplet, $\{\{\tilde{C}_\alpha\}, \tilde{H}, \tilde{\rho}\}$. Any measure of classicality should be defined on equivalent classes of physically equivalent histories. Here the analogy with the statistical mechanics/thermodynamics situation is helpful. A gas is described thermodynamically by listing ingredients and their percentages, pressure, temperature and volume. This thermodynamic description can correspond to an extremely large number of molecular configurations. In practice, one never specifies any particular configuration. Rather, one gets equivalence classes of configurations that support the same average values of kinetic energy, momentum, and intermolecular distances. Statistical laws involving these variables are approximately represented by deterministic laws involving P, V, and T. The statistical account, however, can also support conclusions at variance with thermodynamics, e.g., local or short-time decreases in entropy. Similarly, the G-H project allows for the possibility of quasiclassical systems and also for non-classical systems. It does not supply a criterion for picking out any one of the many possible quasiclassical accounts. Nor does it treat classical reality as a conceptual system, something we will consider in the next chapter.

Dowker and Kent (1995, 1996) criticized the CH interpretation as arbitrary and the implementation just summarized as incomplete. We will separate the issue of the arbitrariness of the CH formulation from the incompleteness of the G-H project. To implement the charge of arbitrariness, Dowker and Kent consider a system whose initial density matrix, ρ_i is given along with the normal complement of Hilbert-space observables. Events are specified by sets, σ_j of orthogonal Hermitian projectors, $P^{(i)}$, characterizing projective decompositions of the identity at definite times. Thus,

$$\sigma_j(t_i) = \{P_I{}^{(i)} : i = 1, 2, \ldots, n_j\}_{t_j}$$

defines a set of projectors obeying Eq. (7.3) at time t_i. Consider a list of sets and time sequences. The histories given by choosing one projection from each set in all possible ways are an exhaustive and exclusive set of alternatives, $\mathcal{S}$. Dowker and Kent impose the Gell-Mann–Hartle medium decoherent consistency conditions, restrict their considerations to exactly countable sets, consider consistent extensions of $\mathcal{S}$, $\mathcal{S}'$, and then ask how many consistent sets a finite Hilbert space supports. The answer is a very large number. This prompts two interrelated questions. How is one set picked out as the physically relevant set? What sort of reality can be attributed to the collection of sets?

Griffiths (1998) countered that these extended sets are meaningless. Their construction leads to histories that could not be assigned probabilities. To make the difficulty more concrete consider the simplest idealized realization of the Dowker-Kent *Ansatz*, a silver atom passing through a Stern-Gerlach (SG) magnet. We will use the simplified notation, X, Y, and Z, for spin in these directions. At t_1 there are three families:

$$X_+(t_1), X_-(t_1) \qquad Y_+(t_1), Y_-(t_1) \qquad Z_+(t_1) Z_-(t_1)$$

The passage from t_1 to t_n allows of 6^{2n} possible histories. For the simple point we wish to make we consider 6 of the 36 possible histories leading form t_1 to t_2

$$(a)X_+(t_1)X_+(t_2)\ (c)X_+(t_1)Y_+(t_2)\ (e)X_+(t_1)Z_+(t_2)$$
$$(b)X_+(t_1)X_-(t_2)\ (d)X_+(t_1)Y_-(t_2)\ (f)X_+(t_1)Z_-(t_2)$$

The formalism does not assign probabilities to these histories. Here the appropriate experimental context would be successive SG magnets with various orientations. Suppose that the atom passes through an SG magnet with a X orientation at t_1 and one with a Z orientation at t_2, then only (e) and (f) can have non-zero probabilities. The selection of histories as meaningful is determined by the questions put to nature in the form of actual or idealized experimental setups. The fact that the formalism does not make the selection is not a shortcoming.

The final objection we will consider is the Dowker-Kent claim that the G-H project cannot demonstrate the preservation of a quasiclassical order. This, I believe, is true. If one accepts the G-H project as a deductive system, then this is a serious, or fatal, objection. If one thinks of the project as a special case or reverse engineering, then the objection is not fatal. The classical order and its perseverence is accepted as something to be explained. The formalism allows for the evolution of large equivalence classes only some of which are quasi-classical. The formalism does not supply a selection principle. The G-H project was never presented as a deductive theory. The goal was to see whether the acceptance of QM as the fundamental science of physical reality allowed for an explanation of the large-scale deterministic laws characterizing classical physics, a reverse engineering project that might eventually lead to a more formal theory.

Consider a hacker trying to reverse engineer a computer game of shooting down alien invaders and assume that he has developed a machine language formulation that accommodates the distinctive features of the alien game at a certain stage of the action. Any such machine language formulation admits of an indefinitely large number of extensions, only a minute fraction of which would preserve ‘quasialienality’. This is not an impediment. The hacker is guided by a goal, reproducing a functioning game, rather than by the unlimited possibilities of extending machine-language code. The G-H project has shown the possibility of programmatically reproducing basic features of the deterministic laws of classical physics. To achieve this goal the project relies on decoherence and various approximations. It is misleading to treat the result as if it were an exact solution capable of indefinite extension. I take the G-H project as a demonstration that it is possible to accept QM as foundational, distinctively QM properties as characterizing basic reality, and outline a schematic program for reproducing distinctive features of classical reality.

Before trying to relate these we should consider some consequences of accepting complementary representations as a basis for a descriptive account. This implies that any attempt to develop an ontology through an analysis of QM or QFT is misguided. This can only give an ontology based on a conceptual foundation known to be inadequate to the reality treated. One might try to find some category, such as ‘event’. ‘process’, ‘haeccitas’, or ‘trope’ that could be shaped to include ‘particle’ and ‘field’. This would obscure the role these concepts play in supporting inferences. The conclusion that should be drawn, in my opinion, is that categories rooted in ordinary language cannot be extended to the quantum realm in a way that supports ontological inferences. Quantum reality is not only stranger than we imagine, it is also stranger than anything we can imagine in an intuitive representation or adequately represent in a conceptual system. The need for complementary accounts shows that the quantum realm is beyond the limits of language.

References

Bassi, Angelo, and Gian Carlo Ghirardi. 1999. Decoherent Histories and Realism. *arXiv:quant-ph/912031.*

Bjorken, James, and Sidney Drell. 1964. *Relatiivistic Quantum Mechanics*. New York, NY: McGraw-Hill.

Bub, Jeffrey. 1997. *Interpreting the Quantum World*. Cambridge: Cambridge University Press.

CMS: The Computing Project; Technical Design Report. 2005. CERN-LHCC-2005-022.

Cowan, G. A., David Pines, and David Meltzer. 1999. *Complexity Metaphors, Models, and Reality*. Cambridge, MA: Perseus Books.

De Witt, Bryce, and Neill Graham. 1973. *The Many-Worlds Interpretation of Quantum Mechanics*. Princeton, NJ: Princeton University Press.

D'Espagnat, Bernard. 1995. *Veiled Reality: An Analysis of Present-Day Quantum Mechanical Concepts*. Reading, MA: Addison-Wesley.

Dowker, Fay, and Adrian Kent. 1995. Properties of Consistent Histories. *Physical Review Letters, 75*, 3038–3041.

Dowker, Fay, and Adrian Kent. 1996. On the Consistent Histories Approach to Quantum Mechanics. *Journal of Statistical Physics, 82*, 1575–1646.

Everett, Hugh. 1957. Relative State Formulation of Quantum Mechanics. *Reviews of Modern Physics, 29*, 454–462.

Foldy, Leslie, and Siegfried Wouthuysen. 1950. On the Dirac Theory of Spin 1/2 Particles and Its Non-Relativistic Limit. *The Physical Review, 76*, 29–36.

Galison, Peter. 1987. *How Experiments End*. Chicago, IL: University of Chicago Press.

Galison, Peter. 1997. Pure and Hybrid Detectors: Mark I and the Psi. In Lillian Hoddeson, et al. (eds.), *The Rise of the Standard Model* (pp. 308–337). Cambridge: Cambridge University Press.

Gell-Mann, Murray, and James Hartle. 1993. Classical Equations for Quantum Systems. *Physical Review D, 47*, 3345–3382.

Gell-Mann, Murray. 1994. *The Quark and the Jaguar: Adventures in the Simple and the Complex*. New York, NY: W. H. Freeman and Company.

Gell-Mann, Murray, and James B. Hartle. 1995. *Strong Decoherence*. arXiv:gr-qc/9509054 v4

Gell-Mann, Murray, and James B. Hartle. 1996. *Equivalent Sets of Histories and Quasiclassical Realms*. arXiv:gr-qc/9404013 v3

Gell-Mann, Murray. 1997. Quarks, Colors, and QCD. In Lillian Hoddeson, Laurie Brown, Michael Riordan, and Max Dresden (eds.), *The Rise of the Standard Model* (pp. 625–633). Cambridge: Cambridge University Press.

Gell-Mann, Murray. 1999. Complex Adaptive Systems. In G. A. Cowan, Pines David, and David Meltzer (eds.), *Complexity Metaphors, Models, and Reality* (pp. 17–46). Cambridge, MA: Perseus Books.

Gell-Mann, Murray, and James Hartle. 1990. Quantum Mechanics in the Light of Quantum Cosmology. In W. H. Zurek (ed.), *Complexity, Entropy, and the Physics of Information*. Reading, MA: Addison-Wesley.

Ghirardi, G., and T. Rimini, A. Weber. 1986. Unified Dynamics for Microscopic and Macroscopic Systems. *Physical Review D, 34*, 470–476.

Goldstein, Sheldon. 1998. Quantum Theory Without Observers–Part One. *Physics Today, 51*, 42–47.

Griffiths, Robert. 1984. Consistent Histories and the Interpretation of Quantum Mechanics. *Journal of Statistical Physics, 36*, 219–272.
Griffiths, Robert. 1986. Correlations in Separated Quantum Systems. *American Journal of Physics, 55*, 11–18.
Griffiths, Robert B. 1996. Consistent Histories and Quantum Reasoning. *arXiv:quant-ph.*
Griffiths, Robert, and James Hartle. 1997. Comment on "Consistent Sets Yield Contrary Inferences in Quantum Theory." *arXiv:gr-qc/9710025 v1.*
Griffiths, Robert. 1998. Choice of Consistent Family and Quantum Incompatibility. *Physical Review A, 57*, 1604.
Griffiths, Robert B. 2002a. *The Nature and Location of Quantum Information.* Physical Review A, *66.*
Griffiths, Robert B. (ed.) 2002b. *Consistent Quantum Theory.* Cambridge: Cambridge University Press.
Halvorson, Hans, and Rob Clifton. 2002. No Place for Particles in Relativistic Quantum Mechanics. *Philosophy of Science, 69*, 1–28.
Hartle, James B. 1993. *The Reduction of the State Vector and Limitations on Measurement in the Quantum Theory.* Cambridge: Cambridge University Press.
Hartle, James B. 2002a. Theories of Everything and Hawking's Wave Function of the Universe. In G. Gibons, P. Shellard, and S. Rankin (eds.), *The Future of Theoretical Physics and Cosmology.* Cambridge: Cambridge University Press.
Hartle, James B. 2002b. The State of the Universe. *arXiv:gr-qc/02090476 v1.*
Healey, Richard. 2004. Gauge Theories and Holism. *Studies in History and Philosophy of Modern Physics, 35*, 619–642.
Kent, Adrian. 1998. Consistent Sets and Contrary Inferences: Reply to Griffiths and Hartle. *arXiv:gr-qc/9808016v2.*
Kent, Adrian. 2000. Quantum Histories and Their Implications. *arXiv:gr-qc/9607073v4.*
Kirk, G. S., and J. E. Raven 1962. *The Pre-Socratic Philosophers.* Cambridge, UK: Cambridge University Press.
Lincoln, Don. 2009. *The Quantum Frontier: The Large Hadron Collider.* Baltimore, MD: The Johns Hopkins University Press.
MacKinnon, Edward. 2009a. The Consistent Histories Interpretation of Quantum Mechanics. *PhilSciArchcives*, # 4549
MacKinnon, Edward. 2009b. The New Reductionism. *The Philosophical Forum, 34*, 439–462.
Omnès, Roland. 1994. *The Interpretation of Quantum Mechanics.* Princeton, NJ: Princeton University Press.
Omnès, Roland. 1999. *Understanding Quantum Mechanics.* Princeton, NJ: Princeton University Press.
Susskind, Leonard. 2006. The Cosmic Landscape: String Theory and the Illusion of Intelligent Design. New York, NY: Back Bay Books.
't Hooft, Gerard. 1997. *In Search of the Ultimate Building Blocks.* Cambridge: Cambridge University Press.
Tegmark, Max. 1997. The Interpretation of Quantum Mechanics: Many Worlds of Many Words? *arXiv:quant-ph/9709132v1.*
Teller, Paul. 1995. *An Interpretive Introduction to Quantum Field Theory.* Princeton, NJ: Princeton University Press.
Wallace, David. 2006. Epistemology Quantized: Circumstances in Which We Should Come to Believe in the Everett Interpretation. *The British Journal for the Philosophy of Science, 57*, 655–689.
Wheeler, John. 1983. Law Without Law. In J. Wheeler, and W. Zurek (eds.), *Quantum Theory and Measurement* (pp. 182–213). Princeton, NJ: Princeton University Press.
Wilczek, Frank. 2008. *The Lightness of Being: Mass, Ether and the Unification of Forces.* New York, NY: Basic Books.
Zee, A. 2003. *Quantum Field Theory in a Nutshell.* Princeton, NJ: Princeton University Press.

Chapter 8
Realism and Reductionism

Metaphysics is the finding of bad reasons for what we believe on instinct, but to find these reasons is no less an instinct.
F. H. Bradley, Appearance and Reality: An Essay in Metaphysics, p. x.

This chapter concludes a long journey through the main stream of physics. As promised, we viewed this stream from the vantage point of a bottom feeder, concentrating more on implicit presuppositions and interpretative perspectives than on outstanding achievements. This path began with mythological thought and the process of demythologizing that launched early Greek drama and philosophy. It led from lived-world semantics though the coupling of mathematics to physical accounts to the scientific revolution and eventually to the standard model. The role of presuppositions emerged with considerations of informal inferences, inferential networks and interpretative perspectives. This generates the larger problem of how conflicting perspectives fit together.

8.1 Physics in Perspective

The basic differences between the overall view of physics presented here and more or less standard presuppositions can be presented in a schematic fashion.

1. **The continuity of physics.** Since Kuhn, it is customary to interpret the development of physics in terms of separate discrete units: conceptual revolutions, successions of theories, paradigms, research programs, and problem-solving methodologies. I have focused on an underlying continuity, based on the development of the language of physics. The contrast could merely signal a difference in emphasis, since no one would deny some underlying linguistic continuity. However, the contrast can be accorded a deeper philosophical significance when the role of language is linked to disputed philosophical issues.
2. **The Status of Classical Physics.** In philosophical accounts classical physics is customarily presented as a collection of separately interpreted theories, primarily

E. MacKinnon, *Interpreting Physics*, Boston Studies in the Philosophy of Science 289,
DOI 10.1007/978-94-007-2369-6_8,

mechanics, thermodynamics, and electrodynamics. Our historical survey traced the path from attempts to develop atomistic mechanism to an informal unification of classical physics in which mechanical concepts played a central role. After the development of quantum mechanics an idealized version of classical physics came to play a role complementary to quantum physics. This idealized classical physics is a phenomenological account relative to the to-be-developed depth account. The representation of reality proper to classical physics is classical reality.

3. **The Role of Measurement.** Physics emerged from the matrix of medieval Aristotelianism through the pivotal role played by the concept of the quantity of a quality. As physics advanced the emphasis shifted from the ontological underpinning of quantitative concepts to mathematical formulations of systems of such concepts and eventually to the conditions for attaching numbers plus dimensions to quantities in a consistent fashion.
4. **The Interpretation of Quantum Mechanics.** This can be reconstructed in terms of stages. The first stage is the measurement interpretation, an austere version of the orthodox or Copenhagen interpretation of quantum mechanics. This effectively treats the mathematical formalism of quantum mechanics as an operational tool rooted in classical physics and in rules for the extension and limitation of classical concepts in the quantum realm. Here interpretation focuses on the experimental basis of QM and the conditions of the possibility of unambiguous communication of information, not on QM as a theory. This is a minimal interpretation and is clearly inadequate to contemporary developments in quantum physics and the philosophy of science. Yet, a proper understanding of the relation of classical to quantum physics requires a recognition of the role this stage played in the development of QM.
5. **Foundations of Quantum Mechanics.** QM should have its own proper foundation. This has not yet been achieved. In the preceding chapter I indicated why I took the Consistent Histories formulation/interpretation as the most promising new approach. The basic reason for this stems from the theme of this book. If one takes the mathematical formulation as the foundation of an interpretation, then the many worlds interpretation or some variant of foundationalism may seem preferable. If one focuses on the role of language and on meeting the conditions for the unambiguous communication of experimental information, then the preferred approach should be downwardly compatible with the measurement interpretation. At present only the CH formulation meets this requirement.

Much of this is tentative, provisional, subject to future revision and possible rejection. However, I believe that two aspects of the CH strategy are necessary features of any future revision. First, the basic non-classical features of QM systems must be accepted as characteristic of reality at a deeper level. Second, a consequence of this, classical reality must be regarded as an emergent system. This is the sharpest difference from the overt classicality of the measurement interpretation and the covert classicality of hidden variable theories or realist versions of foundationalism, that seek to eliminate superposition and interference from the basic representation of reality.

8.2 Continuity and Rationality

Kuhn's original stress on the incommensurability of old and new paradigms supported the accusation that paradigm shifts were irrational. Kuhn gradually weakened the significance he accorded 'incommensurability' in response to adverse criticism and historical counter-examples. Yet, the basic theme of relating ontologies to successive paradigms that differ from the ontology of ordinary language, and presumably differ from the ontology of future theories fostered a profound skepticism on ontological issues and also contributed to the denigration of physics in the science wars. The issue that emerged as central was the rationality of scientific change. This brings up the broader issue of what is meant by rationality.

We begin with the Kuhnian view of rationality as behavior sanctioned by a paradigm in the practice of normal science. Here normal science supplies the general standards for deciding what is reasonable. The normal theoretical physicist is the person who has mastered the textbooks and can solve the assigned problems, not the guy scratching his head and muttering "Hmm, that's funny!" Now switch from normal science to a crisis situation, where the normal reasoning sanctioned by a paradigm leads to anomalies, false conclusions, or contradictions. Reasoning within a paradigm no longer supplies an adequate guide for rational scientific behavior. Is there any other guide?

Plato, Hegel, Marx, Sellars, and many lesser philosophers have presented dialectics as the highest form of critical rationality. In a dialectical development one probes presuppositions, considers alternative viewpoints, checks the consequences of wild hypotheses, and reflects on one's own values. This process can get out of control. Yet, in Plato's dialogs, Hegel's search for a synthesis from the clash of a thesis and antithesis, or Sellars' practice or getting at an issue by beating around the neighboring bushes, the to-be-explained difficulty that launches the dialectic supplies a point of convergence. The scientific dialectic that often mediates the transition from an old to a new paradigm often manifest the same traits. The underlying continuity hinges on the language of discourse and the problems being considered, not from the theories being developed.

Rather than rehashing the developments considered to bring out the underlying conceptual continuity, we will simply refer to the two physicists who exhibited the most probing concern with the status of foundational concepts. In his *Treatise* Maxwell demonstrated in great detail that field theory and distance theory, though mutually incompatible, were empirically equivalent. This meant that the basic laws of electrodynamics did not require either foundation. This paved the way for Hertz's evaluation: Maxwell's theory is Maxwell's set of equations. When electrodynamics is treated as a phenomenological account anchored by measurements of field strengths and charges, then the incommensurability of field and distance concepts is not an impediment to intercommunication.

For the second example we consider the transition from the Bohr-Sommerfeld atomic model to quantum mechanics. The old model had been very successful in explaining atomic structure, spectral lines, and the effect of electrical and magnetic fields on these lines. Yet, there were outstanding problems, like the anomalous

Zeeman effect, and inconsistencies in the assignment and interpretation of quantum numbers. Sommerfeld, and his Munich students, exhibited the practice of normal science, focusing on the problems that could be treated within the model and slighting the fringe issues that went beyond the model's capacity. Bohr and his associates, chiefly Heisenberg, Pauli, and Kramers, focused on the difficulties and began a critical probing of basic assumptions. Others outside Bohr's inner circle were also probing presuppositions and seeking alternatives. Landé tried a semi-classical approach with his vector model. De Broglie introduced a wave-particle duality. Schrödinger abandoned the particle model and developed a wave formulation. Born sought new mathematical formulations. To label this transition irrational would be a semantic perversion. It is still rightly regarded as one of the high points in the development of physics as a rational explanation of nature.

The underlying philosophical difficulties stem from confining 'rationality' to a narrow framework: theories, paradigms, research programs, or problem-solving methodologies treated as explanatory units. We will cover them by a broad use of 'theory'. When we focus on the language of physics and include a distinction between a relatively phenomenological and a depth level then the overall development of physics is evolution through punctuated equilibrium. On a phenomenological level the language of physics advanced by including new terms and attaching special meanings to old terms for use in mechanics, thermodynamics, and electrodynamics. This functioning rests on the assignment of numbers to properties, not on an ontological clarification of the bearers of these properties. The units for force, charge, and temperature were set by a clarification of the process of measuring these quantities. This reliance on measurement requires a fundamental continuity between the language used in physics and ordinary language. An experimenter has to refer to his instruments, activities, presuppositions, and decisions as well as the results of the experiment. This continuity was established through the co-evolution of the language of physics and mathematics. These new and adapted terms fit the network model. Terms fit in a conceptual network that supported informal inferences and linguistic structures that were isomorphic to mathematical structures. This supported the dual inference system that characterizes functioning physics and the dialog between experimenters and theoreticians. The implicit functional ontology of EOL features public objects, objects that speakers can refer to on the assumption that informed listeners will also recognize them.

In the new physics these public objects essentially played a presuppositional role as the bearers of properties. The mathematical systematization applied to properties. Physics supported idiontology rather than ontology. The conceptual system supporting informal inferences and supplying a foundation for mathematical formulations had to meet a basic consistency requirement. If the conceptual core of EOL generated contradictions then no informal inferences were reliable. The contradictions that inevitably arose were generally handled by the Quinean strategy of pushing them from the core to the periphery and then resolving or taming them by one means or another. The extension of EOL across the classical/quantum divide generated repeated contradictions that resisted such Quinean solutions. The remedy that resolved such difficulties was to preserve the established meanings of the trouble-making terms but to restrict their usage to a framework, or a descriptive account anchored in the spatio-temporal location of a particular measuring apparatus.

A non-ontological aspect of this general development should be mentioned. Physics advanced by coupling physical accounts to mathematical formulas.[1] This gradually led to the inclusion of more and more mathematical forms: complex numbers, vectors, tensors, quaternions, statistics, probability theory, group theory, non-Euclidean geometry, abstract spaces, computer simulations, and the complex mathematics of relativity and quantum theory. This stimulated a rethinking of the role of mathematics in physics. Earlier developments presupposed a math-world correspondence. A number with dimensions designates the quantity a property possesses. In practice, numbers relate to physical reality through various types of measurements.

When we turn from a phenomenological level, or EOL, to theories as units, then we do encounter conceptual discontinuities. In a punctuated equilibrium account new theories, like new species, can rapidly replace their predecessors. Here it is important to evaluate theory replacement in physics by the norms proper to physics, rather than those adapted from philosophical theories. Newton's *Principia* initiated a decisive change in the status of physics. Newton himself recognized the charge of irrationality as a serious difficulty that could block acceptance of his work. Aristotelian natural philosophy and its successors, Cartesian and Leibniz-Wolff natural philosophy, accorded a foundational role to principles established by conceptual analysis, aka a priori principles. This was the common basis for the shared contention that action at a distance cannot be incorporated into any system of natural philosophy. Huygens and Leibniz explicitly cited this as a basis for rejecting Newton's account of gravity. Newton, qua natural philosopher, agreed that this principle is unintelligible. Yet, it played a fundamental role in his treatment of gravity.

The way Newton and his successors handled this problem effectively changed the status of physics as a science. There were two generally accepted methods of developing and defending principles basic to natural philosophy. The first was the a priori method of conceptual analysis exemplified by Descartes' proof that extension is the only essential property of matter or by Leibniz's argument for the identity of indiscernibles. The second was the method of introducing hypotheses and testing their consequences, as Kepler had done with the hypothesis that the five regular geometric solids supply a basis for explaining planetary orbits. In the first edition of the *Principia* Newton had three hypotheses, at the beginning of Book III, which were transformed into three Rules of Reasoning in the second (1713) and third (1726) editions. In the second edition, written when Newton was aware of Leibniz's criticism, he also introduced a new Fourth Rule of Reasoning (p. 444):

---

[1] Penrose (2004) gives a very detailed account of the interplay of physical ideas and mathematical formulations in the development of physics and a brief summary of its significance in section 34.2.

> In experimental philosophy we are to look upon propositions inferred by general induction from phenomena as accurately or very nearly true, notwithstanding any contrary hypotheses that may be imagined, till such time as other phenomena occur, by which they may either be made more accurate, or liable to exceptions.

A natural philosophy based on a priori principles is open only to incidental revision, drawing further consequences from the principles. Newton's new rule makes physics, now effectively separated from natural philosophy, open to essential revision based on observation and experimentation. In this context the inverse square law came to be regarded as the discovery of a basic law of nature rather than an irrational principle. It was established by induction from bodies near the earth, the earth-moon system, the relation between the sun and each of the planets, and the moons of Jupiter. Given the state of astronomical knowledge, this was adequate inductive base. The idea that acceptance of competing paradigms leads to separate scientific communities seems to be illustrated by the situation in the early Eighteenth century, where Aristotelian, Cartesian, Wolffian, and Newtonian natural philosophies competed with each other. Here, however, the charge of irrationality hinges on a holistic view of paradigms. If the meanings of key terms are implicitly defined by the way they function in a paradigm, then the members of the separate communities talk past each other. What is required here is a recognition of the semi-independent role of language and the way technical terms are extended from an ordinary language basis. This leads to the account of the meanings of key Newtonian terms given earlier, an extension of Newton's own account. Though terms like 'force', 'inertia', and 'quantity of matter' were given distinctive meanings in the *Principia*, Cartesian and Leibnizian critics were able to understand Newton's physics. Euler and du Châtelet were able to combine Newtonian physics with Cartesian epistemology and Leibnizian metaphysics.

The division of pertinent scientific communities in Newtonian, Aristotelian, Cartesian, and Wolffian, as well as the other paradigm-based community divisions Kuhn treats, was a loose division with fuzzy borders. The Catholic/Protestant division presents an example of more sharply divided communities with distinct and opposed traditions. The burgeoning of the Ecumenical movement in the 1950s illustrates one way in which community-based meaning differences can be overcome. Should Luther, Calvin, Zwingli, and Knox be referred to as 'reformers' or 'heretics'? The neutral term 'religious leaders' proved mutually acceptable. It didn't prejudge whether they led in the right or wrong direction. Should doctrines about Jesus's mother be labeled 'Mariology', a subdivision of theology, or 'Mariolatry', a subdivision of idolatry? Again a neutral term 'Marian studies' fostered dialog. Even fundamental non-theological terms had acquired somewhat different uses. As Karl Barth put it, 'and' is a Catholic term: God and man; faith and reason; scripture and tradition. 'Alone' is a Protestant term: God alone, faith alone, scripture alone. The underlying difficulty comes from treating theories or religious traditions as units of meaning. Then linguistic considerations concern individual words or sentences. The ongoing dialog of the ecumenical movement put the issues of meanings in a broader context treating religious terms as an extension ordinary language terms. Similarly, a recognition of language as a component of science with its own structure, implicit rules, and methods of handling contradictions, and of continued critical dialog as a means of resolving conflicts, plays an essential role in coming to recognize theory change as a critical rational process, rather than an irrational choice.

The scientific skepticism generated by accounts of theory change and the pessimistic meta-induction banks heavily on the role attributed to ontology in scientific theories. If the process of scientific advancement through successive theory replacement entails successive replacement of foundational ontology, then there are no reasonable grounds for accepting any theory-based ontology as more than provisional. The historical analysis presented here interprets physics as relying on idiontology and casts doubt on the foundational role attributed to ontology. What role do ontological assumptions play in the normal functioning of theories? To answer this we will turn away from philosophical analyses of individual theories and consider the notion of effective theories. The application of quantum field theory to particle interactions in the standard model involves series expansions with very many terms and forbiddingly difficult calculations. A method of eliminating the most troublesome calculations is to separate terms into types. The first type includes relatively low energy terms, which are calculated. The relatively high energy terms represent interactions that take place on a much smaller distance scale. The key assumption is that the details of these higher order interactions may be ignored. They have the effect of modifying the coupling constants used on the lower level. These lower level calculations, however, rely on measurements, not higher order calculations to determine coupling constants.

This notion was extended to treat physical theories that function on different energy levels. Molecular chemistry treats the structures and interactions of molecules and uses the shapes of atoms as a basis for some of these calculations. What pass as shapes at this level are the configurations that quantum mechanics treats by detailed calculations of probabilities. Chemists assume that the details of these calculations may be ignored. Similarly, on the next lower level atomic physics treats the nucleus as a unit with mass, charge, and spin and ignores the detailed calculations nuclear and particle physicists must make to explain these parameters.

A qualitative extension of the effective theory approach leads to the notion of a tower of theories.[2] Any theory that relies on input parameters can be regarded as an EFT. A particular theory, so interpreted, supplies a basis for describing interactions within a certain energy range, or at a certain distance scale. It is not an effective tool for describing interactions at higher or lower levels of the energy scale. Then our account of the cosmos and its ingredients can be segmented into a tower of theories with different domains of validity: the universe with galaxies or galactic clusters as units; stars and planets as units; people-sized objects as units; cells; atoms; nuclei; the particles treated in the standard model; supersymmetric unification; string theory. What is the base of the tower? Kane suggests a theory of time and space with no input parameters. String theory is presently the leading

[2] Georgi (1993), Manohar (1996) and Kaplan (2005) present general accounts of effective field theory. The tower of theories is explored in Kane (2000, chap. 3). The philosophical significance of effective theories has been treated in Hartmann (2001), and Castellani (2002).

candidate for a foundational theory. Unfortunately, attempts to include cosmic dark energy in string theory lead to 10^{500} possible string theories.

In the theory-replacement scenario the advance of physics entails replacing the ontology of old theories by the ontology proper to a new theory. No ontological classification is secure until we reach the final fundamental theory. In the tower of theories scenario older basic theories are not rejected. The advance of physics clarifies the limits of their valid applicability. Within these limits a theory may still be used to describe interactions within a certain energy level. Such descriptions presuppose public objects, the things one talks about at a certain level: solar systems, molecules, atoms, particles, or strings. This is the *relative ontology* of a theory, This supports the notion of a functional ontology, an inferential system based on the categorial structure implicit in a theory. Atomic physics and the physics of the standard model of particle physics will remain a permanent part of physics. Further advances should explain the coupling constants presupposed in the standard model, but will not replace this model. It is one of the great triumphs of modern physics.

The twentieth-century development of relativity and quantum mechanics did involve conceptual revolutions, a modification of the basic concepts presupposed in the atomistic mechanism of the nineteenth century. However, neither relativity nor quantum mechanics are constructive theories. Both special and general relativity are principle theories and do not postulate new entities. Quantum mechanics is a methodology that applies across the board. The repeated attempts to reformulate QM as a theory with a new ontology have proved abortive. What we should consider, accordingly, are the changes that relativity and QM introduced in basic concepts.

To put this into context we will focus on the concepts that play a foundational role in atomistic mechanism: 'space', 'time', 'force', 'energy', and 'mass'. In classical physics space and time are represented as independent of matter. As Einstein famously put it: in classical physics if all the matter in the universe disappeared space and time would continue; in general relativity if all the matter in the universe disappeared space and time would disappear with them. In classical physics space and time are separate. Special relativity interrelates them. Classical physics originally represented electrical, and gravitational forces by action at a distance. Classical field theory relied on the propagation of changes in field strengths. QED explains all forces by particle exchanges. Classical physics treats mass as an intrinsic property of bodies. Contemporary physics interrelates mass and energy and seeks to explain mass through the Higgs mechanism and the mass equivalent of energy. It does not follow, however, that one can read a basic ontology from depth theories. To clarify this we have to consider the relation between theories and ontology, or the problem of realism, and the interrelation of theories, or the problem of reductionism.

8.3 The Problematic of Realism

Our present concern is with a clarification of the problem of realism, rather than the proposed solutions. We begin with the familiar split between the problem of realism in an ordinary language context and in a scientific context. Within an ordinary language framework we may distinguish two different, though related types of problems, which we will label 'functional realism' and 'critical realism'. Functional realism relates to the problems of realism encountered by the mythical reasonable man.[3] He accepts as unproblematic the real existence of people, dogs, trees, and the familiar objects of the lived world. However, he may have doubts about the real existence of the Loch Ness monster, Iraq's weapons of mass destruction, the authenticity of the dangers depicted on reality TV, and the claims of telemarketers. Here realism is always a local issue. The basic framework of reality is neither criticized nor questioned. The problems that arise concern whether particular objects, events and properties fit into the familiar framework.

How does this functional realism relate to critical realism? The critical problem of realism generally involves a distinction between appearances and reality. The Vedantic account of sensory experience as Maya, or illusions, Parmenides's denial of the reality of motion Plato's comparison of ordinary experience to shadows on the cave wall, the arguments stemming from Galileo and Locke that secondary qualities are mere appearances, the shared rationalist-empiricist distinction between ideas in the mind and the reality to which they might correspond, the Kantian distinction between noumena and phenomena, Bradley's distinction between appearances and reality, and the various embodiments of idealism, all involve a contrast between a discredited functional realism and a depth reality beyond mere appearances. There is an abiding temptation to refute such claims by an appeal to functional realism. Plato thanked Parmenides's disciple, Zeno, for coming all the way from southern Greece to Athens to argue that motion is impossible. Doctor Johnson considered kicking a stone an adequate refutation of Berkeley's idealism. G. E. Moore set the precedent of appealing to what D. Lewis has dubbed 'Moorean facts', things that we know better than any philosophical arguments to the contrary. Such common sense refutations have rarely dissuaded philosophers from an advocacy of the opposed positions. The philosophers under attack generally retort that no one accepting such refutations as convincing understands the problem. Our immediate concern is with a clarification of the status of this type of critical problem.

The dismissal of functional realism as naïve represents, in my opinion, a serious category mistake. Functional realism is being appraised as if it were a theory of reality competing with other philosophical and scientific theories. Functional realism should be regarded as the outcome of a long process of adaptation to the physical and social world, a process involving both evolutionary roots and personal development. The linguistic expression of this functional realism floats, like the tip of an iceberg, over a large mass of submerged physiological, psychological, and linguistic structures. To see how this emerges as a theory we return to a previous example.

---

[3] The 'reasonable man' became established in legal tradition chiefly through the influence of Justice Oliver Wendell Holmes See Menand (2001, p. 343). Liability for damages in tort cases involves presuppositions about the expectations a reasonable man would have and the precautions he would take.

The shirt I am now wearing is yellow. (S1)
This shirt has the property of being yellow (S2)
Color is a property of extended material objects. (S3)

The transition from the observation report, S1, to the ontological claim, S3, hinges on treating the presuppositions of linguistic usage as ontological props. What is at issue here is the functioning of categories and categorical systems in the structuring of language. As noted earlier, in ordinary language usage the normal use of 'true' and 'false' is a surface issue. One can either assert S1, or quote it and predicate 'true' of the quote. Neither entails predicating 'true' of a transcription of implicit linguistic presuppositions into explicit ontological claims.

There is, however, a certain inevitability to such transcriptions. In a culture, or for an individual, where an ordinary language framework supplies the only medium of discourse, there is no alternative. Even when partially competing philosophical and scientific alternatives are available, the functional realism of ordinary language is a normal feature of discourse. The philosophy professor, who explains to students that the shirt is not really yellow, better not try telling a traffic judge that he is not guilty of running a red light, because the light is not really red. Functional realism can be classified as naïve only in a context where it is treated as a theory of reality. To get away from this we should return functional realism to its native habitat.

As a start we will utilize material covered earlier without repeating supporting arguments. Through a long evolutionary process, advanced organisms have developed complex fine-tuned adaptations to their environments. A bird alighting on a branch, a spider spinning an architectural wonder, a cat leaping to a perfect balance on a narrow rail, all represent complex adaptations. Even non-functional behavior, such as a dog chasing a car, reflects adjustments to earlier environments of wolf packs chasing large prey. Humans are a product of such evolutionary processes. They, however, must adapt to a social as well as a physical world. We will use 'Umwelt' to signify this social-physical world. This is the most complex evolutionary adaptation in our planet's history. Yet, its linguistic spin-off is labeled naïve. We should reflect on the nature of the complexity and its relation to functional realism before considering the naiveté.

How is a basic functional representation of reality reflected in language? Earlier we summarized Davidson's idea of triangulation. To understand the speech of another person she and I must be able to refer to public objects that we both accept. Advances in knowledge lead to the acceptance, at least by specialized communities, of further public objects. These anchor the functional ontology of ordinary language and the specialized extensions used in different specializations. We will supplement our previous considerations by summarizing a pertinent recent development, efforts to explain the systematicity of language. The basic idea is that sentences remain meaningful when terms are replaced by terms of a similar kind. Thus, 'yellow' in (S1) may be replaced by 'green', 'blue', and other color terms. Such substitutivity applies to activities as well as properties. The 'saw' in "John saw the ball" may be replaced by 'hit' 'observed', 'kicked', 'picked up' and many more action terms. However, attempts to explain the systematicity of language purely in terms of syntactical and semantical rules are confounded by the ease with which counterexamples may be constructed. Consider a couple of examples featured in a recent study by Johnson (2004).

Alice showed the book. (8.1a)
Alice described the book. (8.1b)
John stowed his gear. (8.2a)
John put his gear down. (8.2b)

Here the (a) and (b) forms are seen as acceptable systematic variations. Now consider the variations.

Alice showed Martha the book. (8.3a)
Alice described Martha the book. (8.3b)
John stowed his gear down. (8.4a)
John put his gear. (8.4b)

The last three are not acceptable systematic variations. Johnson's detailed analysis of further examples serve to bring out a basic point. An account of the systematicity of language requires a recognition of natural kinds, their properties, activities, and relations, and also a recognition of human agency in the Umwelt. The recognition of natural kinds is a recognition of the role natural kind terms play in systematizing language and facilitating informal inferences. There is no account of the systematicity of language that is adequate to the task of distinguishing all acceptable sentences from anomalous systematic variations. Yet, language users routinely make such distinctions. This ability reflects a detailed knowledge of reality structured in a way that supports material inferences as well as normal activity. The transition from automatic rules to content-dependent variations reflects Davidson's account of the introduction of ontological considerations into language. It happens when automatic constructions prove inadequate to the reality treated.

The functional representation of reality and of human activity implicit in normal language usage evolved as a framework, guide, and constraint for human activity in the Umwelt. It did not evolve as an attempt to represent the world as it exists objectively, or independent of our knowledge of it. To relate this to the problematic of realism we could consider the descriptive metaphysics implicit and functional in ordinary language usage as a thematization of the functional realism that guides and structures our adaptation to the Umwelt. Any systematization of this is a second order thematization. The charge of naïveté arises when this second order thematization is treated as a theory of reality competing with other theories.

Philosophers in the analytic, and especially in the phenomenological tradition, have recognized this. In the human realm, realism is not a theory. It is a basic feature of the framework that makes theories about reality possible. Davidson insists that the interpretation of language rests on a vast amorphous collection of claims that are true, really true. He also insists that the idea of an alternative conceptual framework, not translatable into our language, is incoherent.

Husserl considered common sense realism in terms of the natural standpoint and explicitly recognized the need to go beyond this framework. His attempts to go beyond centered on analyses of acts of consciousness and their transcendental objects. His breakaway disciples focused more on a clarification of the natural standpoint than on noemata. Heidegger distinguished between beings available as tools (*Zuhandenheit*) and beings somehow detached from this normal instrumental network (*Vorhandenheit*). Beings available for use fit into a network of relations to other equipment, human goals, and normal practices. One understands a hammer instrumentally by using it to hammer nails. The hammer emerges from this normal context of usage in the lived world when it breaks and thus becomes a problem. There are other ways of detaching an object from the network of instrumental usage, when it becomes an object of contemplation or is decontextualized through a philosophical perspective. Many aspects of his later philosophy engender serious doubts and even severe criticism. However, one feature of *Being and Time* had a massive impact and should condition any future developments. That is Heidegger's almost anguished insistence that our way of being in the world is prior to and presupposed by any further knowledge of reality. As with Wittgenstein, his philosophical message is the necessity of retrieving and paying critical attention to the features of life and thought that are presupposed in activity, but forgotten or rendered invisible in systematizations of thought and reality. In the *Zuhanden* perspective realism is not a philosophical problem. Coping with reality is.

None of these authors wish to dismiss physics. Yet, the interpretative perspectives inevitably discount any deep ontological significance to physics. The lived world represents our basic reality. The ordinary language conceptual framework admits no alternative conceptual framework. John Searle (1998, chap. 1) champions the Enlightenment vision that there is an external reality that is completely intelligible and that we are capable of understanding it. The default position on external realism centers on the claims: there is a real world independent of our knowledge of it; we have direct perceptual access to it; words typically have clear meanings and can be used to refer; our statements are typically true or false depending on how things are; and causation is a real relation among object and events. In this context, various forms of antirealism, considered as competing accounts, are set up and knocked down with one punch apiece. This effectively treats functioning realism as if it were a theory competing with other philosophical or physical theories.

These demeaning evaluations are supported by some accepted philosophical interpretations of physics. The Continental *Naturwissenschaften/Geisteswissenschaften* distinction was traditionally coupled to a positivistic gloss on the natural sciences. Heidegger, Merleau-Ponty, Gadamer, and others have expressed the position that the physical sciences treat reality externally.[4] For many phenomenologists only a probing of immediate consciousness linked to a critical awareness of our bodily presence in the lived world leads to the problem of being. Merleau-Ponty has

[4] *A Companion to Continental Philosophy* (Critchley et al. 1998) has 56 articles surveying the overall development. None of these treat physics or the type of information physics supplies.

given this position its clearest expression: "They have done well in removing any ontological significance from the principles of science and leaving only a methodological significance, a reservation introducing no essential change in philosophy since the only thinking being resists being defined by the methods of science" (Merleau-Ponty 1945, p. 67. [my translation]).

In the post Quine/Sellars analytic tradition, ordinary language analysis is sharply separated from philosophy of physics. The separation, and the common neglect of physics by analytic philosophers, can be justified if one accepts the schematization of physics as a collection of separately interpreted theories and the contention that the ontological significance of a theory is determined by an analysis of the theory's mathematical formulation. In this context the language of a theory plays no interpretative role. Then the method of ontological analysis in theoretical contexts is sharply separated from methods proper to analysis and phenomenology. If these considerations are supplemented with Kuhnian relativism and Laudan's pessimistic induction, then the neglect of physics seems justified.

A recognition that the language that plays an integrating role in physics could liberate analysis from its present restrictive constraints and make methods of analysis available to philosophers of physics. As a point of departure we extend the two Davidsonian claims just considered. A shared interpretation of scientific discourse depends on the mutual acceptance of a vast, but not so amorphous, collection of claims as true. The second Davidsonian principle concerns triangulation and the role of public objects. For the starkest illustration of the role these play in scientific discourse, consider the most ambitious attempt to extend this discourse, the search for alien intelligence elsewhere in the galaxy. The SETI project beams messages towards potential listeners and listens for corresponding messages beamed to us. The encoding and decipherment of such messages presupposes a process of triangulation. Our minds and alien minds must both recognize public objects and structures that supply common referents. The operative assumption of the SETI messages is that aliens capable of receiving and deciphering messages would be familiar with the distinction between stars, planets, and moons; with atoms and the periodic table of elements; with electrodynamics; and with the fundamental law of arithmetic.

We communicate by imposing structures on electromagnetic radiation and assume that the aliens can detach such structures from electromagnetic radiation. The imposed structures are, a matter of practical necessity, serially ordered units. The law that any number is uniquely factorable into primes supplies an unambiguous basis for assembling structures by factoring a long sequence into two prime numbers, n and m, and then reducing the sequence to an $n \times m$ array. Atoms, and their periodic properties, as well as stars, planets, and moons are represented by iconographic symbols in the array. We assume that intelligent aliens would recognize the role of symbols in communication. These are common sense assumptions of practicing scientists, not consequences of a philosophical account of theories. Do intelligent aliens share such assumptions? The answer to this may have to wait till amiable aliens beam down some version of *Intergalactic for Dummies.* Our immediate concern is with these presumably shared objects and structures. The operative assumption here is that Maxwell's laws of electrodynamics and the periodic table of elements represent objective discoveries. They should be present, in some form, in the science of any sufficiently advanced culture.

Philosophical debates about scientific realism have generally focused on the status of theoretical entities. This is coupled to analyses of whether an acceptance of particular theories entails an acceptance of the entities the theory is a theory of. Antirealism rarely involves arguments that the theoretical entities do not exist. The common contention is that the inference from theory acceptance to entity acceptance is not compelling.[5] Empirical adequacy suffices. Attempts to counter the pessimistic meta-induction have occasioned a shift from realistic claims about entities to realistic claims about structures. The role of language is not treated as a significant factor in either set of arguments and counterarguments. To clarify the role of language in functioning scientific realism we must situate the debate in a larger perspective, scientific attempts to explain all of reality.

Here, fortunately, Steven *Weinberg's 2008 Cosmology* comes to the rescue. This was written after Weinberg had completed his authoritative three volume treatise on quantum field theory and an earlier treatise on general relativity. His professed aim is to present a self-contained account that relies as much as possible on physical arguments and analytic mathematics (See p. vii, 257). When appropriate he compares his results to computer calculations. It is a definitive treatise, not a popularization. Implementing this program requires a detailed knowledge of current astronomical research, a mastery of all the main branches of physics, and an incredible amount of calculation. Very few physicists, and no philosophers, meet these exacting standards. What I wish to siphon off from this massive work are some methodological considerations on the integrating role of language and something of a dual-inference treatment of particles. I will summarize pertinent parts of the book on the assumption that few philosophers will make a detailed study of this long difficult work.

A brief sketch of the overall structure anchors these considerations in particular settings. The first half of the book, chaps. 1–4, assumes that on a sufficiently large scale both the distribution of matter and of background radiation is isotropic and homogeneous. He presupposes a familiarity with descriptive accounts of the main stages: big bang,; inflation; particle creation; uncoupling of gravitational, strong, and weak forces; radiation-dominated era; recombination and the matter-dominated era; formation of galaxies and stars. These are analyzed in a reverse temporal order, beginning with the present era and working backwards. The second half of the book is concerned with small departures from isotropy and their use in inferring structures. With this descriptive background particular problems are generally treated by beginning with a physical account relying on presuppositions and plausibility arguments to select the factors that require analysis. This determines the theories or formulas needed. A detailed working out of conclusions is then given a

[5] Hacking (1983) shifted the realism debates from a focus on theories to a focus on experiments and the use of entities as tools. I believe that in functional scientific realism the emphasis should be on entities accepted as public objects, some of which may serve as experimental tools. Franklin (1981, 1983, 1986) and Galison (1987) emphasize the role of experimental traditions in setting up and interpreting experiments.

physical interpretation. This is the general pattern of mixed formal and informal inferences we discussed earlier in developing the dual-inference model of scientific explanation.

We begin with the treatment of particles in the radiation-dominated era. As Weinberg notes (p. 201) treatment of earlier stages rely on speculation and have uncertainty on some crucial points. The present dominance of matter over antimatter must have been preceded by an asymmetry between particles and antiparticles, now weakly reflected in violations of CP (charge conjugation and parity inversion) symmetry. However, the form this took at the earlier stage is not at all clear. The homogeneity and isotropy of cosmic microwave background radiation (CMB) requires an earlier inflation, but the precise form this took is debatable. After electron-positron annihilation ceased the universe consisted of particles in equilibrium. This simplifies the considerations in two ways. First, a physical treatment of equilibrium does not depend on how equilibrium was achieved. The uncertainties considering the earlier era are not an impediment to a physical account of late developments. Second, the account of particles in equilibrium relies on well established principles of physics.

Before considering these principles we should situate their application. The expanding space-time is represented by a Robertson-Walker metric with a parameter, a(t), characterizing the expansion (chap. 1). This relates to the red shift, z, by the formula, $1 + z = a(t_0)/a(t_1)$, where $a(t_0)$ is the expansion coefficient when the light was emitted and $a(t_1)$ is the coefficient when the light is observed. The radiation-dominated era covers a time from about 1.08 s after the big bang to around 300,000 years, when the background radiation became decoupled. During this era the particle collision rate was so much higher than the expansion rate that there was equilibrium. After this time the radiation still fit the Planck formula quite precisely, but with the temperature term dropping as $1/(a_t)$ from 30,000 to 2.725 K. The production of black-body radiation requires an equilibrium condition between matter and radiation at the time of production. We will consider the treatment of particles in the equilibrium preceding decoupling and in the process of decoupling.

At the beginning of this era when the temperature was 10^{11} K electron-positron pairs were still being produced. Neutrinos and perhaps even dark matter were in equilibrium with radiation. (p. 149). However, neutrinos decoupled before photons. Dark matter, which does not relate to electromagnetic forces, effectively ceased to interact. In equilibrium the laws of thermodynamics hold. In our earlier terminology leptons, baryons, and photons are being treated as *particles*$_c$. The baryon/photon ratio is so low $(1/10^8)$ that the baryon contribution to equilibrium may be ignored. It will be treated separately. Then the count for the particles in equilibrium through collisions is: photons, with two spin states; 3 types of neutrinos and antineutrinos, each with one spin state; and electrons and positrons, each with two spin states. In considering equilibrium the energy density of photons is given by the Stefan-Boltzmann law. The leptons, following the particle count, contribute 7/8 $(6 + 4) =$ 35/4 as much energy, where the 7/8 is a factor for fermions. When these ingredients are treated as a collection of particles in equilibrium, then thermodynamics supplies a basis for determining the entropy density, energy density, and pressure of the total gas and of each constituent. The number of baryons is too small to be a factor in this equilibrium. At the start of this era the number of neutrons is about the same as the number of protons. The problem is to account for the present distribution of baryonic matter. At the initial high temperature neutron decay ($n \rightarrow p + e^- + \nu$) is much slower than the two body collisions ($n + \nu \leftrightarrow p + e^-$, $n + e^+ \leftrightarrow p + \overline{\nu}$). These two-body collisions became relatively negligible when $e^- - e^+$ pairs vanish between 3×10^{10} and 10^{10} K. After that neutron decay competed with the formation of complex nuclei, in which neutrons are stable. Hence the ratios of deuterium and helium to hydrogen supply a basis for inferring stages in the transition era. Deuterium, which is weakly bound, only becomes stable at a much lower temperature than helium, which is tightly bound. This representative example illustrates the way functioning physics makes inferences about unobserved reality. The organizing framework is a descriptive account of stages in cosmic evolution and the processes and particles involved. This account presupposes a vast background knowledge of physics. This combination of a descriptive account and background presuppositions supplies a basis for informal inferences concerning what processes are dominant at a particular stage, such as expansion, collisions, particle production, and particle decay, and what ingredients may be ignored, such as baryons in the equilibrium state, dark matter and neutrinos when the temperature decreases, and rest masses of particles, when the kinetic energy is much greater than the rest mass. This in turn supplies a basis for selecting the applicable mathematical formulations, such as general relativity for the expansion, thermodynamics for the equilibrium state, particle physics for the processes, and Planck's formula for the uncoupled radiation. As a second representative problem we consider recombination and the uncoupling of radiation. This leads to the matter-dominated era, the emergence of galaxies, stars, and planets, and the cooling down of the CMB. The misleading term 'recombination' derives from atomic physics and is still used for the capture of electrons by protons and helium nuclei. Prima facie this might seem o be a simple example of theory application. The ionization energy for hydrogen is 13.6 ev. When the background radiation is below this level, then electrons can be captured in the ground state. Here again, however, theory application depends on a descriptive account of the overall situation and informal inferences about which factors are significant and different energy levels. The expanding plasma prior to recombination consisted of baryons (73% H, 27% He, and a small number of light nuclei), electrons, and 10^8 photons for every lepton. An electron captured by a proton would form a hydrogen atom in an excited state. Further developments depended on the balance between three processes. A hydrogen atom in an excited state emits photons and cascades down the lower energy levels.

Figure 8.1 represents the basic energy levels and some representative transitions, when the fine structure, represented in Fig. 6.1, is ignored. The energy of level n is $-13.6\,\text{ev}/n^2$. The second process, reionization or excitation to a higher energy level, is caused by collision with a CMB photon or a cascade photon from another atom. The third process is cosmic expansion, gradually lowering the energy of all the photons. At 300,000 years, when recombination is virtually complete the CMB photons peak at 0.25 ev, far too low to excite an H atom in the ground state. This atom could be ionized only by a photon produced by a transition to the ground state,

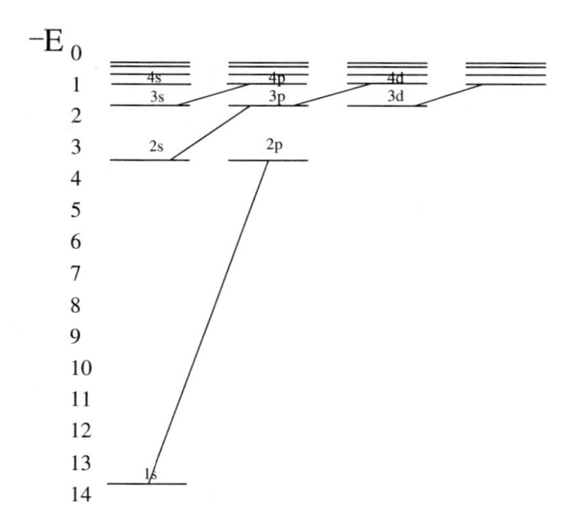

Fig. 8.1 Hydrogen atom energy levels

provided the photon transit time is short enough to avoid significant red-shifting. Thus, even though the transition to the ground state is relatively slow and inefficient, it eventually becomes stabilized.

Recombination and gravitational attraction led to the emergence of large-scale structures. Here we consider only one novel prediction that depends on informal inferences concerning the behavior of particles. Earlier accounts of the growth of large-scale structures, developed before the discovery of dark matter, assumed that galactic clusters emerged first, then individual galaxies. The new assumption (p. 349) that gravitational field perturbations are dominated by dark matter density led to a new scenario. In regions where the average matter density is above average gravitational attraction could cause the matter to collapse into a large spherical region. According to a theorem of Birkhoff the metric and equations of motion of a particle within this sphere are independent of what is happening outside the sphere. (p. 421). Baryons lose energy through radiation. If the region is large enough the atoms could collapse into galaxies and stars, while the dark matter remained in a spherical halo. If the region is not large enough for baryonic collapse, then all the matter would remain in a spherical halo. These clumps of matter could be detected only through gravitational lensing. (sect. 9.3)

Until recently philosophical discussions of scientific realism have centered on the status of theoretical entities. Realists argued that the acceptance of theories as fundamental and explanatory supplies a rational basis for accepting as real the entities posited or presupposed by the theories. Anti-realists counter that such arguments rely on dubious props, such as inference to the best explanation. I am concerned with the status of the philosophical problematic. To make it more specific consider the particles basic to the preceding parts of this chapter, the particles treated in the standard model. Consider the contrasting claims. These particles really exist, or have an existence independent of our knowledge of them. (R). These particles do not exist as such independent of our knowledge. (AR) Does the acceptance of either R or AR contribute to our understanding of reality?

We can attempt an answer in two stages. First, does the acceptance of either position add something to what we learn from physics? I assume that within physics idealism and solipsism are non-starters. There is a reality independent of our knowledge of it. The issue is what kind of knowledge about fundamental reality physics supplies. The most complete account relies on complementarity descriptions. In experimental and engineering contexts fundamental particles are treated as $particles_c$. They move in trajectories and experience collisions. This carries over to descriptive accounts of stages in cosmic evolution. Physicists adapt the thermodynamics developed for gases, treated as collections of colliding molecules, to fundamental particles in equilibrium conditions. In theoretical contexts, fields are basic. QFT does not support sharply localized particles. Attempts to develop an ontology based on QFT suggest that basic reality should be categorized as fields (Auyang 1995),[6] or as quanta (Teller 1995), or by developing new categories (Seibt 2002). The fact that basic physics requires complementarity, i.e., reliance on mutually incompatible categorizations, indicates that our categorial systems are not adequate to fundamental reality. In this context no substantive significance can be attached to either R or AR.

As a second stage, can either R or AR be considered as making a substantive claim beyond the knowledge physics supplies. It could if there were a science of metaphysics that supplies principles of reality independent of physics. As I have repeatedly indicated, I do not believe that there is any such credible science. Rather than repeating the reasons why physics has replaced theology and philosophy as the court of last appeal in matters physical, I will simply indicate something of a philosophical convergence on the position that statements like R and AR do not make a substantive contribution beyond physics.

The trend is best illustrated by Putnam's progress.[7] The Realism he originally championed featured the cutting distinction (cutting nature at the joints). There is a clear distinction between the properties things have in themselves and the properties we project on things. Physics aims at theories that tell us what thing are in themselves. A major difficulty he encountered stems from the Löwenheim-Skolem theorem. A theory expressed in first order logic admits of unintended interpretations. As a logician he extends this to conclude that a theory, considered as a separate unit, cannot fix its own references. Putnam gradually changed from Realism to realism to internal realism to direct realism. In internal realism, the cutting distinction applies only relative to a description. Direct realism is concerned with problems of perception. Putnam's long and winding road reflects something of a shared consensus. If the problem of scientific realism is identified with the acceptance of theoretical entities as having an observer-independent reality, then the arguments supporting

[6] This simple labeling slights the subtlety of her neo-Kantian analysis. However her analysis, like Kant, slights the role of language in concept formation.

[7] See Putnam (1987, 1990). Conant's introduction to (1990) contains a survey of Putnam's development as does Norris (1999) and Putnam (2004).

scientific realism do not compel assent. Acceptance or rejection of scientific realism is essentially a pragmatic issue related to the acceptance of other positions and values.

This interpretation relates Putnam's realism to Van Fraassen's anti-realism. Thus, Van Fraassen categorizes the basic particle of physics as fictions. Yet, he (Van Fraassen 1991) also works out a detailed exposition of the treatment of Fermi-Dirac and Bose-Einstein particles. This can be interpreted as using, but not affirming, the functional realism of ordinary physics while prohibiting any excursions beyond these limits. The functional realism I am advocating is similar to Putnam's internal realism or Arthur Fine's NOA.[8] Both are willing to accept the language of physics at face value. Neither claims that philosophy can go beyond the language of physics in making ontological claims. My position differs from these chiefly through its stress on the role of language. Language supplies the overall framework supporting descriptive accounts that specify which theories and principles are applicable to a particular problem or energy level. An acceptance of the role of language requires a recognition of the limits of meaningful language.

To complete this sketchy survey we will consider the recent discussions of structural realism and relate it to the general issues just treated. Worrall (1989) revived this concept, attributed to Poincaré and many others, to deal with Laudan's pessimistic induction. He shifted the realism debates from entities to structures. The basic argument granted the point that scientific revolutions often do involve the replacement of one ontological basis by another. However, successive theories often support the same equations. What theories get right, Worrall contended, are structures, not entities. This approach soon split into two branches. Epistemic structural realism claims that successful theories inform us about structures in reality, but not about the ontological foundation of these structures. Ontological structuralism (Ladyman 1998) dispenses with epistemologically inaccessible entities supporting structures to emphasize structures as the basic reality. Psillos (2000) and Van Fraassen (2002) have argued that the idea of structures without supporting content is incoherent. My interpretation of physics as supporting idiontology supports a form of epistemic structural realism. The basic claim is that physics focuses more on properties and relations between properties, than on the bearers of these properties. It also accepts claims for entities as public objects. However, it is set in a different interpretative perspective. The effect this has on altering both the defense and criticism of epistemic structural realism can be clarified by beginning with an example treated in Chapter Three.

The theory of heat relied on the assumption of caloric atoms and supported formulas governing heat exchange. The new theory, developed by Helmholtz and others, dispensed with caloric atoms, treated heat as a form of energy, and supported the same equations. Does this imply that the equations for heat exchange correspond to a structure in reality? The carry over of the formulas depends on conservation principles. The older theory claimed that overt heat is a manifestation of free caloric

---

[8] Fine's NOA was developed in his (1986) and elsewhere.

atoms and that caloric atoms are conserved. The newer theory claimed that heat is a form of energy and energy is conserved.

The status of conservation principles has a distinct bearing on the problematic of structural realism. Both the defenders and the opponents of structural realism set up the problem in terms of two components, mathematically formulated theories and observations. The key claim of the epistemological structural realism argument is that equations that carry over from replaced to replacing theories should be explained by a correspondence between a mathematical structure and a structure in reality. Bas Van Fraassen criticizes these arguments as circular. Yet, he relies on the same two component framework: "Science represents the empirical phenomena solely as embeddable in certain abstract structures (theoretical models), and these abstract structures are describable only up to a structural isomorphism." (Van Fraassen 2002, p. 32).

The phenomena physics treats are conceptually structured facts. The conceptual structures generally precede the mathematically formulated theories that systematize them and can survive the demise of these theories. These conceptual structures are embedded in the developed language of physics. An adequate formulation of the problem of structural realism must include four basic components: observed phenomena; structures in the language used to report phenomena; theories; and general principles that transcend particular theories. We should comment on the significance of conceptually structured facts and general principles.

Prior to caloric theory there was a protracted process of conceptual structuring that we considered previously. It led to distinctions between heat and temperature, overt and latent heat, specific heat, specific heat at constant pressure, specific heat at constant volume. We can supplement this by a more recent example of the development of conceptual structures for reporting facts shaping the subsequent formulation of theories. In the era of particle discovery in the 1960s experimenters discovered the law of associated production. The new strange particles always seemed to be produced in pairs. To explain this Gell-Mann and Nishijima independently postulated a strangeness quantum number that is conserved in strong and electromagnetic interactions. Earlier we considered the introduction of isotopic spin following the mathematical rules proper to angular momentum. This led to the conclusion that particles, with isotopic spin I came in multiplets with 2I + 1 members. Gell-Mann used the assumption that interactions can be divided in strong, electromagnetic, and weak to introduce an approximate treatment. In the first approximation electromagnetic and weak forces are shut off and only the strong force is considered. Then the 2I + 1 particle states are symmetric. Switching on the other interactions leads to symmetry breaking and mass differences. This led to the Gell-Mann–Okubo mass formula,

$$M_F = M_0 + M_1 Y + M_2(I(I+1) - Y^2/4), MO) \tag{8.1}$$

where M_0 is the mass in the first approximation, Y is B + S. B is the baryon number and S is the strangeness number. M_1 and M_2 are symmetry-breaking coefficients whose values are determined experimentally. This guided experimenters in the search for and interpretation of new particles. Experimenters, in turn, aided theorists by determining the coefficients needed. The conceptual structuring of particle data in terms of isotopic spin, strangeness quantum numbers, and baryon quantum numbers soon blossomed into the Eightfold Way, developed by Gell-Mann and Ne'eman. This, in turn, led to the development of the standard model. One cannot get at the structures basic to physics by an exclusive concentration on the mathematical formulation of theories.

Conservation and symmetry principles transcend particular theories. Historically, they have supported more inferences from mathematical forms to reality than have theories. The symmetry between particles and antiparticles led to the postulation of new anti-particles for every fermion discovered. As noted earlier, Pauli's exclusion principle led to the introduction of color and the postulation of eight colored gluons. Symmetry principles had a more problematic status. The CPT theorem was already established when Lee and Yang introduced the hypothesis of parity non-conservation for weak interactions. This led to an increasing study of symmetries and experimental tests of which symmetries apply to which interactions. Two types of symmetry arguments played a role in inferring new entities, precisely formulated symmetries, and nebulous symmetries lacking theoretical justification. In quantum field theory the precisely formulated symmetries concern symmetries of the action and of the Lagrangian. The strongest ontological consequences of symmetry principles might lie in the immediate future. If the Large Hadron Collider, or some other future machine, establishes the existence of **a** supersymmetric particle, then symmetry arguments lead to the whole family of supersymmetric particles.

The most important fuzzy symmetry was rooted in the growing realization that there is a basic symmetry between the lepton and quark families of particles. The enlargement of the family of leptons from the original 3 (e, ν, μ) to 4 (e, ν_e, μ, ν_μ) suggested that the family of quarks should be expanded from the original 3 (u, d, s) to 4 (u, d, s, c), a suggestion that was developed and confirmed. Martin Perl's 1976 discovery of the τ lepton was a surprise. The yet-to-be explained symmetry led to the suggestion that there must be 3 more particles: one lepton, ν_τ, and 2 new quarks, b and t. The bottom quark was quickly detected. Eighteen years of intensive labor eventually led to the establishment of the top quark. There have been attempts to explain this parallel, notably the SU(5) unification proposed by Glashow and Georgi. However, neither the X particles nor the proton decay this theory predicted have yet been observed. A grand unifying theory remains a goal, not an achievement. The symmetry remains fuzzy, but intact.

Symmetry principles provide a stronger base for arguing from mathematical structures to structures in reality. Here again, complications require qualifications. Most of the symmetry principles treated in particle physics are spontaneously broken symmetries. In simple terms what this means is that the principles governing the state of a system are symmetric, but the ground state is not symmetric. Hence the symmetry does not quite apply to reality. It only applies when we impose a basic conceptual structuring on reality. One might contend that all these symmetry principles applied exactly at the originating moment of the big bang. This, however, was an explosively unstable system. The conclusion I draw from this is essentially the same as the conclusion concerning entative realism. Physicists have discovered and increasingly rely on symmetry and conservation principles that apply to reality as it is represented in physics. Many of these principles transcend particular theories. Philosophy does not supply any criterion independent of physics for deciding which structure have objective validity.

8.4 Emergence and Reduction

The emergence-reduction (E-R) debates involve philosophy, physics, chemistry, biology, psychology, neurology, and even theology. Here we will skip or skim over most of the substantive issues and focus on a background presupposition. Aspects of this debate, especially arguments for global reductionism, rely on assumptions about the role of physics. However, it is difficult to relate these assumptions to contemporary physics. In addition to the difficulties related to content, there is a preliminary obstacle concerning the language of the E-R debates. Many of the key terms involved, such as 'emergence', 'reduction', and 'supervenience' are not terms used in physics. Even such common terms as 'cause', 'kind', and 'level' have a special significance in these debates. Following the methodology of reductionism one might try to build bridge rules linking key terms in the E-R debates with terms in physical theories. Bridges, however, require secure bases. A perusal of the E-R literature soon reveals that there are no agreed-upon meanings for the key terms just cited.

We analyzed the language of physics as a specialized extension of an ordinary language core. The language of the E-R debates represents a different specialized extension of an ordinary language core, one that passes through philosophy rather than physics. We will try to sketch the passage of the key terms. In their influential study of causality in the law Hart and Honoré (1959, p. 1) clarify the common notion of causality as human intervention in the order of nature.[9] If Joe pushes Bill off a roof then, in the eyes of the law, Joe's action, rather than gravitational attraction or the inflexibility of concrete, is the cause of Bill's death. Intentional human action supplies the primary instance of causal action. This usage is routinely extended to unexpected physical events that disrupt the ordinary course of nature. Investigators are seeking the cause of the fire. In science it is extended to experimental intervention, such as causing a projectile to hit a target.

Physical theories treat forces, not causes. However, the general notion of causality was extended to physics in two significantly different ways. For the first I rely on Cartwright's (1983, chap. 4) distinction between theoretical and causal explanations. A standard way of explaining a phenomenological account is to subsume it in a more

[9] Pinker (2007, pp. 208–225) analyzes primitive notions of causality reflected in dead or submerged metaphors operative in many different ordinary languages. The conjecture is that the primitive notion involved a conflict between an antagonist using force to change the natural state of an agonist. Hanson (1961, chap. 3) presented an influential argument against the extension of the ordinary notion of causality to scientific explanations.

general theory. However, a common way to explain a phenomenon, such as radiation damping or the Hall effect, is to find the cause of the phenomenon. This is an extension of ordinary language causal accounts. The second way is through the notion of classical determinism. Quantum mechanics replaces determinism with probabilities. It is acausal. Questions of the form: "What caused this particular uranium atom to decay at time T?" are not accepted as meaningful.

To see the significance 'cause' has acquired in the E-R debates we consider a dilemma that Kim (1996, p. 237) regards as the most serious difficulty in the philosophy of mind. Assume a simple case of intentional causality. My intention of waving to a friend causes movements in my arm and fingers. Physiologically, these movements can all be explained in terms of nerve impulses stimulating muscular activity. Here Kim would invoke his own exclusion principle: A sufficient cause of an event excludes any appeal to another cause. If a physiological account provides a sufficient cause for the movements then any appeal to a further intentional cause is otiose. A straightforward extension makes mental causation unacceptable as a principle of explanation and discounts the ontological reality of consciousness and qualia, or sensations as experienced. The dilemma is: accept the causal closure of the physical order and discount the ontological reality of qualia, consciousness, and intentionality, or accept their reality and deny the causal closure of the physical order. Prima facie, this is paradoxical. Intentional human action supplies the paradigm cases grounding the ordinary use of 'cause'. Now a principle of causal closure leads to a denial of intentional causality. Kim regards the causal principle as metaphysical, not analytic. This principle would seem to require some form of classical determinism. An attempt to apply this brings up the issue of levels. The idea of a hierarchical ordering of levels of being was basic to the amorphous philosophical tradition labeled 'the great chain of being'. The E-R debates presuppose levels, but reverse this hierarchical ordering. The presupposition is that different levels treat different kind of things with different properties. In a strict reductionist position the causal powers normally attributed to the kinds of entities treated by the special sciences are reduced to the properties of the fundamental physical entities. In their checkered careers both 'kind' and 'property' have accrued a variety of philosophical uses. The term 'property' derives from ordinary language usage and its extension in different substance-property ontologies. Ian Hacking (1991, 2009) has argued that there is neither a strict nor a vague class of classifications that may be called 'natural kinds' and that supports the inductive inferences proper to science. At the other extreme is the term 'supervenience'. As Kim (2002) notes, this is a purely philosophical term with no ordinary language usage. It means whatever philosophers decide it should mean.

With this background we can list the three issues to be treated. The first is the role of theories, and the special role of fundamental theories. The second issue is the classical/quantum divide. This figures in the debates in two key ways. First, the phenomena to be explained are presented in classical terms while the fundamental theories that supply the supporting ontology are in the quantum realm. The second way involves the appeal to quantum properties, such as entanglement, to supply a possible basis for explaining consciousness. For the third issue we switch from the shortcomings of the philosophers to the short-comings of the physicists. The Gell-Mann–Hartle account, treated in the last chapter, accepts the quantum realm as fundamental and attempts to explain the emergence of a quasiclassical realm. This realm involves large-scale statistical regularities that approximate the deterministic laws of classical physics. The further assumption is that studies of complexity will eventually supply a basis for treating intentionality as an emergent property of IGUSes. This leaves a glaring gap. The human order depends on collective intentionality. John Searle (1998, chaps. 4, 5 and 1983) summarizes this in terms of a formula, *X counts as Y in C*. A piece of paper counts as money, or a ticket to a game, or a parking citation, or a diploma in normal circumstances. A person counts as the president, or a felon, or a priest, or a symphony conductor. All such cases presuppose a collective intentionality. The physical reality involved does not explain the functions conferred by the status. A consideration of status functions involves a complex boot-strapping operation of status functions building on status functions. The reality and causal efficacy of such functions depends on an institutionalized collective acceptance of such statuses and functions. This presupposes, but is not reducible to physical reality.

The dialog of developing physics is part of this collective intentional order. How does this human order (Merleau-Ponty's term) relate to reductive accounts based on the primacy of the quantum realm? It doesn't. Neither side includes a consideration of the other. The Gell-Mann–Hartle project leads to the quasiclassical realm which approximates the deterministic laws of classical physics. The conceptual core of classical physics, the system that supports these deterministic laws, is a streamlined extension of the conceptual core of our ordinary language. Though the human order does not figure in the G-H reductive program, it supplies the matrix for any epistemological analysis of scientific knowledge. Far on the other side of this divide the Continental philosophical tradition presents a nuanced analysis of the human order. It is generally developed in a terminology, and with operative presuppositions, that renders questions about any form of global reductionism irrelevant and even meaningless. Searle is one of the few in the analytic tradition who considers ontological as well as epistemological reductionism. Yet he handles it by dismissal rather than any attempt to interrelate these disparate traditions. The present stress on the role of language in physics and on the special problems involved in meaningful extension of language across the classical/quantum divide supplies a kind of tent in which these diverse developments can be juxtaposed.

8.4.1 Reductionism and Physical Theories

To relate physics to the tradition of global reductionism it helps to begin with a consideration of the reasons why physics effectively dropped out of this tradition. The older E-R debates were fairly clear on the role of physics. The classical Oppenheim-Putnam (1958) paper on the unity of science as a working hypothesis listed six major levels: Elementary particles, Atoms, Molecules, Cells, (Multicellular) living things, Social groups. The operative assumption was that the reductive relations between these ontological levels corresponded to some kind of a reductive relation between the theories that treat these levels. The reductionism developed by Feigl (1958), Nagel (1961, chaps. 11 and 12), and Smart (1963, chap. III) banked on a reduction of higher level laws to the laws of physics. Smart insisted that only the laws of physics and chemistry count as scientific laws. In principle this reductive schematism included the physics of elementary particles and a treatment of the human order. Dean Wooldridge concluded The Machinery of Life with the claim: "... the regular and predictable operation of a single body of physical law is sufficient, without supplementation by any form of extra-scientific or 'vitalistic' principle to account for all aspects of human experience" (Wooldridge 1966, p. 204).

Many older defenses of emergentism criticized physics as inadequate because it did not support their version of emergence. Physical accounts were judged incomplete because they do not treat: the goal-driven factor labeled 'entelechy' (Driesch); or becoming (Bergson); or non-physical forces (Broad)ındexBroad, C.; or inner knowledge and radial energy (Teilhard de Chardin); or tacit knowledge (Polanyi); or the primacy of iconic models (Harré). These claims involve a tacit acceptance of the primary role to be accorded a proper physics.

An appreciation of the significance of multiple realizability changed the presuppositional role accorded physics. The music coming out of a small black box could be produced by a radio, or a tape recording, or a CD, or an MP3 player, or miniature musicians, or by some novel new gadget. Similarly the realization that higher level behavior could be explained by different lower-level agents blocked any simple reduction of higher-level phenomena to the theories proper to a particular lower level. Supervenience supplied a noncommittal tool for bridging the gap.[10] Three further considerations favored a strategy of detaching global reductionism from current physics. First, Kim revised the reductionist strategy with the suggestion that a reduction between two levels could rely on a disjunction of heterogeneous lower-level realizers. This deemphasized physical accounts of individual cases. The second factor is the realization that particular reductive accounts based on physics encounter serious difficulties. Such accounts supply a flaccid springboard for a leap from particular accounts to global reductionism. The final factor is Kuhnian relativism. If present physical theories are as likely to be replaced as their predecessors, then it seems appropriate to focus on anticipations of ideal future theories.

The net effect of these trends was to switch the defense of global reductionism from a reliance on physics to a reliance on metaphysics. This metaphysics, however, differs from traditional metaphysics. It is not developed as an independent system. It is intended as a complement to physics, rather than a competitor. The metaphysical principles invoked are principles that should be operative in a final or idealized physics. A few examples may illustrate the trend. Instead of laws of physics one could appeal to laws of nature. These rely on nomic necessity, rather than developing accounts of particles and forces. Nomic necessity concerns laws or relations that must obtain in all possible worlds, or at least in a suitable subset. Following

[10] Surveys of these developments may be found in Kim (1990), Moser and Trout (1995) and Block et al. (1997).

David Lewis (1999, pp. 291–325) it is assumed that such nomic necessity would appear in the formulation of contingent generalizations that play the role of axioms or theorems in future, or ideal, physical theories. Kim gave the most influential defense of reductionism as a metaphysical claim. Linguistic accounts, showing that the language of the mental is irreducible to the language of the physical do not treat the metaphysical problem (Kim 1998). The basic metaphysical principle he relies on is the causal closure of the physical order. Every event that is explainable should be explainable on the basis of causal powers of ultimate physical entities.

To take a simple example, consider a TV meteorologist reporting that an incoming high pressure front will cause a rise in the daily temperature. The real causal activity involves collisions and electromagnetic activities at a molecular level. However, it would be utterly impossible to base weather predictions on attempts to integrate over all molecular activities. Meteorology relies on empirical generalizations and computer simulations. An invocation of a high-pressure front is not an appeal to downward causality, or of a cause in addition to the causal activities of billions of molecules. It is a phenomenological account expressed in the causal terms developed in the familiar pattern of explaining phenomena through causes. In a reductive account this causal explanation is reinterpreted as an oblique reference to the base level causal realizers.

In a simple psychological analogy, nerve cells replace molecules as base units. Nerve cells come in various forms and have a complex structure. In a typical event a stimulus affecting a nerve's dendrites induces a sequence of electrochemical events in the cell, and also environmental interactions involving the transmission of sodium ions, and ultimately to a change in the cell's axons. This can stimulate the dendrites of another nerve cell. For present purposes, we may ignore cell complexity and simply think in terms of a sequence of impulses passing from cell to cell and ultimately stimulating a muscle cell. The normal human brain and nervous system contains approximately ten billion nerve cells. Even if the mechanisms of transmission were better understood, it would clearly be impossible to explain a particular bodily motion by an appropriate integration over all the nerve firings. We might describe the resulting motion by saying: "William raised his hand to signal the waiter". Does this intention exert downward causality? Or, can it be reduced to the ultimate physical realizers of the nervous system, nerve cells?

To analyze this application of causality, and the closely related issue of supervenience, consider the analogy of a dot matrix picture. As a picture, it has properties, e.g., representing Elvis Presley, which the dots do not have. Yet, the qualities of the picture supervene on the properties of the dots and their configuration in an unproblematic way. Now, consider the same dot matrix picture on a computer monitor. A program, like Adobe Photoshop, introduces a higher level of causality, one that changes the quality of the picture by modifying collections of dots. The properties of the dots still supply a sufficient cause for the properties of the picture on the monitor. However, a higher order causal principle is operative in determining the properties of collections of dots. This could be interpreted in terms of the distinction between a phenomenological and a depth level developed in the last section. On a phenomenological level the picture on the monitor is controlled by two separate programs, one assigning values to the properties of dots in a matrix, the other modifying the properties of groups of dots. On a depth level, all is controlled by '1s' and '0s' passing through a CPU. However, these ultimate realizers only relate to picture properties through the mediation of higher level programs.

Conscious control of bodily movements should, the most plausible hypothesis, be something holistic, modifying the activities of collections of nerve cells. Consciousness itself is experienced as a unification. No one is prepared to present a nerve-cell account of bodily activities that has a pretense of being causally sufficient to explain the unity of consciousness. Radical reductionism skips the intermediate levels and focus on the ultimate realizers. Properties and activities of the ultimate entities constituting a human body should explain such higher order properties as consciousness and qualia. Here is where presuppositions play a crucial role. The ultimate entities are not known. Present physics with its layers of effective theories does not support an explanatory interconnection of ultimate entities and mental phenomena. Hence, the reliance on metaphysical principles thought proper to ultimate entities.

Gillett (2007) systematizes, without endorsing, the Kim-inspired metaphysics with a reductionism that accommodates multiple realization. This metaphysics, he insists is a scientific metaphysics, an abstract investigation of ontological issues as they arise within the sciences. Reductive explanations are mechanistic. In a reductive explanation of properties the ultimate realizers are the properties of the basic entities of physics. In his survey of physicalism Stoljar (2001) advocates using 'physicalism' rather than 'materialism' because the latter term is too restricted and: "It is also to emphasize a connection to physics and the physical sciences. Indeed, physicalism is unusual among metaphysical doctrines in being associated historically with a commitment both to the sciences and to a particular branch of science, namely physics". In spite of this assertion, the survey contains no discussion of contemporary physics. His list of almost one hundred references contains no physics texts or articles.

I will briefly examine some central claims of this metaphysics based on an idealized physics, while prescinding from the differences characterizing the spectrum of positions on reductionism.[11] I believe that these metaphysical principles are best interpreted as idealizations of classical physics and semi-classical atomism. The underlying intuition is a matter of taking evolution seriously. All living beings are considered outcomes of evolutionary processes that did not intend life, sentience, or intelligence as goals. Taking evolution seriously entails disallowing any appeal to non-physical factors, such as special creation, souls, entelechies, or divine intervention. Atheism entails, but is not entailed by, this dismissal of extra-physical factors.[12] I accept this general orientation. The E-R debates should be conducted within a naturalistic framework. In explaining the relation between two levels, e.g.,

---

[11] A more detailed account including these differences can be found in MacKinnon (2009).

[12] Surprisingly, Thomas Aquinas set the methodology for conducting such arguments in an unprejudical fashion. In his methodology natural philosophy begins with the assumption that all beings are material beings and postulates non-material beings only if this proves inadequate. The reasons he found this assumption inadequate in explaining motion and concept formation are no longer viable.

atoms and molecules, simple molecules and complex molecules, reductionism is the defeasible default position. In the E-R tradition, as in the old great chain of being tradition, these levels are spoken of as levels of being. Whether affirmed or denied, the position that different kind of entities proper to different levels have different characteristic properties relies on terminology developed in the ontological tradition. This inter-level reductionism may be extended to global reductionism through a quasi-metaphysical principle

Everything real is physical. (P)

This principle may be given a non-controversial negative interpretation. It rejects spiritual souls, angels, and demons as not real. The positive interpretation admits of variations. The strongest reductive interpretation is that all facts supervene on basic physical facts. Then this can be related to a second quasi-metaphysical principle,

The properties attributed to base entities are objectively real. (M)

This is vague enough to accommodate both property monism, which relies on a type-type reduction of higher order properties (e.g., pain is really nothing by nerve fiber excitations); or a Kim-type property dualism that admits higher properties but insists their causal powers ride piggy-back on lower order properties and ultimately on properties of base entities.

Radical reductionism requires a strong, but ambiguous, dependence on the theory-entity inference. To clarify this dependence it is necessary to make a clear distinction between two types of reduction. Mereological reduction aims at explaining a complex whole in terms of its constituent parts. Arguments supporting mereological reduction often presuppose idealized theories and a methodology of theory reduction. One assumes that each level, L_i is characterized by an appropriate theory, T_i and postulates basic theoretical entities, E_i. The assumption that the sequence of theories, T_1 $T_2, \ldots, T_{final}$ is characterized by the transitivity of theoretical reduction supports a corresponding mereological reduction from E_1 to E_{final}. Within this general framework we can give a minimal condition for ontological reduction. For any token state described by a high-level theory there is a token physical state at the base level. We can also give a minimal condition for ontological emergence. Higher order properties can exert downwards causation. Mental states, to take the key issue, can produce physical results.

The reductive assumption is that the base level entities are those posited by the most basic physical theory. Since such a theory is not yet known and presently accepted theories may be replaced, the base entities and their properties are rarely specified. There is, however, an implicit assumption that the ontological terminology of entities with characteristic properties and related causal powers can be meaningfully applied to whatever base entities future physics relies on (see Kim 1993).

The final quasi-metaphysical principle we will consider is,

The physical order is causally closed. (C)

The problematic feature of this claim, as previously noted, is the significance to be attached to 'causally'. It banks on some form of determinism that is not sanctioned by quantum physics. In different ways Gillett (2007) and Glymour (2007) develop a closure that is compositional, rather than causal. Here we focus on Gillett's exposition of, but not a defense of, compositional reductionism. To accommodate multiple realizability Gillett assumes that different packages of individuals with different powers, properties, and processes may together compose qualitatively different higher powers in the entities studied by the special sciences. This supports an ontological reductionism without a semantic reductionism of higher order predicates. This ontological reductionism is mechanistic. "The truth of physicalism allows compositional reductionism, if successful, to show that the entities of microphysics are the only entities". (Gillett, p. 206).

(P), (C), and (M) are metaphysical principles in a loose sense. They are regarded as general principles that should apply independent of any particular system of metaphysics. They are, accordingly, given a loose informal formulation. Such assumptions result in a kind of status ambiguity. Radical reductionism is generally presented as the tough-minded option, the stance based on an acceptance of physics as the court of last appeal. These physics-inspired metaphysical principles actually appeal to an idealization based on classical physics and semiclassical atomism. Before considering this we will consider the other side of the debate.

8.4.2 Emergence

'Emergence', like 'reduction', is an accordion term. By the proper stretching and squeezing one can grind out many melodies. One can speak of the emergence of laws, entities, or properties. In the present context the crucial issue is ontological emergence. A weak doctrine of emergence holds that complexes have properties that are not properties of proper parts of the complexes. Since this is a ubiquitous phenomenon, something more is needed about the nature of the higher order properties. The previous laser-jet example illustrates the problem. The higher-order properties of the picture are not ontologically emergent properties. The picture strongly supervenes on the dots. There can be no change in the properties of the picture without a change in the dots. Mental states supervene on brain states. The disputed issue concerns whether this is strong or weak supervenience. Strong supervenience entails that two individuals could not differ in their mental states if their physical states were identical. Ontological emergence insists that higher-order entities have properties that cannot be explained through such strong supervenience.

O'Connor (1994) listed four conditions for a property, P, of an object, O, to be ontologically emergent:

1. P supervenes on properties of the parts of Q;
2. P is not had by any of the object's parts;
3. P is distinct from structural properties of O; and
4. P has a direct ('downward') determinative influence on the pattern of behavior involving O's parts.

Silberstein and McGeever (1999) have "Ontologically emergent features are features of systems or wholes that possess causal capacities not reducible to any of the intrinsic causal capacities of the parts nor to the (reducible) relations between the parts". A key emergentist assumption is that an ontologically emergent property of a complex entity is not explained through supervenience on the properties of base entities (O'Connor and Jacobs 2003). Humphrey (1997) developed a logic for this in the framework of a system of levels. Assume entities on an i-level with i-level properties. If there is a fusion operations (*), different from concatenation, joining i-level properties to produce an $i + 1$ level property with casual properties, then this is an ontologically emergent property.

We finish the survey by considering a related position, non-reductive physicalism, which Kim considers equivalent to emergentism FAPP. Davidson's influential development of this position stressed two key points. First, intentions are causes of actions. Second, he developed a position labeled 'anomalous monism' (Davidson 1970). Though the mental interacts with the physical, there are no strict laws governing this interaction. Non-reductive physicalism often attempts to finesse the controverted issues. The non-a priori aspect is manifested in a methodology of holding positions P and C, but not using them as a basis of argumentation. The idea is that, instead of invoking metaphysical principles, it is better to leave the science to the scientists. This is exemplified by John Searle's repeated claim, "The brain causes consciousness, but don't ask me how". These versions of non-reductive physicalism lead to a consequence that is sharply at variance with reductive accounts of mind. Reductivists argue that if two persons are identical in all physical respects, then they are identical simpliciter. Davidson explicitly rejects this (Davidson 2001, p. 33). In the Searle-Chalmers debate[13] Chalmers defended and Searle denied the thesis that it is possible to imagine a world exactly like ours except that complex physical organisms physically identical to earthly animals lack consciousness or mental states. These arguments have focused on individuals and on the issue of whether two individuals could be identical in their physical states and yet differ in mental states. It sidesteps the fact that a total specification of a person's physical state is impossible. The accounts of quantum measurement discussed earlier lead to the conclusion that any attempt to gain a complete knowledge of a person's physical state would destroy life and any mental states. Here again the philosophical debates refer to an ideal unrealizable physics.

Which of these positions are compatible with the interpretation of physics developed here? The radical reductionism summarized is simply incompatible with presently accepted physics and its foreseeable extensions. This evaluation allows reductionists an easy escape route of simply rejecting the interpretation of physics presented here. To put this evaluation on a broader basis I would like to show that the traditional global reductionism implicitly relies on the general principles of classical

[13] John Searle's New York Review article of March 6, 1997 led to an exchange between Chalmers and Searle, ibid, May 15, 1997, pp. 60–61.

physics and their extension to semiclassical physics. It is not compatible with any interpretation of physics that takes QM as the fundamental science of reality.

To develop this it is necessary to clarify the conceptual divide between semi-classical and quantum physics. The semiclassical physics of the Bohr-Sommerfeld atomic model relied on the principles of classical mechanics and electrodynamics supplemented by restrictions on their applicability in quantum contexts and by the introduction of quantum hypotheses concerning energy levels and transitions. Semi-classical models of atoms are still extensively used, especially by chemists explaining how spatial properties of atoms determine molecular configurations. This is part of an effective theory that can be used to describe interactions within a certain energy range, roughly from 10^{-5} to 1 ev. This justification does not work for the program of global reductionism. This program must rely on properties actually possessed by base entities, not on properties only appropriate to a model with a limited range of applicability. Real atoms are atoms as described within quantum physics. This presents three major problems for the type of global reductionism we have been considering.

The first is the reliance on levels buttressed by the implicit assumptions that these may be treated as a sequence of roughly equivalent steps supported, at least in principle, by some sort of theory reduction linking successive steps. The classical/quantum transition is the major conceptual break in the development of physics. There is a radical difference in the methods and presupposition on either side of the divide. Classical determinism supports a reliance on casual accounts. Quantum indeterminism does not. An adequate treatment of the reduction-emergence problem requires a consideration of reductionism within the classical realm and the relation of the classical realm to the quantum realm. It is meaningful to treat levels within a realm, but not the same type of levels across realms.

The second difficulty concerns general methodological principles. Kim effectively broke the multiple realizability gap by allowing mereological reduction between two levels based on a disjunction of heterogeneous lower-level realizers. This works well in explaining levels above the C/Q divide. A particular type of pain can be realized by different brain states. It does not work very well across the divide. Thus attributions of definite size and shape to different types of atoms plays a role in accounts in which these are the base realizers. From a quantum perspective atoms with definite sizes and shapes are decoherent atoms. Decoherence relies on the effect of an indefinitely large number of virtual processes. As we have seen in the treatment of the Lamb shift, basic properties of atoms, such as energy levels, are also shaped by an indefinitely large number of virtual processes. An expansion of mereological reduction to include a potentially infinite number of virtual processes vitiates the intended goal.

The third difficulty concerns the implicit reliance on theory reduction. This does not fit the present practice of physics or its foreseeable extensions. As we have seen, physicists regard different levels, such as atomic physics, nuclear physics, and particle physics as different energy levels treated by different effective theories. String theory represents a forseeable extension. The goal is not to have string theory replace the properties of the particles in the standard model. The hope is that string theory may explain the parameters that the standard model uses but does not explain. Regardless of whether string theory succeeds, it illustrates one crucial point. Any ultimate theory, i.e., a theory that functions at the Planck level, has features, such as 10-dimensional space, branes, and supersymmetrical exchange forces radically different from anything in the classical perspective. The goal is to use such features to deduce parameters of the next highest level. Global reductionism attributes to ultimate entities the properties needed to explain away mental causality. What this fits is not the perspective of contemporary physics, but that of Augustus de Morgan's poem:

> Great fleas have little fleas upon their backs to bite 'em
> And little fleas have lesser fleas, and so on ad infinitum.

Emergentism and non-reductive physicalism are quite compatible with the interpretation of physics presented here. The reasons are chiefly negative. Both rely on a rejection of strong reductionism. Neither banks on anticipations of the form of future or idealized physics. These considerations lead to the following appraisal. Reductionism remains the defeasible default position for relating any two levels. Physics does not support the claims of radical reductionism either that the overall collection of levels can be explained reductively or that higher order properties must be explained in terms of causal properties of ultimate entities. The reduction of the mental to the physical presents unresolved problems. The older reductionist attempt to explain mental properties by reduction to the properties of semi-classical atoms proved hopeless. Real atoms, however, are quantum entities. Do quantum considerations offer any hope of bridging the gap?

Entanglement has been demonstrated for particles or systems that were related and then separated. These demonstrations work only in limited contrived conditions, where measurement destroys entanglement, but supports an inference to a prior entanglement. More complex forms of entanglement and superposition are much more difficult to detect. However, one can always have recourse to the George W. Bush principle. *Absence of evidence is not evidence of absence*. It is plausible to assume that relatively dense complex systems can have quantum holistic properties similar to entanglement. If such properties have adaptive value, then it is plausible that evolutionary developments would select and enhance them. The intuitive appeal is that an extension of this notion might provide a conceptual tool for explaining the unifying role of conscious awareness. I will summarize some attempts to treat the problem of consciousness in distinctively quantum perspective.

The early stage in the development of QM occasioned discussions of psycho-physical parallelism and the role of consciousness. von Neumann (1955 [1927], chap. vi) London and Bauer (1983 [1939]). Wigner (1967, sect. III) expanded this psychological duality into his account of two types of reality. The existence of my consciousness is absolute. The existence of everything else is relative. The culmination of this trend is, in my opinion, Wheeler's doctrine of observer participancy. It begins with the Bohrian observation that no phenomenon is a phenomenon until it is an observed phenomenon. It is extended into an account of the indispensable role of observer participancy in bringing the universe into being. The key assumption behind these strong versions of psycho-physical parallelism is that an act of consciousness is required to reduce a superposition to a mixture. In this perspective consciousness must be accepted as something fundamental and objectively real. I am assuming that environmental decoherence introduces a similar reduction FAPP. So, this appeal to the causal role of consciousness is not needed.

The second quantum approach to the mind-body problem, developed by Henry Stapp (1996, 2007), was also influenced by von Neumann's account of measurement. Stapp schematizes measurement in terms of three processes. Process 1 is the experimenter's choice of the setup, or the question put to nature. Process 2 is nature's response to this experimental question. This, unlike process 1, is covered by normal quantum mechanics and involves the spreading of the initial state. Process 3 (the Dirac choice) replaces the quantum superposition of states by a unique classical choice, e.g., a spot on a photographic plate. Here Stapp adds a novel element, an adaptation of the quantum Zeno effect. Consider an atom with ground state |g>, an excited state, |e>, and a radiation field that can force transitions from |e> to |g>. The probability of a decay in this simple case is given by an exponential decay law, with a factor, $e^{(-i(E_{|e>}-E_{|g>})t/S}$, expressing the time dependence of the decay probability. A recording of the decay collapses this wave function. An observation of the excited state would also collapse the wave function, effectively resetting it to 0 in the decay probability. Repeated observations would keep resetting the decay law to $t = 0$. In this idealized version, continuous observation would prohibit decay. The experimental situation is complicated by the fact that observations take time, require a relaxation time between observation, and also by the fact that the detector interacts with the environment.

To adapt this, Stapp considers a person choosing to perform some physical action and assumes that this choice is implemented through neuronal activity. This activity involves chemical and ionic processes that require quantum mechanical accounts. Treat this as a quantum observation. If the choice is instantaneous, then there is an immediate jump to the Dirac process and quantum effects are washed out. William James proposed an alternative to such instantaneous choice. He argued that making a choice involves holding something in conscious awareness. Recent empirical studies of the process of choosing offer empirical support to this Jamesian hypothesis. If quantum mechanics governs these processes, then the fixation of consciousness is similar to the continuous observation of the quantum Zeno effect. This implies that processes 2 and 3, involving superpositions and their replacement by mixtures, are also modified. There may be templates, or patterns of fixating attention to achieve particular results, that are routinely used in normal activity. This might provide an account of a link between mental and physical activity at variance with classical determinism.

The cosmologist, Roger Penrose, developed the third account with the assistance of the anesthesiologist, Stuart (Hameroff and Penrose 1996). Following the discovery that neurons have a microtubular structure was the further discovery that each tubulin molecule has two slightly different configurations that are determined by the position of a special electron in one of two stable locations. Hameroff noted a correlation between the position of this special electron and the effect of various anesthetics on consciousness. The idea that a special electron functions like a molecular on-off switch suggests the possibility that neural structures form the basis for a kind of quantum computer. Penrose's elaboration of this involves quantum gravity and Gödel's incompleteness theorem. Here again, decoherence presents a serious problem that this hypothesis must overcome.

These hypotheses should be treated as first tentative steps in a largely unexplored area with an intuitive appeal.[14] Quantum entanglement introduces holistic accounts. To see the intuitive appeal we can jump from tentative steps to an even more tentative goal. What would it be like to have an explanation of consciousness in which quantum entanglement plays an integrating role? Quantum information theory attempts to exploit multiple entanglements. This has led to studies of how mixed network states can generate singlet states through entanglement. In a singlet state one can attribute properties only to the state, not to its composite parts. Maximally entangled singlet states can be created between arbitrary points in a network, with a probability that is independent of the distance between them (Broadfoot et al. 2009). The basic network in the brain consists of axons coated with myelin sheaths. A normal 20 year old male human has about 176,000 km. of myelinated axons. Suppose that the action of transmitting signals continually generates short-lived quantum entanglements. This would entail billions of entanglements that would be like effervescence for a fluid. Individual bubbles are constantly emerging and vanishing. The effervescence perseveres. An effervescence of quantum entanglements might make subsystems parts of a unified system in a definite quantum state. This could be coupled to the conjecture that complexity of the right sort causes the emergence of consciousness. The introduction of quantum entanglement sharply limits the type of structures that might serve. This wild conjecture is presented to indicate the type of argument that might be developed utilizing quantum entanglement.

On a more basic level the E-R debates ultimately hinge on a clash of intuitions. The generating intuition behind reductionism is the acceptance of the natural order in place of any reliance on non-natural or supernatural factors. Since life, consciousness, and human intelligence evolved from a material basis, they should be explained through that basis. However, the acceptance of the natural order and evolution does not entail the acceptance of wholesale mereological reductionism and a sequence of inter-theory reductions. Evolutionary accounts of emergent properties require long periods of development with branching into new species and the inclusion of accidental factors. Synchronic reductionism is ahistorical. I find a denial of intentional causality totally implausible. In treating individuals and even opposed to the practice of physics. Accounts of experimental physics and of the role of models in theoretical physics require a consideration of the choices made by physicists. When we switch from an individual to a collective basis, then the problem is more extreme. The human order, or Karl Popper's third world, is grounded in collective intentionality. Language as a means of communication presupposes the

---

[14] See Tarlaci (2010) for an historical survey of attempts to invoke quantum mechanics in explaining brain processes.

attribution of beliefs and intentions. The institutions of property, marriage, morality, religion, and law, the sharing of a culture, the ongoing dialog of philosophy, physics, and other specializations, presuppose a collective intentionality. The idea that these institutions could also function the same way in a world comprised of individuals physically identical to us, but lacking mental states, represents a desperate attempt to save a flawed philosophical thesis.

An acceptance of emergent properties or of non-reductive physicalism need not be seen as a position in the E-R debates. It can also be interpreted as a rejection of the manner in which the problem has been articulated and of implicit presuppositions on both sides.[15] In physics the acceptance of the effective theories scenario militates against theory reduction. By stressing the functional role of theories in systematizing the interactions proper to some energy level, it undercuts the inference from theories to entities and the mereological reduction of higher-level entities to lower-level entities. The acceptance of the human order as a functioning level in which individual intentions can produce physical results and in which collective intentionality plays a constitutive role is obviously compatible with emergentism or non-reductive physicalism. It is also compatible with the thrust of complexity, the development of mathematical descriptions for various forms of individual and collective behavior. Then the general issue is one of explaining how the human order relates to the other three units that we have considered: the quantum order, the quasiclassical order, and the classical order. Here the switch from ontological terms like 'realm' and 'reality' to the more neutral term 'order' signals a switch from ontological to epistemological considerations. How should these four conceptual systems be related?

8.5 The Four Orders

T. S. Elliot's *Four Quartets* pivots around the theme: My end is my beginning. We began with the human order and attempted to show how the development of physics emerged from this order. We ended with the quantum realm and are attempting to see how this relates back to the human order. The interrelation of the human order, classical reality, the quasiclassical realm, and the quantum realm, does not fit any of the common philosophical programs of ontological reduction, epistemological reduction, paradigm replacement, or theory replacement. An extended analogy may help to clarify the novel aspects of this problem.

Music had a well developed tradition of systematization prior to any understanding of sound in terms of wave motion. The organization of music in terms of notes and scales grew out of music as experienced. For the purposes of the present analogy we will call this *The Musical Order* and takes notes as basic units. A note as heard is characterized by pitch, quality, and loudness. Within the musical order there is no theoretical explanation of these properties. However, even an amateur can easily distinguish a $C^\sharp$ on a piano, violin, or flute. Professionals organize music

---

[15] Such a position has been developed in Bitbol (2007).

into many types from cantatas to hip hop. Notes as represented in musical notation only symbolize pitch and duration. Supplemental information is required to indicate quality and loudness. Using this as a basis we can construct The Fourier Order. Here each sound is represented by a collection of numbers representing the fundamental frequency, the overtones, and the relative intensities of each component. We assume that it is possible to give and implement a complete numerical specification of each musical sound.

The reductive physicalist claims that all sounds are really nothing but phenomenological manifestations of molecules in motion. Suppose we grant that. On this basis, however, it is difficult to explain the difference between a $C^\sharp$ on a piano and a violin, and impossible to explain the difference between an aria and a blues-song, or between a mediocre and a brilliant performance of either. So the physicalist studies some developments in philosophy of science and then restructures his questions. Musical sounds as part of The Musical Order are supervenient on molecular motions. Supervenience holds strictly. There can be no change in the phenomena without a change in the subvenient base. A reliance on supervenience obviates the necessity of giving detailed molecular accounts of each particular sound. One need only show that it is possible in principle. This, however, does not supply any basis for treating types, a concerto vs. a sonata, or perceived qualities, such as an inspired vs. a mechanical performance. To address such questions our developing physicalist ascends from treatments of individual molecular motions and perceived sounds to the systems in which they are systematized.

Now the issue of reductionism takes a new form. How can we use the resources of *The Molecular Order* to construct *The Quasi-Fourier Order*? This would be an approximate molecular analog of The Fourier Order. Before undertaking such a task we should clarify the goal. The Fourier Order is much more precise than sounds in the musical order, because it is an idealization. The quasi-Fourier Order should produce structures that are approximately isomorphic to structures in the Fourier Order. It cannot be expected to have the absolute precision of numbers and relations in the Fourier Order. Our physicalist, however, insists that The Quasi-Fourier Order represents reality as it exists objectively, while The Fourier Order is an idealized outgrowth of musical order relations.

The first step in the reconstruction is to replace molecules in motion by hydrodynamic variables. In simpler terms we treat air, or other media that transmit sound, as continuous rather than discrete. Further coarse-graining is needed to isolate the components of molecular motion that contribute to the vibrations from random molecular motion. Then one can treat air as a fluid having an equilibrium pressure, $\mathbf{p}$, a displacement from equilibrium, $\mathbf{r}$, and the velocity associated with this displacement, $\mathbf{v}$. The intensity is given by a formula, $\mathbf{p} \bullet \mathbf{v}$. The intensity level is measured in units of $10 \times \log$ (intensity in microwatts/cm^2) $+ 100$, with $1\,\text{watt/cm}^2 = 160\,\text{dB}$. We still have to account for frequencies and overtones. Here our physicalist makes a strategic switch. These issues are best handled by a causal analysis. The frequencies associated with molecular motions in the air should match the frequencies of the sources producing the vibrations. Musical instruments rely on three basic sources of sound production, vibrating strings, vibrating columns of air, and surface vibrations of percussion instruments, such as drum heads. This suffices for an initial analysis. More complex sources, such as the human voice, are postponed to future dissertation topics. The motion, u, of the basic sources is covered by the Helmholtz equation, $\nabla^2 u + k^2 u = 0$, where k is related to the angular frequency proper to the source vibrations, ω, and the frequency c by $k = \omega/c$. The solution of this equation for vibrating strings with fixed end points, vibrating air columns with one end closed, and vibrating drum heads with the outer circular edge of the membrane fixed, leads to various fundamental frequencies and different overtones with varying intensities. It supplies a basis for more detailed analysis of particular sources.

At this point our physicalist declares, “Mission accomplished”. He has shown that it is possible in principle to use the resources of The Molecular Order to produce a Quasi-Fourier Order that supports structures and relations that are approximately isomorphic to structures and relations in The Fourier Order. What of the distinctively phenomenological qualities characterizing music in The Musical Order? The physicalist does not deny the reality of such properties. Explaining them, however, is not a physical problem. So he appeals to supervenience and leaves further details to musicians, psychologists, physiologists, and other interested parties.

We are concerned with the interrelation of four orders: the human, the classical, the quasi-classical, and the quantum. In an initial characterization we may say that *the human order* is epistemologically foundational while *the quantum order* is ontologically foundational. The other two are derivative idealizations. Explaining the human order presents many formidable problems, which I will not treat. I am concerned with clarifying the problematic of how the human order fits into this larger perspective. We can get at a crucial aspect of this problematic by contrasting two opposed presentations of the problem.

The first involves attempts to develop scientific explanations of the mind and of mental processes. Patricia Churchland (1986) has clarified the philosophical problematic. The ultimate goal is a unified theory of neuronal processes and psychological states. This is a much more limited goal than global reductionism and relates to actual science, rather than a metaphysical extrapolation from classical physics. It would not be a simple reduction of the mental to the physical, but a relation of two theories. Within the reducing theory, T_B, presumably an empirically based neuronal theory, one could build an analog, $T_R{}^*$, of the laws of the to-be-reduced psychological theory, T_R. The development of such an analog theory circumvents meaning problems generated by simple cross-theoretical identification, e.g., a pain is nothing but an excitation of a C-fiber. Since we are focusing on the problematic I will grant that some types of psychological theories, such as those concerning perceptions or conditioned reflexes could be paired with neuronal theories. If Joe and Jim hear the same C note, it is plausible to assume that a neuronal explanation could cover a neuronal analog to both experiences. Suppose they also both believe that the Sharks will win the Stanley Cup. Joe believes this because he has studied the players and the records of the competing teams. Jim believes it because his interpretation of astrology charts supports this. There is no plausibility to the hypothesis that a neuronal account would cover both cases. One way of handling such problems is to dismiss psychological explanations based on beliefs, intentions, and desires as **folk psychology**, a topic requiring further analysis.

We are all accustomed to giving and hearing explanations of the form, she threw the glass because she was angry. He is not guilty of first degree murder because he did not intend to kill her. A quasi-theoretical generalization from such examples leads to explanations based on the role of perceptions, beliefs, desires, intentions, and even of the faculties involved. A systematization of such explanatory procedures can be interpreted as a primitive form of psychological theory. Folk psychology, so interpreted, can fit in the general schema of progress through theory replacement. In Paul Churchland's influential evaluation,[16] folk psychology is an empirical theory that posits mental and perceptual states. It is a very old theory that has not made any significant progress in some three thousand years. It should be rejected as a false theory. Rejecting the theory entails rejecting its ontology, the inner mental states the theory posits. Analytic philosophy, in the Churchlands' evaluation, smuggles in folk psychology in the form of a priori meaning relations (Churchland and Churchland 1998).

I think that folk psychology should not be treated as a primitive scientific theory, but a wedge into non-reductive explanations involving the human order. Here accounts presuppose individuals embedded in and fashioned by a social order. Since the social order cannot be explained in ontological terms, any attempt at an ontological reduction of beliefs, intentions, and desires is a category mistake. To get at this we begin with analytic accounts of behavior. Analysts rely on attributions of mental states. Kenny (1992, chap. 1) argues that explaining behavior through mental states should not be considered a causal account. He distinguishes between: symptoms of mental states, where the inference from behavior to states is a matter of empirical discovery; and criteria of mental states, where the connection is something that must be grasped to interpret the behavior. Thus speaking French is a criterion for knowing French. In many cases behavior can be identified as behavior of a certain kind only if it is seen as proceeding from a certain type of belief or desire. To recognize particular behavior as an instance of buying, selling, promising, marrying, lying, or telling a story, requires attributing the requisite intentions to the agent. This is not a marginal issue. It is basic to our social institutions and our legal and literary traditions. The differences between murder, manslaughter, and accidental homicide, or rape and consensual sex, depends on the intentions accorded agents. The intentions attributed to Don Quixote, Hamlet, Faust, Raskolnikov, and Willy Lohman are more significant that their overt actions. The primary function of the concept of a mental state is one of enabling us to interpret and understand the behavior of other persons. This connection is built into the learning and normal usage of these concepts.

There is a point that John Searle has developed in detail in various publications (See Searle 1998, chap. 4). Intentions with propositional content have conditions of satisfaction. We assimilate such intentional specifications through examples

---

[16] Churchland (1981), reproduced in many anthologies. When confronted with criticisms, such as the inconsistency involved in defending a belief that beliefs do not exist, he has broadened the status of folk psychology to a pervasive framework.

featuring such connections, not through mastering a primitive theory of mental states. Davidson's accounts of radical translation and of triangulation rely on attributions of beliefs, intentions, and desires. His account of anomalous monism blocks any attempt to relate explanations through psychological causes to scientific explanations based on neurophysiology. A fortiori, neither analysis nor phenomenology develops a relation between the human order and the physical order, whether interpreted classically or quantum-mechanically.

This clash seems to put us in something like Kim's dilemma. If we accept the reality of the human order with its presuppositions of free will, agent causality, a causal role for mental states, and moral responsibility, then we forgo the possibility of fitting the human order into a scientific world view. If we accept science as the court of last appeal in judging what is real or true, then we dismiss the casual efficacy of the intentional order. This dilemma relies on an oversimplification of both positions. Explanations that presuppose the reality of the human order are not ontological explanations. The proposed scientific explanations presuppose some form of synchronic reduction. Any attempted reduction of the human order requires a consideration of the evolution of the human order. We will get at this by beginning with a simple form of intentionality and then moving on to accounts of behavior.

Following Dennett (2009) we can adopt an *intentional stance*. This is a strategy of interpreting the behavior of an entity (a person, IGUS, animal, robot, whatever) *as if* it were a rational agent whose freely chosen actions are often determined by a consideration of beliefs, intentions, and desires. This simplifies and streamlines accounts of behavior. I have a computer program that plays bridge. I would like to anticipate how it would respond to a move I am considering, e.g., leading a low card from a suit in which I have the King. If I were to attempt a prediction by adopting a *physical stance* I would have to base the prediction on a consideration of all the electrical impulses passing through a CPU and the electrical states of 'memory' storage systems, a forbiddingly difficult task. It would be simpler to skip this and explain its behavior through a *design stance*. If I had the bridge program in a computer language that I understood then, with a considerable amount of effort, I could predict how it would respond to my move. In the intentional stance I assume that the program knows the rules, intends to win, and follows reasonable strategies. Then most predictions are automatic. The program will follow suit, will trump when possible. If the program also believes that I would not lead from a King it would play the Ace on my low card lead rather than the Queen. I try the low lead. If it works I learn something about the program's beliefs about my beliefs. Fortunately I can safely assume that I am a cognitive step above the program on this point. I am capable of learning from experience and adjusting my beliefs, though some of my associates doubt this. The simple bridge program I have can not learn from experience.

The attribution of beliefs, intentions, and desires to other persons supplies a basis for normal human interactions. It is remarkably successful in predicting routine behavior. In driving down a crowded California freeway I assume that other drivers know the basic rule of the road, believe that others also know these rules, and intend to avoid accidents. If I am making a non-routine move, e.g., changing lanes or attempting a U-turn, then I have to anticipate the likely response of other drivers. This anticipation is based on the belief that other drivers will also make non-routine moves, such as swerving or braking, to avoid an accident. There is no doubt about the fact that an intentional stance supplies a good guide for normal behavior in the Umwelt.

To put this in a broader perspective we should consider the evolution of behavior (See Dennett 2003). One cannot understand the type of behavior proper to any entity simply by considering the type of entity it is. Behavior evolves to meet different sorts of challenges. In a group there is competition for food, mating, and a place in the pecking order. The group faces challenges from harsh or changing environments and from prey-predator relations. The evolution of the behavior of members of the group is also shaped by the competition between free-loaders, who have a short-term advantage, and altruists, who contribute to the long-term survival of the group. The net result is evolved behavior that is beyond the cognitive capacity of the individual practicing the behavior. Orgel's second rule is: *Evolution is cleverer than you are*. Migrating birds and salmon returning to spawn are guided by subtle navigational tools without a concept of navigation. Bees, bears, and bobcats prepare for the winter without, it seems, any cognitive awareness of seasonal changes.

Humans evolved in groups. The development of language enabled the development of more complex forms of individual and group behavior. We will focus on one stage of this evolution, the one considered in chapters 1 and 2. Terms referring to overt behavior supplied a basis for a metaphorical extension of these terms to refer to mental acts and states. The music analogy supplies a parallel. People played and experienced music before, or independent of, the systematization of musical notation initiated by medieval monks. However, after standard musical notation was accepted and assimilated it became an essential tool for learning and performing music. It also made possible an extension of music from one, or a few, instruments to large orchestras playing complex works. The introduction of terms referring to mental acts or states allowed for behavior that is not possible without such usage. It allowed people to make promises, enter into agreements, deliberately implant and shape beliefs in other people, and develop the collective intentionality manifested in institutions such as money, marriage, property, the recognition of rights and obligations. Since these linguistic extensions made these practices possible, an analysis of these practices reveals an implicit commitment to the reality and causal efficacy of the mental states referred to. Does this suffice to determine the ontological status of these states?

Human behavior evolved by adapting to physical and increasingly complex social challenges. Mentalistic term were, the most plausible hypothesis, developed as tools for controlling behavior and adapting behavior to more complex social customs and institutions. If *mind* is regarded as the inner source of behavior then 'mind' is an essentially social concept. The mind manifest an understanding of tribal practices and social structures that the individual need not and generally does not comprehend. Our evolved behavior is often cleverer than we are. The brain can be studied by considering an individual isolated from a physical and social environment. The mind cannot. This interpretation of 'mind' supports Davidson's anomalous monism. Beliefs, intentions, and desires are related to physical processes, but the relations are not lawlike. We should study such mental states by learning how they develop and function.

An individual learns the use of mentalistic terms by indoctrination and assimilation, not by learning a theory. In recent years Piaget's analysis of children's cognitive development has been put on a more critical basis by analyses that have expanded and have been subjected to more experimental tests. Striano and Tomasello (2001) One striking finding was that new born infants (less than 48 hours old) reliably imitated adult behavior directed at them, such a opening a mouth or sticking out a tongue. The infant perceives adult faces visually, but has only an incipient proprioceptive perception of her own face. A key point in these development studies is that a growing infant learns behavioral control by adapting to the behavior of others.

An interesting new development here is the study of *motherese*. Across cultures mothers have a distinctive way of speaking to their babies. They speak in a high pitch, or a whisper, have a sing-song style of speaking, exaggerated prosodic variations, and simplified content. Dean Falk contends that motherese played a decisive role in the evolution of language, a contention that has provoked some controversy.[17] Our concern is with the development of mind reading, rather than with the origins of language. The infant's initial learning is conditioned by her mother's actions. Through facial expressions and verbal tones, she highlights basic features of language that the child can assimilate, and she rewards success. For the child to respond, she must have some reading of her mother's behavior, as directed at the baby, as encouraging or discouraging. A similar shaping of responses continues at later stages through the shaping and articulation of a child's emotions. (Whee! Isn't this fun? Oh, you're sad). Anyone who works with retarded children, as I do, recognizes the need to articulate a child's emotions. When you teach a child that she is crying because she is sad, or that she is throwing things because she is angry, then she gradually learns to interpret her own states through public terms. The shaping and articulation of mental and emotional states is dependent upon sustained mutual interactions. This process of learning to recognize and articulate one's own mental states through a care giver's identification and reinforcement carries over to higher order processes. We teach children to make good decisions, have the approved beliefs, recognize disturbing symptoms, and avoid dangerous pleasures. Most of their social learning, however, is through interaction, assimilation, and adaption. Mind reading is not a magical trick. It is an essential part of the process of assimilating and adapting to a social order. Since we all do this, the skills acquired are not conspicuous. They are skills we take for granted in our normal reading of behavior. When Shakespeare's Caesar declares "yon Cassius has a lean and hungry look",

[17] Her published article is available online at http://www.bbsonline.org/Preprints/Falk/Referees. It contains an extensive bibliography. Initial reactions are summarized in Scientific American, 291 (Aug. 2003, 30–32).

his audience knows what intentions to infer. The absence of such skills, as in some autistic children, is more noticeable. The difficulty involved in reconstructing them gives some measure of their complexity. It is not difficult to reconstruct the skills involved in catching a baseball or balancing on a bike. As Grice's rules of conversational implicature, and their very limited success, show, it is extremely difficult to reconstruct the skills involved in a normal conversation.

Accepting the normal functioning of the human order entails accepting persons as agents who act in accord with beliefs, intentions, desires, and a functional recognition of accepted norms of behavior. Such acceptance fits an intentional stance. We can predict the behavior of persons or computer bridge programs by treating them *as if* they were agents whose actions are determined by beliefs, intentions, and desires. Even reductionists are compelled to act this way if they wish to fit into normal society. Does this also entail accepting the reality and causal efficacy of free choices? It clearly does not for the bridge program. If we take a design stance then the 'choices' made can be explained by a computer program, which shows that they are not really free choices. If one takes a physical stance then it is meaningless to talk of a bridge program making free choices. In the case of human behavior the physical stance is unrealizable. If we take a design stance, then we need an evolutionary account of human behavior. This relies on accounts of branching points and the emergence of complex structures and adaptive behaviors, but not on intelligent design. One might still argue, following Patricia Churchland, that a theory of neural states would have structures corresponding to the rules implicitly governing the prediction of behavior based on mental states. However, this would be a token reduction (or analog) not a type reduction. We have to ascend to a broader level. How do we understand the relation between the human order and the quantum order?

In our adaption of the Gell-Mann–Hartle project the classical and quasiclassical orders are theoretical constructs that mediate the relation between the human and the quantum orders. This serves to explain broad features of the classical order on a reductive basis. The classical order could explain the human order only if it is supplemented by a thesis of universal determinism, something incompatible with the quantum foundation. Without this thesis the classical order is understood as something derived from the human order by abstraction, simplification, and mathematical representation. It cannot explain the features omitted through the process of abstraction.

We are left with the human order, which is epistemologically foundational, and the quantum order, which is ontologically foundational. Neither can be reduced to the other. The human order presupposes a physical basis. But it cannot be explained by strong supervenience on that basis. It presupposes such non-physical features as language, actions based on intentions, desires, and beliefs, and institutions based on collective intentionality. On an individual basis the arguments we considered supported consciousness and free choice as emergent properties. On a collective basis they also support the human order as something emergent. In both cases emergence is defended on a limited and basically negative basis. Arguments based on global reduction and/or strong supervenience fail. The basis for the human order is man, the riddle, jest, and glory of the world.

References

Auyang, Sunny. 1995. *How is Quantum Field Theory Possible?* New York, NY: Oxford University Press.

Bitbol, Michel. 2007. Ontology, matter and Emergence. *Phenomenology and the Cognitive Sciences, 6*, 293–300.

Block, Ned, Owen Flanagan, and Guven Guzeldere. 1997. *The Nature of Consciousness: Philosophical Debates*. Cambridge: MIT Press.

Broadfoot, S., U. Dorner, and D. Jaksh. 2009. Mixed State Entanglement Percolation. *arXiv:0906/1622v1*.

Cartwright, Nancy. 1983. *How the Laws of Physics Lie*. Oxford: Clarendon Press.

Castellani, E. 2002. Reductionism, Emergence, and Effective Field Theories. *Studies in History and Philosophy of Modern Physics, B33*, 251–267.

Chalmers, David, and John R. Searle. 1997. Consciousness and the Philosophers: An Exchange. *The New York Review of Books, 44*.

Churchland, Patricia Smith. 1986. *Neurophilosophy*. Cambridge, MA: MIT Press.

Churchland, Paul. 1981. Eliminative Materialism and Propositional Attitudes. *The Journal of Philosophy, 78*, 67–90.

Churchland, Paul, and Patricia Churchland. 1998. *On the Contrary Critical Essays, 1987–1997*. Cambridge, MA: MIT Press.

Critchley, Simon, William Ralph Schroeder, and Jay Bernstein. 1998. *A Companion to Continental Philosophy*. Malden, MA: Blackwell.

Davidson, Donald. 1970. Mental events. In L. Foster, and J. Swanson (eds.), *Experience and Theory. Humanities Press* (pp. 79–101). Oxford: Reprinted in Essays on Action and Events (Oxford University Press, 1980).

Davidson, D. 2001. *Subjective, Intersubjective, Objective*. Oxford: Calarendon Press.

Dennett, Daniel. 2003. *Freedom Evolves*. New York, NY: Viking.

Dennett, Daniel. 2009. Intentional Systems Theory. In B. McLaughlin, A. Beckermann, and S. Walter, (eds.), *Oxford Handbook of the Philosophy of Mind*. Oxford: Oxford University Press.

Falk, Dean. 2003. Prelinguistic Evolution in Early Homins: Whence motherese? *Behaviorall and Brain Science, 27*, 491–534.

Feigl, Herbert. 1958. The 'Mental' and the 'Physical'. In Herbert Feigl, Michael Scriven, and Grover Maxwell (eds.), *Minnesoat Studies in the Philosophy of Science* (pp. 370–497). Minneapolis: University of Minnesota Press.

Fine, Arthur. 1986. *The Shaky Game*. Chicago, IL: University of Chicago Press.

Franklin, Alan. 1981 What Makes a Good Experiment. *British Journal for Philosophy of Science, 32*, 367–374.

Franklin, Allan. 1983. The Discovery and Acceptance of CP Violation. *Historical Studies in the Physical Sciences, 13*, 207–238.

Franklin, Alan. 1986. *The Neglect of Experiment*. Cambridge: Cambridge University Press.

Galison, Peter. 1987. *How Experiments End*. Chicago, IL: University of Chicago Press.

Georgi, Howard. 1993. Effective Field Theory. *Annual Review of Nuclear Physics, 43*, 209–252.

Gillett, Carl. 2007. The New Reductionism. *The Journal of Philosophy, 104*, 193–216.

Glymour, Clark. 2007. When is a Brain Like the Planet? *Philosophy of Science, 74*, 330–346.

Hacking, Ian. 1983. *Representing and Intervening: Introductory Topics in the Philosophy of Natural Science*. Cambridge: Cambridge University Press.

Hacking, Ian. 1991. A Tradition of Natural Kinds. *Philosophical Studies, 51*, 109–126.

Hacking, Ian. 2009. Natural Kinds: Rosy Dawn, Scholastic Twilight. *Address to the Bay Area Philosophy of Science Association*.

Hameroff, S. R., and R. Penrose 1996. Conscious Events as Orchestrated Space-Time Selections. *Journal of Consciousness Studies, 3*, 36–63.

Hanson, Norwood Russell. 1961. *Patterns of Discovery: An Inquiry into the Conceptual Foundations of Science*. Cambridge Eng.: Cambridge University Press.
Hart, H. L. A., and Antony Maurice Honoré. 1959. *Causation in the Law*. Oxford: Clarendon Press.
Hartmann, S. 2001. Effective Field Theories, Reductionism and Scientific Explanation. *Studies in History and Philosophy of Modern Physics, 32B*, 267–304.
Humphrey, Paul. 1997. How Properties Emerge. *Philosophy of Science, 64*, 1–17.
Johnson, Kent. 2004. On the Systematicity of Language and Thought. *The Journal of Philosophy, 101*, 111–139.
Kane, Gordon. 2000. *Supersymmetry*. Cambridge, MA: Perseus Publishing.
Kaplan, David. 2005. Five Lectures on Effective Field Theory. *arXiv:nucl-th/0510023v1*, arXiv:nucl-th/0510023.
Kenny, Anthony. 1992. *The Metaphysics of Mind*. Oxford, England; New York, NY: Oxford University Press; Clarendon Press.
Kim, Jaegwon. 1982. Psychophysical Supervenience. *Philosophical Studies, 41*, 51–70.
Kim, Jaegwon. 1990. Supervenience as a Philosophical Concept. *Metaphilosophy, 21*, 1–27.
Kim, Jaegwon. 1993. *Supervenience and Mind: Selected Philosophical Essays*. Cambridge, England, New York, NY: Cambridge University Press.
Kim, Jaegwon. 1996. *Philosophy of Mind*. Boulder, CO: Westview Press.
Kim, Jaegwon. 1998. *Mind in a Physical World an Essay on the Mind-Body Problem and Mental Causation*. Cambridge, MA: MIT Press.
Kim, Jaegwon. 2002. *Supervenience*. Aldershot, Hants, England; Brookfield, VT: Ashgate; Dartmouth.
Ladyman, James. 1998. Structural Realism: Epistemology or Metaphysics? *Studies in the History and Philosophy of Science, 29*, 409–424.
Lewis, David. 1999. Psychophysical and Theoretical Identifications. In David Lewis (ed.), *Papers in Metaphysics and Epistemology* (pp. 248–261). Cambridge: Cambridge University Press.
London, Fritz, and Edmund Bauer. 1983 [1939]. The Theory of Observation in Quantum Mechanics. In J. Wheeler, and W. Zurek (eds.), *Quantum Theory and Measurement* (pp. 217–259). Princeton, NJ: Princeton University Press.
MacKinnon, Edward. 2009. The New Reductionism. *The Philosophical Forum, 34*, 439–462.
Manohar, Aneesh. 1996. Effective Field Theories. *arXiv:hep-ph/96061222*.
Menand, Louis. 2001. *The Metaphysical Club*. New York, NY: Farrar, Straus, and Giroux.
Merleau-Ponty, Maurice. 1945. *Phénoménologie de la perception*. Paris: Librairie Gallimard.
Moser, Paul, and J. D. Trout. 1995. General Introduction: Contemporary Materialism. In Moser Paul, and J. D. Trout (eds.), *Contemporary Materilism* (pp. 1–35). London: Routledge.
Nagel, Ernest. 1961. *The Structure of Science: Problems in the Logic of Scientific Explanation*. New York, NY: Harcourt, Brace & World.
Norris, Christopher. 1999. Putnam's Progress. *The Philosophical Forum, 30*, 61–90.
O'Connor, Timothy. 1994. Emergent Properties. *American Philosophical Quarterly, 31*, 91–104.
O'Connor, Timothy, and Jonathan Jacobs. 2003. Emergent Individuals. *The Philosophical Quarterly, 33*, 540–555.
Oppenheim, Paul, and Hilary Putnam. 1958. Unity of Science as a Working Hypothesis. In Herbert Feigl, Michael Scriven, and Grover Maxwell (eds.), *Minnesota Studies in the Philosophy of Science: Volume II. Concepts, Theories, and the Mind-Body Problem* (pp. 3–36). Minneapolis: University of Minnesota Press.
Penrose, Roger. 2004. *The Road to Reality: A Complete Guide to the Laws of the Universe*. New York, NY: Knopf.
Pinker, Steven. 2007. *The Stuff of Thought: Language as a Window into Human Nature*. New York, NY: Viking.
Psillos, Stathis. 2000. The Present State of the Scientific Realism Debate. *British Journal for the Philosophy of Science, 51*, 705–728.
Putnam, Hilary. 1987. *The Many Faces of Realism*. La Salle, IL: Open Court.
Putnam, Hilary. 1990. *Realism with a Human Face*. Cambridge, MA: Harvard University press.

Putnam, Hilary. 2004. *Ethics Without Ontology*. Cambridge, MA: Harvard University Press.
Searle, John R. 1983. *Intentionality: An Essay in the Philosophy of Mind*. Cambridge: Cambridge University Press.
Searle, John R. 1997. Consciousness and the Philosophers: An Exchange. *The New York Review of Books, 44*.
Searle, John R. 1998. *Mind, Language and Society: Philosophy in the Real World*. New York, NY: Basic Books.
Seibt, Johanna. 2002. Quanta, Tropes, or Processes: Ontologies for QFT Beyond the Myth of Substance. In Meinrad Kuhlmann, et al. (eds.), *Ontological Aspects of Quantum Field Theory*. Singapore: World Scientific.
Silberstein, Michael, and John McGeever. 1999. The Search for Ontological Emergence. *The Philosophical Quarterly, 49*, 182–200.
Smart, J. J. C. 1963. *Philosophy and Scientific Realism*. London: Routledge and Kegan Paul.
Stapp, Henry. 1996. The Hard Problem: A Quantum Approach. *Lawrence Berkeley Lab*, LBL-37163.
Stapp, Henry. 2007. Quantum Approaches to Consciousness. In Max Velmans, and Susan Schneider (eds.), *The Blackwell Companion to Consciousness*. Malden, MA: Blackwell.
Stoljar, Daniel. 2001. Physicalism. *Stanford Encyclopedia of Philosophy*. Stanford, CA: Stanford University Press.
Striano, T., and M. Tomasello. 2001. Infant Development: Physical and Social Cognition. In N. J. Smelsen, and P. B. Bate (eds.), *International Encyclopedia of the Social and Behavioral Sciences* (pp. 7410–7414). Oxford: Pergamon.
Tarlaci, Sultan. 2010. A Historical View of the Relation Between Quantum Mechanics and the Brain. *Neuroquantology, 8*, 120–136.
Teller, Paul. 1995. *An Interpretive Introduction to Quantum Field Theory*. Princeton, NJ: Princeton University Press.
Van Fraassen, Bas. 1991. *Quantum Mechanics: An Empiricist View*. Oxford: Clarendon Press.
Van Fraassen, Bas. 2002. Structure: Its Shadow and Substance. *Philosophy of Science Archives, 63*.
von Neumann, John. 1955 [1927]. *Mathematical Foundations of Quantum Mechanics*, trans. Robert T. Beyer. Princeton, NJ: Princeton University Press.
Weinberg, Steven. 2008. *Cosmology*. Oxford: Oxford University Press.
Wigner, Eugene P. 1967. *Symmetries and Reflections: Scientific Essays*. Bloomington: Indiana University Press.
Wooldridge, Dean E. 1966. *The Machinery of Life*. New York, NY: McGraw-Hill.
Worrall, John. 1989. Structural Realism: The Best of Both Worlds. *Dialectica, 43*, 99–124.

Index

E. MacKinnon, *Interpreting Physics*, Boston Studies in the Philosophy of Science 289,
DOI 10.1007/978-94-007-2369-6,

PGSTL